JOURNAL OF REPRODUCTION AND FERTILITY

SUPPLEMENT No. 34

REPRODUCTION
IN
DOMESTIC RUMINANTS

Proceedings of a Symposium held at

Ithaca, New York,

U.S.A.

July 1987

EDITED BY

G. D. Niswender, D. T. Baird, J. K. Findlay

and

Barbara J. Weir

Journal of Reproduction & Fertility

1987

First published 1987

ISSN 0449-3087
ISBN 0 906545 12 9

Published by **The Journals of Reproduction and Fertility Ltd.**

Agent for distribution: **The Biochemical Society Book Depot, P.O. Box 32, Commerce Way, Whitehall Industrial Estate, Colchester, CO2 8HP, Essex, U.K.**

Printed in Great Britain by
Henry Ling Ltd., at
The Dorset Press, Dorchester, Dorset

CONTENTS

J. Reprod. Fert., Suppl. **34** (1987), v

Printed in Great Britain

Preface

The Second International Symposium on *Reproduction in Domestic Ruminants* was held in Ithaca, New York, on 8–11 July 1986.

There were 130 participants, and the meeting was supported by numerous companies. Direct financial contributions were received from Ferring Laboratories, Inc. (Ridgewood, New Jersey), Merck Sharp & Dohme Research Laboratories (Rahway, New Jersey), Monsanto Company (St Louis, Missouri), Sandoz Research Institute (East Hanover, New Jersey), SmithKline Beckman Corporation (Philadelphia, Pennsylvania) and the Upjohn Company (Kalamazoo, Michigan).

ORGANIZING COMMITTEE

D. T. Baird
J. K. Findlay
D. L. Foster

W. Hansel
K. Inskeep
G. D. Niswender (*Chairman*)

Participants

J. P. Advis (Rutgers, NJ, U.S.A.)
U. Aladin (Cornell, NY, U.S.A.)
H. W. Alila (Cornell, NY, U.S.A.)
R. P. Amann (Colorado State, CO, U.S.A.)
G. B. Anderson (California-Davis, CA, U.S.A.)
K. J. Betteridge (Guelph, Ontario, Canada)
E. L. Bittman (Massachusetts, MA, U.S.A.)
A. von Brackel (Ithaca, NY, U.S.A.)
J. H. Britt (North Carolina State, NC, U.S.A.)
R. L. Butcher (West Virginia, WV, U.S.A.)
W. E. Butler (Cornell, NY, U.S.A.)
R. Canfield (Cornell, NY, U.S.A.)
Y. Chandrasekhar (London, Ontario, Canada)
J. R. Chenault (Kalamazoo, MI, U.S.A.)
R. K. Christenson (Clay Center, NE, U.S.A.)
I. J. Clarke (Melbourne, Australia)
L. Claypool (Michigan, MI, U.S.A.)
P. W. Concannon (Cornell, NY, U.S.A.)
B. Condon (New Hampshire, NH, U.S.A.)
R. I. Cox (Blacktown, NSW, Australia)
M. Crowder (Colorado State, CO, U.S.A.)
R. A. Dailey (West Virginia, WV, U.S.A.)
B. N. Day (Missouri, MO, U.S.A.)
M. Day (Ohio State, OH, U.S.A.)
D. R. Deaver (Pennsylvania State, PA, U.S.A.)
J. R. Diehl (Clemson, SC, U.S.A.)
H. Dobson (Liverpool, U.K.)
T. G. Dunn (Wyoming, WY, U.S.A.)
F. Eblin (Michigan, MI, U.S.A.)
H. Engelhardt (Western Ontario, Ontario, Canada)
C. Farin (Colorado State, CO, U.S.A.)
M. Fields (Florida, FL, U.S.A.)
J. P. Figueroa (Cornell, NY, U.S.A.)
J. K. Findlay (Melbourne, Victoria, Australia)
N. L. First (Wisconsin, WI, U.S.A.)
T. A. Fitz (Maryland, MD, U.S.A.)
G. L. Foley (Cornell, NY, U.S.A.)
R. H. Foote (Cornell, NY, U.S.A.)
R. G. Foraage (East Roseville, NSW, Australia)
J. E. Fortune (Cornell, NY, U.S.A.)
D. L. Foster (Michigan, MI, U.S.A.)
S. L. Fu (Cornell, NY, U.S.A.)
H. A. Garverick (Missouri, MO, U.S.A.)
A. K. Goff (Montréal, Quebec, Canada)
H. D. Hafs (Merck Sharp and Dohme, NJ, U.S.A.)
J. Hall (West Virginia, WV, U.S.A.)
R. H. Hammerstedt (Penn State, PA, U.S.A.)
W. Hansel (Cornell, NY, U.S.A.)
M. T. Hochereau-de Reviers (INRA, Nouzilly, France)
M. K. Holland (Vanderbilt, TN, U.S.A.)
P. B. Hoyer (Arizona, AZ, U.S.A.)
A. G. Hunter (Minnesota, MN, U.S.A.)
M. G. Hunter (Nottingham, U.K.)
K. Inskeep (West Virginia, WV, U.S.A.)
J. J. Ireland (Yale, CT, U.S.A.)
A. L. Johnson (Rutgers, NJ, U.S.A.)
M. T. Kaproth (EAIC, NY, U.S.A.)
C. L. Keefer (Georgia, GA, U.S.A.)
D. H. Keisler (Missouri, MO, U.S.A.)
D. J. Kennaway (Adelaide, South Australia, Australia)
M. W. Khalil (Western Ontario, Ontario, Canada)
D. J. Kiehm (London, Ontario, Canada)
G. Killian (Pennsylvania State, PA, U.S.A.)
J. E. Kinder (Nebraska, NE, U.S.A.)

A. King (Montréal, Quebec, Canada)
G. King (Guelph, Ontario, Canada)
T. Kiser (Georgia, GA, U.S.A.)
J. J. Knickerbocker (Colorado State, CO, U.S.A.)
G. E. Lamming (Nottingham, U.K.)
R. B. Land (ABRO, Edinburgh, U.K.)
B. Lasley (California-Davis, CA, U.S.A.)
J. W. Lauderdale (Upjohn, MI, U.S.A.)
R. A. Lepley (Kalamazoo, MI, U.S.A.)
S. Leshin (Georgia, GA, U.S.A.)
P. E. Lewis (West Virginia, WV, U.S.A.)
D. M. Linkie (Ferring Labs Inc., NJ, U.S.A.)
S. P. Lonton (ABSD, WI, U.S.A.)
J. Lussier (Saskatchewan, Saskatchewan, Canada)
P. Malven (Lafayette, IN, U.S.A.)
J. G. Manns (Saskatchewan, Saskatchewan, Canada)
G. A. Massmann-Figueroa (Cornell, NY, U.S.A.)
M. Mellado (Ithaca, NY, U.S.A.)
S. Meredith (Lincoln, MO, U.S.A.)
R. Milvae (Cornell, NY, U.S.A.)
G. E. Moss (Wyoming, WY, U.S.A.)
L. Munson (Cornell, NY, U.S.A.)
C. D. Nancarrow (Blacktown, NSW, Australia)
T. Nett (Colorado State, Co, U.S.A.)
J. H. Nilson (Case Western Reserve, OH, U.S.A.)
G. D. Niswender (Colorado State, CO, U.S.A.)
R. J. Orts (3M Center, MN, U.S.A.)
J. D. O'Shea (Melbourne, Victoria, Australia)
J. Parks (Cornell, NY, U.S.A.)
J. Parrish (Wisconsin, WI, U.S.A.)
J. Pelletier (INRA, Nouzilly, France)
A. R. Peters (MLC, U.K.)
M. G. Pimental (Cornell, NY, U.S.A.)
J. Raeside (Guelph, Ontario, Canada)
R. Randel (Texas A&M, TX, U.S.A.)
L. Rhodes (Cornell, NY, U.S.A.)
H. A. Robertson (Victoria, BC, Canada)
J. F. Roche (University College, Dublin, Ireland)
D. Sadowsky (Cornell, NY, U.S.A.)
S. Samaras (Cornell, NY, U.S.A.)
G. R. Sasser (Idaho, ID, U.S.A.)
H. R. Sawyer (Colorado State, CO, U.S.A.)
D. Schams (Munich, Western Germany)
B. D. Schanbacher (Clay Center, NE, U.S.A.)
D. Schlafer (Cornell, NY, U.S.A.)
G. E. Seidel, Jr (Colorado State, CO, U.S.A.)
E. L. Sheldrick (Cambridge, U.K.)
M. Shemesh (Kimron VI, Beit Dagan, Israel)
M-A. Sirard (Quebec, Canada)
M. F. Smith (Missouri, MO, U.S.A.)
R. D. Smith (Cornell, NY, U.S.A.)
M. R. Temple (Cornell, NY, U.S.A.)
W. W. Thatcher (Florida, FL, U.S.A.)
A. Tsafriri (Weizmann Institute, Rehovot, Israel)
P. C. Tsang (Cornell, NY, U.S.A.)
W. C. Wagner (Urbana, IN, U.S.A.)
J. S. Walton (Guelph, Ontario, Canada)
C. D. Wathes (Bristol, U.K.)
C. W. Weems (Hawaii-Manoa, HI, U.S.A.)
R. P. Wettemann (Oklahoma State, OK, U.S.A.)
J. E. Wheaton (Minnesota, MN, U.S.A.)
S. Whisnant (West Virginia, WV, U.S.A.)
G. L. Williams (Texas A&M, TX, U.S.A.)
M. E. Wise (Arizona, AZ, U.S.A.)

J. Reprod. Fert., Suppl. **34** (1987), 1–8

Control of GnRH secretion

I. J. Clarke

*Medical Research Centre, Prince Henry's Hospital, St Kilda Road, Melbourne, Victoria 3004,
Australia*

Introduction

During a discussion at the 1st International Symposium on Reproduction in Domestic Ruminants following Dr Gerald Lincoln's presentation, it became apparent that we needed a method to monitor gonadotrophin-releasing hormone (GnRH) secretion in sheep. Within the 6 years that have elapsed, three models have been developed for this purpose and it is now possible to monitor accurately the secretion of GnRH from the median eminence and to relate this to LH and FSH release. We are therefore able to conduct meaningful experiments to ascertain the roles of the various neural systems and feedback effects that might regulate GnRH secretion.

This paper will review the progress that has been made in measuring the secretion of GnRH, particularly in sheep, and consider steroidal feedback effects. Finally, brief consideration will be given to some of the various neural systems that might be involved in regulating GnRH secretion.

Models for the measurement of GnRH secretion

Two systems have been developed for the direct sampling of hypophysial portal blood in conscious sheep. One involves the accessing of the anterior face of the pituitary gland by unilateral transnasal transsphenoidal surgery (Clarke & Cummins, 1982). An artificial ethmoid sinus is created anterior to the gland and two 12-gauge needles are implanted through the non-operative side to enter the artificial sinus. One needle is directed at the portal vessels that course down the anterior face of the pituitary gland in this species, and this is later used as a guide needle for a stillette that punctures the vessels. The other needle is used for sample collection.

After surgery, sheep are allowed to recover for up to 3 days before they are heparinized and their portal vessels are punctured and sampled. This method has many advantages.

(1) It allows the direct sampling of portal blood from conscious animals whilst not totally compromising pituitary function; thus GnRH and LH secretion may be monitored contemporaneously.

(2) The surgical preparation time is less than 2 h so that 3–4 sheep can be prepared early in a week then sampled during the remainder of the week.

(3) The surgical operation does not disturb the brain or the pituitary gland.

(4) The portal blood collection rate is generally 0·2–0·4 ml/min, providing adequate volumes of blood in 5–10-min intervals for GnRH assay.

(5) Samples can be collected for up to 10 h, without significant discomfort to the animal.

A major disadvantage of this procedure is that an animal may be used only once. Another is that, due to variations in the vascular arrangements, identical lesions cannot be made in a series of sheep and this limits interpretations of differences in GnRH pulse amplitude between animals.

The push–pull perfusion system, which was originally developed by Levine & Ramirez (1980) in rats, has also been adapted for use in sheep (Levine *et al.*, 1982). This method requires the introduction of concentric needles through the brain and into the median eminence. The surgical technique requires X-ray equipment to guide the cannulae into position. Animals are allowed to recover and GnRH secretion is monitored in perfusates of the median eminence that are pushed through the inner needle and pulled through the external needle. Accurate calibration of the 'push' and 'pull' pumps is essential to prevent tissue damage. Like the method of Clarke & Cummins (1982), this 'push–pull' perfusion may be performed on conscious animals with minimal trauma and GnRH secretion and LH secretion may be monitored simultaneously. Drawbacks with the 'push–pull' system are that X-ray equipment is necessary for cannula placement and tissue damage occurs at the time of sampling. Other disadvantages are that GnRH is not always measurable in the perfusates and, since one is not measuring GnRH in portal blood, one cannot be absolutely certain that the 'pulses' seen in the perfusates also appear in portal blood. Finally, this technique generally requires some histological examination of the median eminence to identify the site of cannula tip placement. Caraty *et al.* (1982) have also described a 'push–pull' perfusion system for sheep but in their model portal blood was sampled; it was not made clear whether the samples also contained cerebrospinal fluid.

In conscious monkeys, it is possible to monitor the pulsatile release of GnRH by the 'push–pull' perfusion (Levine *et al.*, 1985) or by the withdrawal of cerebrospinal fluid from the third ventricle (Van Vugt *et al.*, 1986). The latter system has the considerable advantage of enabling samples to be taken from animals on more than one occasion, without perturbation of the hypothalamo–pituitary anatomy. One problem with this procedure is that GnRH and LH pulses do not always appear synchronously; the appearance of some LH pulses without concomitant GnRH pulses suggests that GnRH pulses that are presumably present in portal blood do not occur in the cerebrospinal fluid. In sheep, the measurement of GnRH in cerebrospinal fluid has not provided satisfactory results (R. S. Carson, personal communication).

Feedback control of GnRH secretion by ovarian steroids

To adopt a systematic approach to the study of steroidal effects on GnRH secretion we have defined three types of feedback effect. These are best defined in ovariectomized animals and in terms of the changes that occur after oestrogen treatment of ovariectomized sheep (Clarke *et al.*, 1982). The first and immediate effect is a fall in plasma LH concentrations; this may be called a short-term negative feedback effect. Following this, plasma LH concentrations rise well above preinjection values (Clarke *et al.*, 1982) by what may be called a positive feedback mechanism. Thereafter, and with continued oestrogen treatment, plasma LH concentrations are held below those of untreated ovariectomized animals by a long-term (tonic) negative feedback effect (Diekman & Malven, 1973). The positive feedback effect may also be obtained in seasonally anoestrous ewes (Goding *et al.*, 1969).

This classification may be applied to normal cyclic events. The short-term negative feedback effect might operate in the late follicular phase of the sheep oestrous cycle as suggested by Thomas (1983) and by the data of Scaramuzzi & Radford (1983). The positive feedback mechanism that is seen in oestrogen-treated ovariectomized ewes, and in anoestrous ewes, has been considered similar to that which generates the normal preovulatory LH surge although this may not be the case (see below). Finally, long-term tonic negative feedback is pertinent in terms of the patterns of gonadotrophin secretion that are seen during the luteal phase of the cycle and during anoestrous periods.

Short-term negative feedback

Sakar & Fink (1980) found that an intravenous injection of 1 µg oestradiol caused an acute reduction in GnRH secretion in ovariectomized rats but in another study (Sherwood & Fink, 1980), the subcutaneous injection of 50 µg oestradiol benzoate had no acute effect. In both studies the doses used may be regarded as high. The intravenous injection of 1 μg oestradiol had no immediate effect on portal GnRH concentration in ovariectomized monkeys (Carmel *et al.*, 1976).

In sheep, it was found that the secretion of GnRH continued after a 50 μg intramuscular injection of oestradiol benzoate (Clarke & Cummins, 1985); this could occur when the secretion of LH ceased. Schillo *et al.* (1985), using similarly treated ewes but sampling by 'push–pull' perfusion of the median eminence, also found that secretion of GnRH could continue during the 'pre-surge' period, whilst LH secretion ceased. This led to the conclusion that short-term negative feedback is the result of a pituitary action of oestrogen and support for this notion was gained by studies in ovariectomized ewes with hypothalamo–pituitary disconnection given oestrogen during pulsatile GnRH treatment. It was clearly seen that the initial effect of oestrogen was virtually to eliminate pituitary responses to GnRH (Clarke & Cummins, 1984).

Thus, with the exception of one study in rats, the evidence clearly indicates that there is no short-term negative feedback effect of oestrogen at the hypothalamic level to limit GnRH secretion, but that the effect observed in hypothalamo–pituitary intact ewes (Clarke *et al.*, 1982) is the result of pituitary action of oestrogen.

Positive feedback

The preovulatory surge in LH secretion has fascinated neuroendocrinologists since it first became apparent that a signal emanating from the brain was responsible for ovulation and that this signal was triggered by coitus (in reflex ovulators) or by ovarian steroids. By virtue of this mechanism, the pituitary gland in sheep may lose up to 90% of its stores of LH within a few hours (Roche *et al.*, 1970).

It is clearly established that the responsiveness of the pituitary gland to GnRH is increased at the time of the normal cyclic preovulatory LH surge (Reeves *et al.*, 1971) and at the time of an oestrogen-induced surge in ovariectomized ewes (Coppings & Malven, 1976). This increased responsiveness accounts for a 2–3-fold rise in plasma LH concentrations (Clarke & Cummins, 1984). To achieve the much higher levels that are seen during the preovulatory LH surge, an additional factor is required. This is most likely to be a rise in the level of GnRH secretion or an alteration in the pattern of its secretion. Studies of rats (Sarkar *et al.*, 1976), women (Miyake *et al.*, 1980) and monkeys (Neill *et al.*, 1977) indicate that the LH surge is associated with a rise in GnRH secretion. In sheep the primary signal for the generation of a surge in LH secretion appears to be a rise in ovarian oestrogen secretion (Goding *et al.*, 1969). In some species, e.g. humans (Liu & Yen, 1983), the positive feedback trigger may involve a combination of oestrogen and progesterone effects. This does not appear to be the case in sheep since there is no rise in plasma progesterone at this time (Baird & Scaramuzzi, 1976). If, in sheep, oestrogen alone is the trigger, how is the rise in GnRH secretion manifest and is there an increase in GnRH pulse frequency, amplitude or both?

To investigate this issue, the GnRH concentrations in portal blood were monitored in oestrogen-treated ovariectomized ewes before and during the LH surge (Clarke & Cummins, 1985). At the time of these studies we were attempting to sample the upper reaches of the portal network and, on some occasions this resulted in blood being sampled from the median eminence, leading to artefactually high GnRH readings in some of the controls and treated animals. In spite of this, an increase in GnRH pulse frequency was detected at the time of the LH surge; mean (± s.e.m.) inter-pulse intervals were 26·8 ± 9·8 min during the LH surge compared to 53·5 ± 8·7 min in controls. A striking example of this increase in the secretion of GnRH at the time of the LH surge is seen in Fig. 1(a). Schillo *et al.* (1985) found variable patterns of GnRH secretion associated with oestrogen-induced surges of LH secretion in ovariectomized ewes. GnRH was undetectable in the perfusates of 5 of 11 sheep sampled during the surge and, in those with detectable GnRH, 3 patterns of secretion were seen. In some ewes a large GnRH pulse was detected before the surge and a sustained rise was seen during the surge; in others there was a gradual rise in GnRH output and in some there was little change during the surge. Figure 1(b) shows results from one ewe in this study,

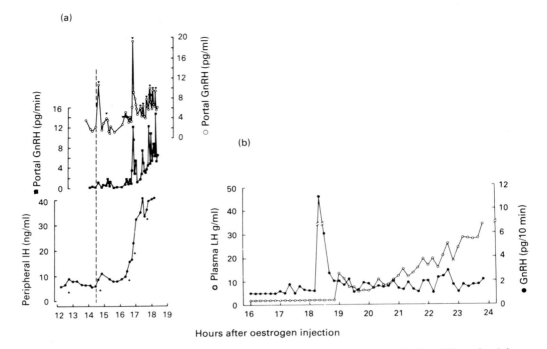

(a)

(b)

Hours after oestrogen injection

Fig. 1. GnRH secretion during the LH surge induced by oestrogen injection (50 µg, i.m.) in ovariectomized ewes. (a) Hypophysial portal plasma GnRH shown as pg/min (■) or pg/ml (○) and jugular venous LH (●) after the start of the LH surge (vertical broken line). GnRH pulses (▼) and LH pulses (▲) are indicated. Taken from Clarke & Cummins (1985), with permission. (b) GnRH (●) in 'push–pull' perfusates of the median eminence and jugular venous LH (○). Taken from Schillo *et al.* (1985) with permission.

illustrating the large GnRH pulse that was sometimes seen at the onset of the LH surge. In ovariectomized monkeys and again using the 'push–pull' technique, Levine *et al.* (1985) also found variable patterns of GnRH secretion associated with the oestrogen-induced LH surge. Is this between animal variation in GnRH secretion at the time of the LH surge an artefact of the collection technique, or can the surge in LH secretion be generated by more than one pattern of GnRH secretion? Preliminary data from cyclic ewes support the latter, for reasons discussed below.

The long-term ovariectomized animal may not be an appropriate model for the study of GnRH secretion during the LH surge because of the chronic deprivation of steroids and the inherently high pulse frequency. A much better model may be the anoestrous ewe, in which both of these problems are overcome, and in which an LH surge may be induced by oestrogen (Goding *et al.*, 1969). Accordingly, GnRH secretion was monitored in 6 anoestrous Corriedale ewes after injections of 50 µg oestradiol benzoate. In 4/6 ewes the LH surge was clearly associated with elevated GnRH secretion. In the remaining animals a large GnRH pulse was seen at the beginning of the LH surge (data not shown). At this point, it was apparent that oestrogen provoked a rise in GnRH secretion but the pattern of response differed in the 2 experimental ovariectomized and anoestrous ewes; in the former there was a rise in GnRH pulse frequency and in the latter a more pronounced 'surge' of secretion was evident. Which of these is more likely to represent the situation in the cyclic ewe?

To monitor GnRH secretion during cyclic LH surges preovulatory events were precipitated by injecting prostaglandin F-2α and causing luteal regression. GnRH secretion during the LH surge has been measured in 6 ewes (data to be published). In one ewe there was a large GnRH pulse at the start of the LH surge and very little activity thereafter, this being reminiscent of the results obtained in some of the animals studied by Schillo *et al.* (1985) and in one of our oestrogen-treated anoestrous ewes. In 2 other ewes the GnRH concentrations and pulse frequencies were similar to those of the follicular phase. In 3 ewes GnRH concentrations during the LH surge were clearly elevated above those in the follicular phase of the cycle. Even in ewes in which there was a clear elevation in GnRH secretion, the values were comparatively modest (up to 30 pg/ml) and well below those seen in 4/6 oestrogen-treated anoestrous ewes (> 100 pg/ml, data not shown). This may mean that the latter is inappropriate as a model for the cyclic LH surge.

Based on these studies of sheep and monkeys and using two different techniques and in 3 laboratories, it is apparent that the pattern of secretion of GnRH that is associated with the positive feedback surge in LH secretion may vary between animals. It is therefore extremely difficult to generalize about the neural mechanisms involved in the positive feedback phenomenon.

Long-term negative feedback

It is well accepted that a combination of oestrogen and progesterone exerts a tonic negative feedback effect on gonadotrophin secretion during the luteal phase of the oestrous cycle. Using ovariectomized Suffolk ewes, Goodman & Karsch (1980) dissected the separate effects of these two steroids and showed that oestrogen reduced LH pulse amplitude and progesterone reduced pulse frequency. In a subsequent study, Goodman *et al.* (1981) used a lower dose of progesterone which, when given alone, had no effect on LH secretion, but in combination with oestrogen was able to reduce pulse frequency. Treatment of ovariectomized Merino sheep with lower dosages of oestrogen and progesterone than those used by Goodman & Karsch (1980) had no effect on LH secretion during the breeding season unless the steroids were given in combination (Martin *et al.*, 1983).

When ovariectomized Corriedale ewes were treated with oestrogen or progesterone in the doses used previously by Goodman & Karsch (1980), either steroid, given either during the anoestrous or breeding season, completely abolished GnRH secretion (Karsch *et al.*, 1987). Two major conclusions were drawn from this. Firstly, both steroids are able to exert powerful negative feedback effects on GnRH secretion. Secondly, since these treatments abolished GnRH/LH secretion in Corriedale ewes but did not eliminate LH secretion in Suffolk ewes (Goodman & Karsch, 1980), the former breed appears more responsive to lower amounts of steroid than the latter. To identify alterations of GnRH frequency and amplitude that might result from either oestrogen or progesterone feedback, it will be necessary to use lower doses in our Corriedale sheep.

Neuromodulation of GnRH secretion

The generation of GnRH pulses depends upon central aminergic systems which have not been as exhaustively studied in the sheep as the rat. Jackson (1975, 1977) has shown that α-adrenergic and dopaminergic systems are involved, but since the aminergic systems have not been mapped in the sheep brain, we are unable to pinpoint the centres involved. In other species (e.g. rat) the aminergic afferent inputs to the hypothalamus arise from the brain stem nuclei and it is highly likely that this is also the case for the sheep.

There is strong evidence to suggest that opioid systems may interact with the aminergic elements that regulate GnRH secretion. It is significant that some endorphin- and dynorphin-containing neurones may concentrate oestrogen in the rat hypothalamus (Morrell *et al.*, 1985), and that opiate receptors are found in areas such as the locus coeruleus (origin of the A6 ascending fibres; Pert *et al.*, 1976), and the medial preoptic area (Hammer, 1985), the site where GnRH cell bodies are found (Lehman *et al.*, 1986).

I. J. Clarke

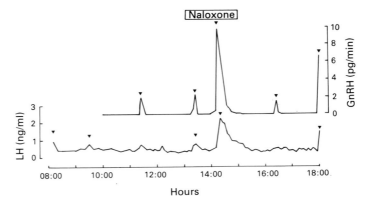

Hours

Fig. 2. GnRH secretion (pg/min) into hypophysial portal blood and jugular venous plasma concentrations of LH in a ewe given naloxone (40 mg/h) during the mid-luteal phase of the oestrous cycle. ▼ Pulses. (R. Horton & I. J. Clarke, unpublished data.)

In male (Schanbacher, 1982; Ebling & Lincoln, 1985) and female (Brooks *et al.*, 1986) sheep, opiate agonists can suppress LH secretion and opiate antagonists increase LH secretion. In females the effect of the antagonist naloxone is most marked during the luteal phase of the oestrous cycle (Brooks *et al.*, 1986), whereas in males the effect depends upon the presence of testosterone (Schanbacher, 1982). Studies in a variety of species (see Van Vugt, 1985, for references) have suggested that steroidal feedback on GnRH involves opioid systems. In the female, this seems most pertinent to the luteal phase of the ovarian cycle (Ferin *et al.*, 1984). There are, however, some problems with this hypothesis. Firstly, treatment with naloxone is not always followed by an LH response (Van Vugt *et al.*, 1983, 1984; Brooks *et al.*, 1986). Secondly, after continued treatment with agonist or antagonist the effect on LH is temporary (Ebling & Lincoln, 1985; Brooks *et al.*, 1986). This may mean that the opioid system is facilitatory rather than obligatory for feedback regulation.

Using the portal access model we have monitored the GnRH secretion that results from the treatment of mid-luteal phase ewes with naloxone. An example of the responses obtained is shown in Fig. 2. It is clear that the naloxone-induced GnRH pulse is of greater amplitude than that normally seen, and that the effect is short-lived. Further studies are now required to determine which sub-class(es) of opiate receptors is/are responsible for this response, and which neural centres are involved.

Conclusions

The development of methods for the measurement of GnRH secretion in conscious sheep has permitted definition of the feedback effects of ovarian steroids at the central level. In particular, the patterns of secretion associated with the preovulatory LH surge have been demonstrated. Further studies are required to determine the exact nature (change in frequency and/or amplitude) of the long-term negative feedback action of steroids. Feedback effects at the level of the pituitary gland have also been characterized using the hypothalamo–pituitary disconnected ewe (see Clarke, 1987). Having defined the sites of the various feedback effects and the resultant changes in GnRH and gonadotrophin secretion, we are now in a position to study the central mechanisms that mediate these feedback effects in sheep.

References

Baird, D.T. & Scaramuzzi, R.J. (1976) Changes in the secretion of ovarian steroids and pituitary luteinizing hormone in the peri-ovulatory period in the ewe: the effect of progesterone. *J. Endocr.* **70**, 237–245.

Brooks, A.N., Lamming, G.E., Lees, P.D. & Haynes, N.B. (1986). Opioid modulation of LH secretion in the ewe. *J. Reprod. Fert.* **76**, 693–708.

Caraty, A., Orgeur, P. & Thiery, J.-C. (1982) Mise en evidence d'une secretion pulsatile du LH-RH du sang porte hypophysaire chez la Brebis par une technique originale de prelevements multiples. *C. r. hebd. Séanc. Acad. Sci. Paris* **295**, 103–106.

Carmel, P.W., Araki, S. & Ferin, M. (1976) Pituitary stalk portal blood collection in rhesus monkeys: evidence for pulsatile release of gonadotrophin-releasing hormone (GnRH). *Endocrinology* **99**, 243–248.

Clarke, I.J. (1987) Ovarian feedback regulation of GnRH secretion and action. In *Endocrinology and Physiology of Reproduction*. Eds D. T. Armstrong, H. G. Friesen & P. C. K. Leung. Plenum, New York (in press).

Clarke, I.J. & Cummins, J.T. (1982) The temporal relationship between gonadotropin releasing hormone (GnRH) and luteinizing hormone (LH) secretion in ovariectomized ewes. *Endocrinology* **111**, 1737–1739.

Clarke, I.J. & Cummins, J.T. (1984) Direct pituitary effects of estrogen and progesterone on gonadotrophin secretion in the ovariectomized ewe. *Neuroendocrinology* **39**, 267–274.

Clarke, I.J. & Cummins, J.T. (1985) Increased GnRH pulse frequency associated with estrogen-induced LH surges in ovariectomized ewes. *Endocrinology* **116**, 2376–2383.

Clarke, I.J., Funder, J.W. & Findlay, J.K. (1982) Relationship between pituitary nuclear oestrogen receptors and the release of LH, FSH and prolactin in the ewe. *J. Reprod. Fert.* **64**, 355–362.

Coppings, R.J. & Malven, P.V. (1976) Biphasic effect of estradiol on mechanisms regulating LH release in ovariectomized sheep. *Neuroendocrinology* **21**, 146–156.

Diekman, M.A. & Malven, P.V. (1973) Effect of ovariectomy and estradiol on LH patterns in ewes. *J. Anim. Sci.* **37**, 562–567.

Ebling, F.J.P. & Lincoln, G.A. (1985) Endogenous opioids and the control of seasonal LH secretion in Soay rams. *J. Endocr.* **107**, 341–353.

Ferin, M., Van Vugt, D. & Wardlaw, S. (1984) The hypothalamic control of the menstrual cycle and the role of endogenous opioid peptides. *Recent Progr. Horm. Res.* **40**, 441–485.

Goding, J.R., Catt, K.J., Brown, J.M., Kaltenbach, C.C., Cumming, I.A. & Mole, B.J. (1969) Radioimmunoassay for ovine luteinizing hormone. Secretion of luteinizing hormone during estrus and following estrogen administration in the sheep. *Endocrinology* **85**, 133–142.

Goodman, R.L. & Karsch, F.J. (1980) Pulsatile secretion of luteinizing hormone: differential suppression by ovarian steroids. *Endocrinology* **107**, 1286–1290.

Goodman, R.L., Bittman, E.L., Foster, D.L. & Karsch, F.J. (1981) The endocrine basis of the synergistic suppression of luteinizing hormone by estradiol and progesterone. *Endocrinology* **109**, 1414–1417.

Hammer, R.P., Jr (1985) The sex hormone-dependent development of opiate receptors in the rat medial preoptic area. *Brain Res.* **360**, 65–74.

Jackson, G.L. (1975) Blockade of estrogen-induced release of luteinizing hormone by reserpine and potentiation of synthetic gonadotropin-releasing hormone-induced release of luteinizing hormone by estrogen in the ovariectomized ewe. *Endocrinology* **97**, 1300–1307.

Jackson, G.L. (1977) Effect of adrenergic blocking drugs on secretion of luteinizing hormone in the ovariectomized ewe. *Biol. Reprod.* **16**, 543–548.

Karsch, F.J., Cummins, J.T., Thomas, G.B. & Clarke, I.J. (1987) Steroid feedback inhibition of the pulsatile secretion of gonadotropin-releasing hormone in OVX ewes. *Biol. Reprod.* (In press).

Lehman, M.N., Robinson, J.E., Karsch, F.J. & Silverman, A.J. (1986) Immunocytochemical localization of luteinizing hormone-releasing hormone (LHRH) pathways in the sheep brain during anestrus and the mid-luteal phase of the estrous cycle. *J. comp. Neurol.* **244**, 19–35.

Levine, J.E. & Ramirez, V.D. (1980) *In vivo* release of luteinizing hormone-releasing hormone estimated with push-pull cannulae from the mediobasal hypothalamus of ovariectomized, steroid-primed rats. *Endocrinology* **107**, 1782–1790.

Levine, J.E., Pau, K.-Y., Ramirez, V.D. & Jackson, G.L. (1982) Simultaneous measurement of luteinizing hormone-releasing hormone and luteinizing hormone release in unanesthetized, ovariectomized sheep. *Endocrinology* **111**, 1449–1455.

Levine, J.E., Norman, R.L., Gliessman, P.M., Oyama, T.T., Bangsberg, D.R. & Spies, H.G. (1985) *In vivo* gonadotropin-releasing hormone release and serum luteinizing hormone measurements in ovariectomized, estrogen-treated rhesus monkeys. *Endocrinology* **117**, 711–721.

Liu, J.H. & Yen, S.S.C. (1983) Induction of midcycle gonadotropin surge by ovarian steroids in women: a critical evaluation. *J. clin. Endocr. Metab.* **37**, 797–802.

Martin, G.B., Scaramuzzi, R.J. & Henstridge, J.D. (1983) Effects of oestradiol, progesterone and androstenedione on the pulsatile secretion of luteinizing hormone in ovariectomized ewes during spring and autumn. *J. Endocr.* **96**, 181–193.

Miyake, A., Kawamura, Y., Aono, T. & Kurachi, K. (1980) Changes in plasma LRH during the normal menstrual cycle in women. *Acta endocr., Copenh.* **93**, 257–263.

Morrell, J.I., McGinty, J.F. & Pfaff, D.W. (1985) A subset of β-endorphin- or dynorphin-containing neurons in the medial basal hypothalamus accumulates estradiol. *Neuroendocrinology* **41**, 417–426.

Neill, J.D., Patton, J.M., Dailey, R.A., Tsou, R.C. & Tindall, G.T. (1977) Luteinizing hormone releasing hormone (LH-RH) in pituitary stalk blood of rhesus monkeys: relationship to levels of LH release. *Endocrinology* **101**, 430–434.

Pert, C.B., Kuhar, M.J. & Snyder, S.H. (1976) Opiate

receptor: autoradiographic localization in rat brain. *Proc. natn. Acad. Sci. U.S.A.* **73**, 3729–3733.

Reeves, J.J., Arimura, A. & Schally, A.V. (1971) Pituitary responsiveness to purified luteinizing hormone-releasing hormone (LH-RH) at various stages of the estrous cycle in sheep. *J. Anim. Sci.* **32**, 123–126.

Roche, J.F., Foster, D.L., Karsch, F.J., Cook, B. & Dziuk, P.J. (1970) Levels of luteinizing hormone in sera and pituitaries of ewes during the estrous cycle and anestrus. *Endocrinology* **86**, 568–572.

Sarkar, D.K. & Fink, G. (1980) Luteinizing hormone releasing factor in pituitary stalk plasma from long term ovariectomized rats: effects of steroids. *J. Endocr.* **86**, 511–524.

Sarkar, D.K., Chiappa, S.A., Fink, G. & Sherwood, N.M. (1976) Gonadotrophin-releasing hormone surge in pro-oestrous rats. *Nature, Lond.* **264**, 461–463.

Scaramuzzi, R.J. & Radford, H.M. (1983) Factors regulating ovulation rate in the ewe. *J. Reprod. Fert.* **69**, 353–367.

Schanbacher, B.D. (1982) Naxolone-provoked LH release in rams, wethers and wethers implanted with testosterone. *J. Androl.* **3**, 41–42.

Schillo, K.K., Leshin, L.S., Kuehl, D. & Jackson, G.L. (1985) Simultaneous measurement of luteinizing hormone-releasing hormone and luteinizing hormone during estradiol-induced luteinizing hormone surges in the ovariectomized ewe. *Biol. Reprod.* **33**, 644–652.

Sherwood, N.M. & Fink, G. (1980) Effect of ovariectomy and adrenalectomy on luteinizing hormone-releasing hormone in pituitary stalk blood from female rats. *Endocrinology* **106**, 363–367.

Thomas, G.B. (1983) A study of the pulsatile secretion of luteinizing hormone in the merino ewe. *Proc. Aust. Soc. Reprod. Biol.* **15**, Abstr. No. 47.

Van Vugt, D.A. (1985) Opioid regulation of prolactin and luteinizing hormone secretion. In *CRC Handbook of Pharmacologic Methodologies for the Study of the Neuroendocrine System* pp. 173–184. Eds R. W. Steger & A. Johns. CRC Press, Boca Raton.

Van Vugt, D.A., Bakst, G., Dyrenfurth, I. & Ferin, M. (1983) Naloxone stimulation of luteinizing hormone secretion in the female monkey: influence of endocrine and experimental conditions. *Endocrinology* **113**, 1858–1864.

Van Vugt, D.A., Lam, N.Y. & Ferin, M. (1984) Reduced frequency of pulsatile luteinizing hormone secretion in the luteal phase of the rhesus monkey involvement of endogenous opiates. *Endocrinology* **115**, 1095–1101.

Van Vugt, D.A., Diefenbach, W.D., Alston, E. & Ferin, M. (1986) Gonadotropin-releasing hormone pulses in third ventricular cerebrospinal fluid of ovariectomized rhesus monkeys: correlation with luteinizing hormone pulses. *Endocrinology* **117**, 1550–1558.

J. Reprod. Fert., Suppl. **34** (1987), 9–16

Searching for an inhibitory action of blood-borne β-endorphin on LH release

P. V. Malven

Department of Animal Sciences, Purdue University, West Lafayette, Indiana 47907, U.S.A.

Summary. Concentrations of β-endorphin were quantified in peripheral blood plasma of sheep by a radioimmunoassay that cross-reacted with β-lipotrophin. Plasma concentrations of β-endorphin increased abruptly after physical confinement, bacteraemia, and electroacupuncture treatment for induction of analgesia. In these experimental situations in which plasma concentrations of β-endorphin increased, plasma concentrations of LH often decreased. To test the hypothesis that increases in blood-borne β-endorphin actually caused the decrease in LH release, naloxone was administered to antagonize the opioid receptors at which blood-borne β-endorphin might act. In no case did administration of naloxone disrupt the temporal correlation between experimentally induced increases in plasma β-endorphin and decreases in plasma LH. It was concluded that the increases in blood-borne β-endorphin did not cause the decrease in LH release. Other research investigated whether β-endorphin might be delivered via blood from pituitary to hypothalamus in locally enriched concentrations. Even when pituitary release of β-endorphin was acutely stimulated, it was not possible to demonstrate retrograde delivery of β-endorphin to the hypothalamus without dilution in the systemic circulation. In conclusion, it is unlikely that blood-borne β-endorphin inhibits the release of LH, and β-endorphin should not be classified as a hormone until blood concentrations of the peptide can be shown to exert some effect at a location distant from its site of secretion.

Introduction

It has been 10 years since Li & Chung (1976) isolated and identified a unique peptide from an extract of pituitary glands from camels and suggested the name β-endorphin for this peptide. This same laboratory had reported earlier (Li *et al.*, 1975) that this extract from 500 camel pituitaries lacked β-lipotrophin which had been abundantly present in pituitary extracts from other species. Li *et al.* (1975) suggested that the camel may be unique with regard to β-lipotrophin or that β-lipotrophin was destroyed before they received the glands from the slaughterhouse in Iraq. Li & Chung (1976) observed that the amount of β-endorphin which they found in their extract of camel pituitaries approximated the amount of β-lipotrophin which one would expect to be present using calculations based on pituitaries from other domestic species. Within the same year, there appeared three reports showing that several identified, as well as unidentified, fragments of β-lipotrophin from camels, pigs and sheep possessed morphine-like activities in bioassays and in radioreceptor assays (Bradbury *et al.*, 1976; Cox *et al.*, 1976; Lazarus *et al.*, 1976). In these studies the peptide corresponding to residues 61 to 91 at the C-terminus of β-lipotrophin produced the greatest effects and, in agreement with the earlier suggestion of Li & Chung (1976), it has come to be universally called β-endorphin. Many advances have been made in our understanding of β-endorphin in the decade since its discovery. These include recognition that (1) β-endorphin is only one of many proteolytic cleavage products of β-lipotrophin and its precursor pro-opiomelanocortin (POMC), (2) β-endorphin is synthesized from POMC in the pars anterior and pars intermedia of the pituitary gland as well as in brain tissue, and (3) β-endorphin is secreted into blood concomitantly with adrenocorticotrophic hormone (ACTH), another product of POMC (Imura *et al.*, 1983; Akil *et al.*, 1984).

Blood-borne β-endorphin

Shortly after the discovery of β-endorphin, several laboratories developed radioimmunoassays to demonstrate its presence in blood (Guillemin *et al.*, 1977; Hollt *et al.*, 1978; Akil *et al.*, 1979). Concentrations of blood-borne β-endorphin have now been measured in many species, and in general they parallel changes in ACTH and/or adrenocortical steroids in blood (Vuolteenaho *et al.*, 1982; Orth *et al.*, 1982; DeSouza & Van Loon, 1983). We have validated methods for quantifying β-endorphin immunoreactivity in blood plasma of rats (Davis *et al.*, 1983), horses (Bossut *et al.*, 1983) and sheep (Leshin & Malven, 1984b). A major problem of β-endorphin quantification is that most antisera to β-endorphin also recognize the β-endorphin sequence when it is still part of the β-lipotrophin molecule. Therefore, a portion of the estimated β-endorphin in plasma represents β-lipotrophin unless a separation step precedes the immunoassay. Although the fluctuations of β-endorphin and β-lipotrophin appear parallel in most cases, the relative contribution of each molecule to the measured concentration is not always known for every plasma sample.

The source of most blood-borne β-endorphin is thought to be the pars anterior of the pituitary gland. Selective removal of the pars anterior leaving the pars intermedia and pars nervosa intact in rats markedly reduced basal concentrations of β-endorphin in plasma (Przewlocki *et al.*, 1982), but these authors also concluded that during stress the pars intermedia may secrete some β-endorphin into blood. However, the cells of the pars anterior and pars intermedia process POMC differently, and the production of β-endorphin appears to be regulated differently in the two tissues (Smyth & Zakarian, 1980; Shiomi & Akil, 1982; Millan *et al.*, 1982; Lim & Funder, 1983). Other possible sources of blood-borne β-endorphin include the brain (Zakarian & Smyth, 1982) and peripheral organs such as the gonads and other tissues (see Autelitano *et al.*, 1986, for review) because β-endorphin is synthesized in a variety of tissues, but their contribution to plasma β-endorphin relative to that of the pituitary gland is probably minor.

Is β-endorphin a hormone?

β-Endorphin produced in neural tissue can act in a neurocrine or paracrine manner to influence opioid receptors in adjacent neural elements. This paper will not address such actions of β-endorphin but will focus on the question of whether β-endorphin is a hormone. By definition, a hormone travels in blood to a site some distance from its site of secretion and modulates a physiological process. β-Endorphin is secreted into blood by the pituitary gland and concentrations of blood-borne β-endorphin fluctuate widely, both acutely and chronically. The only hormonal criterion which has been difficult to satisfy is modulation of a physiological process by blood-borne β-endorphin (Cox & Baizman, 1982). I propose to restrict the remainder of my remarks to this subject using data primarily from our laboratory and focussing mainly on the inhibition of release of pituitary LH. There is considerable evidence that in specific physiological states unidentified endogenous opioid peptides inhibit LH release in ewes (Malven *et al.*, 1984; Brooks *et al.*, 1986a, b; Gregg *et al.*, 1986) and other species (see Malven, 1986, for review). A large part of this evidence has been obtained through use of naloxone, an antagonist of opioid receptors. When administration of naloxone abruptly increases plasma concentrations of LH, it can only be concluded that naloxone antagonized the LH inhibitory action of some unidentified opioid. However, such an effect of naloxone, which readily passes the blood–brain barrier, does not establish whether a blood-borne opioid was antagonized or whether naloxone antagonized a locally produced opioid acting on the pituitary LH cells or on neurones which release LH-releasing hormone (LHRH) or which modulate LHRH neurones (Malven, 1986). To demonstrate an LH-inhibitory action of blood-borne β-endorphin, we have examined several situations in which there is an association between elevated plasma β-endorphin concentrations and depressed plasma LH values (i.e. a temporal correlation). To test whether this temporal correlation was in fact a cause–effect relationship, we have attempted to disrupt the association by administering naloxone. Our rationale was that systemic administration of naloxone would reach all the sites reached by blood-borne β-endorphin and should be able to block the interaction of this β-endorphin with those opioid receptors which mediate the suppression of LH release. If naloxone attenuates the suppression of LH release, we can conclude that blood-borne β-endorphin is a possible mediator of the effect. If naloxone has no effect on the suppressed levels of LH, it seems unlikely that the elevated blood concentrations of β-endorphin are responsible for the inhibition of LH release.

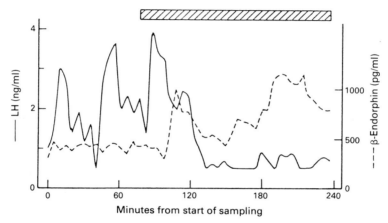

Fig. 1. Plasma profiles of LH and β-endorphin in one ovariectomized ewe. During the period of physical confinement (hatched bar), the ewe was maintained in an enclosed chamber (1·8 × 0·7 × 1·6 m) which moderately restricted movement and produced visual isolation identical to that described by Rasmussen & Malven (1983).

Inhibition of LH during confinement stress and bacteraemia

Episodic secretion of LH in ovariectomized ewes decreased after imposition of physical confinement of the ewes. As ewes habituated to confinement, the frequency and amplitude of LH peaks increased (Rasmussen & Malven, 1983). Although plasma concentrations of β-endorphin were not reported by Rasmussen & Malven (1983), one example, of an ovariectomized ewe exposed to a similar environment is presented in Fig. 1. The profile of plasma LH showed two distinct peaks before the onset of confinement and a third peak soon after confinement. At that point (about 105 min after start of sampling), the plasma concentration of β-endorphin increased abruptly, and there were no more peaks of LH for the remainder of the experiment. In this ewe, a temporal association clearly existed between the confinement-induced increase in β-endorphin and suppressed release of LH.

Another method of increasing plasma concentrations of β-endorphin is by induced bacteraemia. Intravenous administration of bacteria or their endotoxins provokes substantial increases in plasma β-endorphin (Carr *et al.*, 1982; Leshin & Malven, 1984a). Figure 2 shows 4 examples of ovariectomized ewes in which such a bacteria-induced increase of β-endorphin occurred. Plasma LH was fluctuating episodically during each sampling period as expected of ovariectomized ewes.

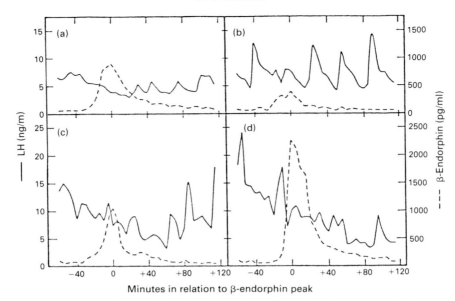

Fig. 2. Plasma profiles of LH and β-endorphin in 4 ovariectomized ewes plotted relative to the peak of β-endorphin in plasma. Each panel presents a sampling sequence wherein a transient increase in plasma β-endorphin occurred probably as a result of bacteraemia (taken from Exp. 1 of Leshin & Malven, 1984a)

The increases in plasma β-endorphin in Figs 2(a) and 2(b) occurred concurrently with lower than usual peaks of plasma LH. Profiles of LH in Figs 2(c) and 2(d) were less regular, but there appeared to be some slight inhibition of LH release during the periods of transiently elevated β-endorphin concentrations and plasma LH values continued to decline steadily during the period of the β-endorphin discharge into blood. These examples in Fig. 2 did not involve a standardized induction of bacteraemia and in some cases the bacteraemia was probably inadvertent (Leshin & Malven, 1984a).

To investigate more thoroughly the temporal association between decreased LH and increased β-endorphin during bacteraemia, Leshin & Malven (1984a) administered a standardized bolus of cultured *E. coli* to ovariectomized ewes. A variety of cardiovascular, thermoregulatory and hormonal effects were noted after *E. coli* infusion, and plasma LH decreased gradually while appearing to become less episodic. Mean plasma LH values during the 180-min period after *E. coli* infusion were lower ($P < 0.05$) than during the 180-min period before infusion (Fig. 3). To determine whether opioid receptors were required for this temporal association, naloxone was infused at a dosage of 2·5 mg/kg/h for the entire 6-h sampling period. Naloxone infusion had no significant effect on the mean LH profile either before or after *E. coli* infusion, and the profile in Fig. 3 represents an average of both saline and naloxone infusions. If our dosage of naloxone was adequate, these results indicate that the bacteria-induced increase in blood-borne β-endorphin was not acting via naloxone-sensitive opioid receptors to produce the concurrent suppression of plasma LH.

Effects of electroacupuncture on opioid–LH interactions

Electroacupuncture treatment of specific loci in male and female sheep confirmed the Chinese veterinary literature by quantifying the resulting analgesia (Bossut *et al.*, 1986). Application of this electroacupuncture treatment to castrated male sheep inhibited plasma LH (Malven *et al.*, 1984)

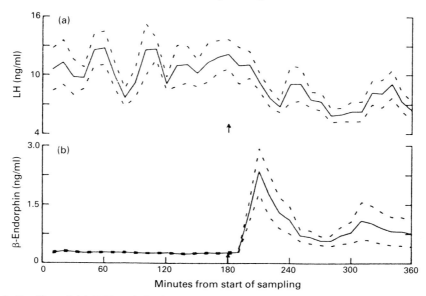

Fig. 3. Profiles of (a) LH and (b) β-endorphin in plasma during 6-h sampling. At 180 min (denoted by arrows), cultured *E. coli* were injected into one jugular vein. Trials in which saline or naloxone (2·5 mg/kg/h) were infused continuously were combined because there were no statistical differences between them. Solid lines depict means (*n* = 10) and broken lines depict the s.e.m. (Adapted from Leshin & Malven, 1984a.)

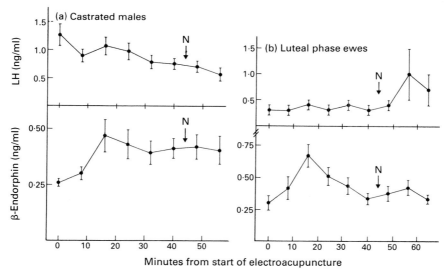

Fig. 4. Plasma profiles of LH and β-endorphin in (a) castrated male sheep and (b) cyclic luteal-phase ewes subjected to electroacupuncture for induction of cutaneous analgesia (redrawn from the data of Malven *et al.* (1984) and of Bossut *et al.* (1986)). Means are presented ± s.e.m. with 16–19 (a) and 8–12 (b) observations per mean. Time of naloxone injection is denoted by the arrow (N). LH values of luteal-phase ewes used as controls for the electroacupuncture treatment and which received the same dosage of naloxone have been presented elsewhere (Malven *et al.*, 1984).

and increased plasma concentrations of β-endorphin (Fig. 4a). This temporal correlation between depressed LH and elevated β-endorphin concentrations in plasma was similar to those observed during confinement and bacteraemia. Injection of naloxone (1·1 mg/kg) during the period of depressed plasma LH values failed to stimulate release of LH (Fig. 4a) suggesting that blood-borne β-endorphin was probably not responsible for the electroacupuncture-induced suppression of LH in those castrated males.

Malven et al. (1984) reported that naloxone administration (1·1 mg/kg) to cyclic female sheep abruptly stimulated the release of LH provided that the ewes were in the luteal phase of the oestrous cycle. This observation, as well as that of Trout & Malven (1984), indicates that during progesterone-induced suppression of LH release an unidentified endogenous opioid was inhibitory to LH release and that a moderate dosage of naloxone was able to antagonize this opioid thereby stimulating LH release. When luteal-phase ewes were treated with electroacupuncture for induction of cutaneous analgesia, injection of naloxone at the effective dosage for control ewes (Malven et al., 1984) did not increase plasma LH significantly (Fig. 4b). Electroacupuncture treatment of these ewes also increased plasma β-endorphin (Fig. 4b), similar to but not as sustained as its effect in castrated males (Fig. 4a). Therefore, electroacupuncture treatment of luteal-phase ewes increased plasma β-endorphin and blocked the ability of naloxone to stimulate the release of LH. It seems unlikely that blood-borne β-endorphin could be responsible for antagonizing the LH-releasing effect of naloxone during electroacupuncture treatment of luteal-phase ewes, but it remains a possibility that the dosage of naloxone, which was adequate to stimulate LH release during sham treatment, subsequently became inadequate during acupuncture-induced elevations of plasma β-endorphin.

Retrograde delivery of pituitary β-endorphin to the brain

The vascular anatomy of the pituitary gland is consistent with retrograde delivery of secreted pituitary hormones to the hypothalamus and brain without dilution in the systemic circulation (Page, 1982). Since opioid receptors are present in large numbers in the brain, there is a strong physiological rationale for delivery to the brain of enriched concentrations of pituitary-derived β-endorphin. Moreover, surgical removal of all or parts of the pituitary gland reduced the concentrations of β-endorphin in hypothalamic tissue (Ogawa et al., 1979; Przewlocki et al., 1982). To investigate the possibility of retrograde delivery of β-endorphin in ewes, Leshin & Malven (1985) implanted catheters for blood sampling into the dorsal longitudinal sagittal sinus very near the place where diencephalic venous effluent entered the sinus. Plasma concentrations of β-endorphin were measured in pairs of samples withdrawn simultaneously from the sagittal sinus and from the carotid artery at regular 20-sec intervals. Our rationale was that retrograde delivery of pituitary-derived β-endorphin to the brain should be reflected in more β-endorphin in the sagittal sinus samples than in the paired carotid artery sample.

Release of β-endorphin into blood was provoked by induced bacteraemia and less frequently by injection of naloxone and pentobarbitone sodium. There were very few series of blood samples over the entire experiment in which concentrations of β-endorphin were consistently greater in the sagittal sinus than in the carotid artery. We concluded that these few series were insufficient to establish retrograde delivery of pituitary-derived β-endorphin to the diencephalon. Rather, they may represent β-endorphin being secreted into blood from the diencephalon. In summary, our failure to demonstrate enriched concentrations of β-endorphin in diencephalic effluent blood of ewes did not support the hypothesis of retrograde delivery, but negative results such as these cannot exclude such a possibility. Since this experiment did not demonstrate locally enriched concentrations of β-endorphin in the vicinity of its CNS receptors, we cannot counteract the contention (Cox & Baizman, 1982) that blood concentrations of β-endorphin are generally much lower (picomolar range) than the levels needed to affect in-vitro opioid binding sites (nanomolar range).

Conclusions

Blood-borne β-endorphin satisfies several of the criteria for being a hormone, but demonstrating an action of β-endorphin on distant sites regulating LH release has not yet been achieved. Temporal correlations between elevated plasma concentrations of β-endorphin and suppressed release of LH were repeatedly demonstrated, but administration of naloxone at dosages which should antagonize all the opioid effects of β-endorphin failed to relieve the suppression of LH release. These results suggest that the temporal correlations were not reflecting a cause and effect relationship. Moreover, retrograde blood-borne delivery of enriched concentrations of β-endorphin to the brain could not be demonstrated. In summary, β-endorphin cannot yet be classified as a hormone on the basis of its effects on LH release because it fails to satisfy the criterion of acting at a location distant from its site of secretion.

This manuscript is published as Journal Paper No. 10,792 of the Indiana Agricultural Experiment Station. I thank L. S. Leshin and D. F. B. Bossut who conducted many of the experiments from which the present data were taken; and S. A. Haglof for technical assistance with all of the experiments.

References

Akil, H., Watson, S.J., Barchas, J.D. & Li, C.H. (1979) β-Endorphin immunoreactivity in rat and human blood: radioimmunoassay, comparative levels and physiological alterations. *Life Sci.* **24,** 1659–1666.

Akil, H., Watson, S.J., Young, E., Lewis, M.E., Khachaturian, H. & Walker, J.M. (1984) Endogenous opioids: biology and function. *Ann. Rev. Neurosci.* 7, 223–255.

Autelitano, D.J., Lolait, S.J., Smith, A.I. & Funder, J.W. (1986) Pregnancy-associated changes in ovarian immunoreactive β-endorphin in rats. *J. Endocr.* **108,** 343–350.

Bossut, D.F.B., Leshin, L.S. & Malven, P.V. (1983) Radioimmunological measurement of beta-endorphin in equine plasma. *Proc. Soc. exp. Biol. Med.* **173,** 454–459.

Bossut, D.F.B., Stromberg, M.W. & Malven, P.V. (1986) Electroacupuncture-induced analgesia in sheep: measurement of cutaneous pain thresholds and plasma concentrations of prolactin and β-endorphin immunoreactivity. *Am. J. vet. Res.* **47,** 669–676.

Bradbury, A.F., Smyth, D.G., Snell, C.R., Birdsall, N.J.M. & Hulme, E.C. (1976) C fragment of lipotrophin has a high affinity for brain opiate receptors. *Nature, Lond.* **260,** 793–795.

Brooks, A.N., Haynes, N.B., Yang, K. & Lamming, G.E. (1986a) Ovarian steroid involvement in endogenous opioid modulation of LH secretion in seasonally anoestrous mature ewes. *J. Reprod. Fert.* **76,** 709–715.

Brooks, A.N., Lamming, G.E., Lees, P.D. & Haynes, N.B. (1986b) Opioid modulation of LH secretion in the ewe. *J. Reprod. Fert.* **76,** 693–708.

Carr, D.B., Bergland, R., Hamilton, A., Blume, H., Kasting, N., Arnold, M. Martin, J.B. & Rosenblatt, M. (1982) Endotoxin-stimulated opioid peptide secretion: Two secretory pools and feedback control in vivo. *Science, N.Y.* **217,** 845–848.

Cox, B.M. & Baizman, E.R. (1982) Physiological functions of endorphins. In *Endorphins: Chemistry,* *Physiology, Pharmacology and Clinical Relevance,* pp. 113–196. Eds J. B. Malick & R. M. S. Bell. Marcel Dekker Inc., New York.

Cox, B.M., Goldstein, A. & Li, C.H. (1976) Opioid activity of a peptide, β-lipotropin (61–91), derived from β-lipotropin. *Proc. natn. Acad. Sci., U.S.A.* **73,** 1821–1823.

Davis, J.M., Lowy, M.T., Yim, G.K.W., Lamb, D.R. & Malven, P.V. (1983) Relationship between plasma concentrations of immunoreactive beta-endorphin and food intake in rats. *Peptides* **4,** 79–83.

DeSouza, E.B. & Van Loon, G.R. (1983) A triphasic pattern of parallel secretion of β-endorphin/β-lipotropin and ACTH after adrenalectomy in rats. *Am. J. Physiol.* **245,** E60–E66.

Gregg, D.W., Moss, G.E., Hudgens, R.E. & Malven, P.V. (1986) Endogenous opioid modulation of luteinizing hormone and prolactin secretion in postpartum ewes and cows. *J. Anim. Sci.* **63,** 838–847.

Guillemin, R., Vargo, T., Rossier, J., Minick, S., Ling, N., Rivier, C., Vale, W. & Bloom, F. (1977) β-Endorphin and adrenocorticotropin are secreted concomitantly by the pituitary gland. *Science, N.Y.* **197,** 1367–1369.

Hollt, V., Przewlocki, R. & Herz, A. (1978) β-Endorphin-like immunoreactivity in plasma, pituitaries and hypothalamus of rats following treatment with opiates. *Life Sci.* **23,** 1057–1066.

Imura, H., Nakai, K., Nakao, K., Oki, S., Tanaka, I., Jingami, H., Yoshimasa, T., Tsukada, T., Ikeda, Y., Suda, M. & Sakamoto, M. (1983) Biosynthesis and distribution of opioid peptides. *J. Endocr. Invest.* **6,** 139–149.

Lazarus, L.H., Ling, N. & Guillemin, R. (1976) β-lipotropin as a prohormone for the morphinomimetic peptides endorphins and enkephalin. *Proc. natn. Acad. Sci., U.S.A.* **73,** 2156–2159.

Leshin, L.S. & Malven, P.V. (1984a) Bacteremia-induced changes in pituitary hormone release and effect of naloxone. *Am. J. Physiol.* **247,** E585–E591.

Leshin, L.S. & Malven, P.V. (1984b) Radioimmunoassay for β-endorphin/β-lipotropin in unextracted plasma from sheep. *Domest. Anim. Endocr.* **1**, 175–188.

Leshin, L.S. & Malven, P.V. (1985) Continuous measurement of cerebral arteriovenous differences of beta-endorphin in sheep. *Neuroendocrinology* **40**, 120–128.

Li, C.H. & Chung, D. (1976) Isolation and structure of an untriakontapeptide with opiate activity from camel pituitary glands. *Proc. natn. Acad. Sci., U.S.A.* **73**, 1145–1148.

Li, C.H., Danho, W.O., Chung, D. & Rao, A.J. (1975) Isolation, characterization, and amino acid sequence of melanotropins from camel pituitary glands. *Biochemistry, N.Y.* **14**, 947–952.

Lim, A.T.W. & Funder, J.W. (1983) Stress-induced changes in plasma, pituitary and hypothalamic immunoreactive β-endorphin: effects of diurnal variation, adrenalectomy, corticosteroids, and opiate agonists and antagonists. *Neuroendocrinology* **36**, 225–234.

Malven, P.V. (1986) Inhibition of pituitary LH release resulting from endogenous opioid peptides. *Domest. Anim. Endocr.* **3**, 135–144.

Malven, P.V., Bossut, D.F.B. & Diekman, M.A. (1984) Effects of naloxone and electroacupuncture treatment on plasma concentrations of LH in sheep. *J. Endocr.* **101**, 75–80.

Millan, M.H., Millan, M.J. & Herz, A. (1982) Selective destruction of the ventral noradrenergic bundle but not of the locus coeruleus elevates plasma levels of β-endorphin immunoreactivity in rats. *Neurosci. Lett.* **29**, 269–273.

Ogawa, N., Panerai, A.E., Lee, S., Forsbach, G., Havlicek, V. & Friesen, H.G. (1979) β-Endorphin concentration in the brain of intact and hypophysectomized rats. *Life Sci.* **25**, 317–326.

Orth, D.N., Holscher, M.A., Wilson, M.G., Nicholson, W.E., Plue, R.E. & Mount, C.D. (1982) Equine Cushing's disease: plasma immunoreactive proopiolipomelanocortin peptide and cortisol levels basally and in response to diagnostic tests. *Endocrinology* **110**, 1430–1441.

Page, R.B. (1982) Pituitary blood flow. *Am. J. Physiol.* **243**, E427–E442.

Przewlocki, R., Millan, M.J., Gramsch, C., Millan, M.H. & Herz, A. (1982) The influence of selective adeno- and neurointermedio-hypophysectomy upon plasma and brain levels of β-endorphin and their response to stress in rats. *Brain Res.* **242**, 107–117.

Rasmussen, D.D. & Malven, P.V. (1983) Effects of confinement stress on episodic secretion of LH in ovariectomized ewes. *Neuroendocrinology* **36**, 392–396.

Shiomi, H. & Akil, H. (1982) Pulse-chase studies of the POMC/beta-endorphin system in the pituitary of acutely and chronically stressed rats. *Life Sci.* **31**, 2185–2188.

Smyth, D.G. & Zakarian, S. (1980) Selective processing of β-endorphin in regions of porcine pituitary. *Nature. Lond.* **288**, 613–615.

Trout, W.E. & Malven, P.V. (1984) Progesterone facilitates a naloxone-sensitive inhibition of LH release by endogenous opioids in sheep. *J. Anim. Sci.* **59** (Suppl. 1), 344, Abstr.

Vuolteenaho, O., Leppaluoto, J. & Mannisto, P. (1982) Rat plasma and hypothalamic β-endorphin levels fluctuate concomitantly with plasma corticosteroids during the day. *Acta physiol. scand.* **115**, 515–516.

Zakarian, S. & Smyth, D.G. (1982) Distribution of β-endorphin related peptides in rat pituitary and brain. *Biochem. J.* **202**, 561–571.

J. Reprod. Fert., Suppl. **34** (1987), 17–26

Neurotransmitter regulation of luteinizing hormone and prolactin secretion

R. A. Dailey*, D. R. Deaver‡ and R. L. Goodman†

Division of Animal and Veterinary Sciences and †Department of Physiology and Biophysics, West Virginia University, Morgantown, West Virginia 26506, and ‡Department of Animal and Dairy Science, The Pennsylvania State University, University Park, PA 16802, U.S.A.

Introduction

The importance of the hypothalamus for the regulation of gonadotrophin secretion has long been recognized (Harris, 1955). Extensive studies, primarily in laboratory rodents, have established that biogenic amines, particularly noradrenaline, dopamine and serotonin, influence secretion of gonadotrophins (LH and FSH) and prolactin (Weiner & Ganong, 1978). Most research on effects of biogenic amines in ruminants has been with sheep because of the relative ease of experiments, the reduced expense and the added dimension of effects of seasonality. Major techniques applied to determine areas of the hypothalamus involved include brain lesions, multiunit electrical recordings and measurements of tissue contents of neurotransmitters and of LHRH. The roles of neurotransmitters and releasing hormones have been assessed by measurement of their concentrations in the portal circulation and by exogenous treatment with these compounds or agonists and antagonists to them.

Pharmacological studies have been utilized to elucidate the neuroendocrine control of anterior pituitary function (Steger & Johns, 1985). Administration of authentic amines or congeners of these compounds provided some of the earliest evidence that the monoamines were involved with the control of LH release. Administration of the adrenergic agonist, dibenamine, blocked ovulation in rabbits and rats (Sawyer *et al.*, 1947; Everett *et al.*, 1949). In numerous studies exogenous dopamine, noradrenaline and serotonin influenced secretion of gonadotrophins and led to the development of various hypotheses regarding the roles of endogenous amines in the control of anterior pituitary function. As emphasized by Deaver & Dailey (1982), interpretation of results by this approach is influenced by the pharmacological agent used, the route of administration and the physiological state of the animal. One concern is the site of action of the biogenic amines, which do not penetrate the blood–brain barrier (Oldendorff, 1971) and therefore may act only at the median eminence or pituitary gland when given systemically. Other pharmacological agents do penetrate this physiological barrier and act centrally. Dose is important because at a low dose the agent may be an agonist and at high doses an antagonist. For example, at low doses bromocriptine is a dopamine agonist acting via D_1 receptors while at high doses it is an antagonist acting upon D_2 receptors (Kebabian & Calne, 1979). Also, a compound may antagonize the actions of more than one amine. Phentolamine, an α1- and α2-antagonist, blocks responses due to serotonin (Weiner, 1980).

The intent of this paper is to review studies that enhance our knowledge of the neuroendocrine regulation of the secretion of gonadotrophins and prolactin in domestic ruminants. We shall examine the experimental protocols used, cite pertinent results, and formulate working hypotheses that might serve for designing future studies. Of the various techniques described, only the pharmacological approach has been used to any extent with ruminants. Because of limited information concerning actions of biogenic amines on secretion of FSH, the review will concentrate on secretion of LH and prolactin.

Effects of biogenic amines on secretion of LH

Dopamine

Effects of dopamine on secretion of LH are dependent on the dose administered, season and whether the animals are ovariectomized or intact. Przekop *et al.* (1975) reported that intraventricular infusion of dopamine did not influence the release of LH in anoestrous ewes. Similarly, Deaver & Dailey (1982) reported that intravenous infusion of dopamine did not affect tonic or LHRH-induced release of LH during seasonal anoestrus. However, in ovariectomized ewes infusion of dopamine at the same time of year resulted in dose-dependent alterations of LH release (Table 1). At doses of 0·66 µg/kg/min increases in LH release were observed, while at a dose of 66·6 µg/kg/min release of LH was inhibited. The results with ovariectomized ewes could be used as evidence to support the contention that dopamine might influence the release of LH, and it is possible that the dopaminergic systems were operating at maximum capacity in intact ewes and that further stimulation of the post-synaptic pathways was simply ineffective. In fact, Meyer & Goodman (1985) provided evidence that dopaminergic systems inhibit the secretion of LH in anoestrous ewes (see below).

Different hypothalamic mechanisms might control the release of LH at different times of the year. During the luteal phase of the ovarian cycle in ewes, dopamine appears to exert primarily inhibitory effects on the secretion of LH. McNeilly (1980) suggested that dopamine might be important as an inhibitor of LH in ewes during the breeding season. Deaver & Dailey (1983), with the intravenous infusion of dopamine were able to inhibit for 30–36 h the increase in secretion of LH that normally occurs after the administration of PGF-2α. Despite this, the timing or magnitude of the preovulatory surge of LH was not affected.

Table 1. Effect of infusion of dopamine on mean concentrations of LH in plasma before and after LHRH (25 µg, i.m.) in ovariectomized ewes (adapted from Deaver & Dailey, 1982)

Dose of dopamine (µg/kg/min)	Mean conc. LH (ng/ml)	
	Before*	After†
0	12·2	220·0
0·06	17·1	307·1
0·66	14·4	124·1
6·66	−7·5	81·6

*Mean of 15-min samples for 2 h (linear, $P < 0.05$, quadratic, $P < 0.05$).

†Mean of 20-min samples for 4 h (linear, $P < 0.05$, quadratic, $P < 0.01$).

Noradrenaline

Intraventricular administration of noradrenaline induced preovulatory surges of LH in ewes late in the anoestrous season (Przekop *et al.*, 1975). However, other investigators have demonstrated that noradrenaline inhibited tonic secretion of LH in males and females. Intravenous infusion of noradrenaline (6·6 µg/kg/min) inhibited the secretion of LH in ovariectomized ewes during the anoestrous and breeding seasons. Release of LH after administration of LHRH was diminished by infusion of noradrenaline in intact anoestrous ewes (Deaver & Dailey, 1982) and in prepubertal beef heifers (Hardin & Randel, 1983). Infusion of noradrenaline reduced the frequencies but not amplitudes of pulses of LH (Table 2) after the induction of luteal regression with

Table 2. Effects of infusion of noradrenaline on pulsatile LH in pro-oestrous ewes (unpublished data)

Dose (μg/kg/min)	No. of ewes	Frequency (pulses/6 h)	Amplitude (ng/ml)
0	4	9·75[a]	1·11
0·06	5	9·60[a,b]	1·00
0·66	5	7·00[b]	0·84
6·66	6	3·00[c]	1·10

a *vs* b, $P < 0.10$; b *vs* c, $P < 0.01$.

PGF-2α (F. E. Barr, D. R. Deaver & R. A. Dailey, unpublished). In the male, intraventricular infusion of noradrenaline lowered the concentrations of LH in peripheral plasma (Riggs & Malven, 1974). In contrast, *in vitro*, noradrenaline stimulated LHRH-induced release of LH from pituitaries collected from wethers, and this effect appeared to be mediated by β-adrenergic receptors (Swartz & Moberg, 1986).

Serotonin

The indoleamines, serotonin and melatonin, have been demonstrated to influence pituitary release of LH. We will address only the effects of serotonin.

Domanski *et al.* (1975) reported that intraventricular administration or infusion of serotonin into the medial basal hypothalamus delayed the onset of the preovulatory release of LH in ewes. However, intravenous infusion of serotonin was shown to exert biphasic effects on the secretion of LH in ovariectomized ewes. In contrast to the biphasic effects of dopamine, low doses of serotonin lowered concentrations of LH and higher doses increased concentrations of LH in ovariectomized ewes (Deaver & Dailey, 1982). However, in intact anoestrous and breeding ewes, intravenous infusion of serotonin appeared only to increase the release of LH. Tonic secretion of LH was decreased by intraventricular administration of serotonin in wethers (Riggs & Malven, 1974) and by intracarotid infusion in ovariectomized cows (Mondragon *et al.*, 1986).

Effects of catecholamine antagonists on LH secretion

There have been few studies of the effects of catecholaminergic antagonists on gonadotrophin secretion in ruminants and all of this work has been done in sheep. Even with this limited information, it is clear that the actions of these antagonists vary markedly with the endocrine status of the animal.

Early work by Jackson (1977) demonstrated that both an α-adrenergic (phenoxybenzamine) and a dopaminergic (pimozide) antagonist suppressed secretion of LH in ovariectomized ewes. In contrast, only pimozide consistently blocked the oestradiol-induced LH surge (Jackson, 1977). The former observation has been confirmed using phenoxybenzamine (Meyer & Goodman, 1986) and fluphenazine, another dopaminergic antagonist (Goodman, 1985). However, relatively high doses of these antagonists were required. Lower doses of pimozide and phenoxybenzamine did not affect LH, even though they produced maximal effects on prolactin and blood pressure (Meyer & Goodman, 1985).

In contrast to their inhibitory effects in ovariectomized ewes, both phenoxybenzamine and pimozide increased pulsatile secretion of LH in intact anoestrous ewes (Meyer & Goodman, 1985). This stimulatory action was observed not only with the same doses that suppressed secretion of LH in ovariectomized animals, but also was evident with 10-fold lower doses. Other α-adrenergic

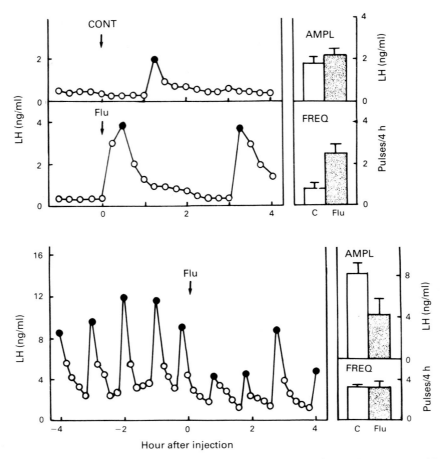

Fig. 1. Effect of a dopamine antagonist on LH secretion in ovary-intact (top two panels) and ovariectomized ewes (bottom panel). Fluphenazine (0·5 mg/kg) was injected i.v. as indicated by arrows, except in control ewes, which received vehicle. Left panels are representative LH pulse patterns. Right panels depict mean (+s.e.m.) LH pulse amplitude (AMPL) and frequency (FREQ). For intact animals data are from injected (shaded bars) and control (open bars) ewes; for ovariectomized animals data are before (open bars) and after (shaded bars) injection.

(dibenamine) and dopaminergic (fluphenazine) antagonists increased tonic secretion of LH in intact anoestrous ewes (Meyer & Goodman, 1986), whereas 6 other neurotransmitter antagonists which act on different receptor types did not (Meyer & Goodman, 1985). These results have led to the hypothesis that catecholaminergic neurones actively suppress LH release in intact anoestrous ewes.

The apparently contradictory actions of catecholamine antagonists in ewes with and without ovaries raised the possibility that at least two different catecholaminergic systems may be involved in regulating episodic secretion of LH. This possibility was strengthened by an analysis of the effects of these antagonists on LH pulse characteristics. As illustrated in Fig. 1, fluphenazine stimulates pulsatile secretion of LH in intact anoestrous ewes by increasing frequency, but inhibits LH in ovariectomized animals by decreasing amplitude. Similar differential effects of phenoxybenzamine on pulse frequency and amplitude in intact and ovariectomized ewes have also been observed (Meyer & Goodman, 1986). A second inference that can be drawn from these results was that the activity of these two systems depends upon the endocrine status of the ewe. For example, since

Table 3. Effect of dopamine antagonists on pulsatile LH secretion in luteal-phase ewes (data taken
from Goodman (1985), Meyer & Goodman (1985,
1986) and unpublished results)

Treatment	Frequency (pulses/4 h)	Amplitude (ng/ml)
Control	0·8	1·7
Pimozide		
0·08 mg/kg	1·4	2·6
0·8 mg/kg	0·4	1·0
Control	1·0	0·8
Domperidone		
0·3 mg/kg	2·4*	1·5*
0·1 mg/kg	2·4*	2·9*
Control	1·2	1·8
Fluphenazine		
0·05 mg/kg	3·2*	2·1
0·5 mg/kg	2·5*	2·2
Trifluoperazine		
0·5 mg/kg	1·6	2·7

*Significantly different from control values, $P < 0.05$.

the catecholaminergic neurones suppressing LH pulse frequency appeared to predominate in anoestrous ewes, an ovarian hormone, probably oestradiol (Goodman *et al.*, 1982), may stimulate the activity of this neural system and thereby suppress LH pulse frequency.

The administration of catecholamine antagonists to intact ewes during the breeding season has yielded confusing and contradictory results. Most of this work has been done during the luteal phase of the ovarian cycle, when LH pulse frequency is slow (Baird, 1978) due to the inhibitory actions of progesterone (Goodman & Karsch, 1980). At this time of the cycle, phenoxybenzamine had no effect on pulsatile secretion of LH (Meyer & Goodman, 1985). The effects of dopamine antagonists, however, were quite variable. Two dopaminergic antagonists, fluphenazine (Goodman, 1985) and domperidone (Deaver *et al.*, 1987) markedly increased LH pulse frequency, whereas others, including pimozide (Meyer & Goodman, 1985), did not (Table 3). The ability of two different dopaminergic antagonists to increase secretion of LH during the luteal phase raises the possibility that dopaminergic neural systems actively inhibit release of LH at this time of the cycle. If this is the case, however, it is unclear why other antagonists were not effective. It is always possible that inadequate doses of the ineffective antagonists were administered. This seems unlikely since, for example, the doses of pimozide used markedly increased prolactin secretion in luteal-phase animals and also increased LH in anoestrous ewes (Meyer & Goodman, 1985). Another possibility is that the variability in responses reflects the selectivity of a subclass of dopamine receptors. For example, the D_1 receptor for dopamine has a higher affinity for fluphenazine than for pimozide (Kebabian & Calne, 1979). However, none of the established dopamine receptors has a selectivity that accounts for the variability in the responses to the four dopamine antagonists tested to date. Further work with antagonists selective for only one of the dopamine receptors may be needed to resolve this issue.

Possible sites and mechanisms of action of biogenic amines

Knowing the site of action of exogenously administered amines is crucial for interpretation of the results from experiments outlined above. Studies using laboratory rodents have demonstrated that turnover rates for various neurotransmitters within discrete hypothalamic nuclei can change in opposite directions during the oestrous cycle (Barraclough & Wise, 1982). Implicit in these findings is the concept that assigning simply a positive or negative role to a particular transmitter may be inappropriate. When infusing catecholamines or indoleamines intravenously one would expect relatively low increases in the hypothalamic content of these substances due to the existence of the blood–brain barrier (Oldendorff, 1971). These biogenic amines could affect the stalk median eminence or any of the other circumventricular organs which have been shown to contain LHRH neurones, since these are outside the blood–brain barrier.

Reductions in LH pulse frequency after administration of biogenic amines are probably indicative of inhibition of LHRH release, which has been shown to be pulsatile in ewes (Levine et al., 1982; Clarke & Cummins, 1982). In vitro, noradrenaline suppressed the release of LHRH from the median eminence of pro-oestrous and ovariectomized cows (Zalesky et al., 1986). Weesner et al. (1986) concluded from in-vitro studies that a neurochemical from the medial basal hypothalamus may exert an inhibitory effect on LHRH release from the median eminence of mature bulls. These observations are further supported by the fact that antagonists to various catecholamines caused decreases in episodic discharges of LH. However, a direct effect of biogenic amines at the level of the pituitary cannot be ruled out.

Previously, investigators concluded that dopamine, noradrenaline and serotonin had no or only minor effects on the secretion of gonadotrophins at the level of the pituitary (McCann, 1983). These conclusions have been drawn mainly from observations that incubation of amines with pituitary cells in vitro failed to affect secretion of LH. In contrast, Dailey et al. (1978) demonstrated that in stalk-sectioned rabbits, dopamine inhibited LHRH-induced release of LH. In addition, concentrations of serotonin were greater in the anterior pituitary than in the median eminence of the cow (Piezzi et al., 1970), and Wheaton et al. (1972) suggested that serotonin might be released from the median eminence at oestrus in sheep. The possibility that amines might act directly on gonadotrophs is supported by identification of receptors for these transmitters in pituitaries from various species (Nunez et al., 1981; Johns et al., 1982).

Prolactin

We have suggested, as a working hypothesis, that gonadotrophin secretion is regulated by stimulatory (LHRH) and inhibitory (neurotransmitters) secretions from the hypothalamus; the effects are modified by hypothalamic changes with season; and neurotransmitters may exert control at the median eminence. A prototype can be drawn from studies on secretion of prolactin in sheep (Fig. 2a).

Concentrations of prolactin were reduced by treatment with 2-bromo-α-ergocryptine (Niswender, 1974) and ergocornine hydrogen maleinate (Louw et al., 1974), both dopamine receptor agonists. Treatment of sheep with phenoxybenzamine, an adrenergic receptor antagonist, or arginine, increased serum concentrations of prolactin (Davis & Borger, 1973). Treatment with both compounds resulted in a greater than additive effect, suggesting separate sites or mechanisms of action. From results in ewes with permanent electrodes placed in the median eminence, Malven (1975) concluded that a prolactin-releasing neurohormone existed in the anterior median eminence and that both this substance and an inhibitory compound were present in the posterior median eminence.

After surgical disconnection of the pituitary from the hypothalamus, there was only a transient increase of prolactin in sheep (Bryant et al., 1971; Clarke et al., 1983; Gust, 1985) and cattle (Anderson et al., 1980). These results contrast with the observations in non-ruminants (Anderson et al., 1982) and suggest that ruminants have a different mechanism for regulation of prolactin secretion. Because infusions of dopamine and noradrenaline but not serotonin (Deaver & Dailey, 1982, 1983; F. E. Barr, D. R. Deaver & R. A. Dailey, unpublished) lowered prolactin concentrations in serum, and dopamine but not noradrenaline (Swartz & Moberg, 1986) inhibited prolactin secretion in vitro, we suggest that dopamine acts directly on the lactotroph and that

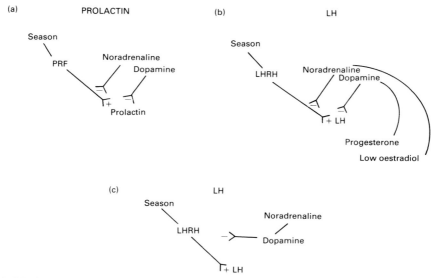

Fig. 2. Models of relationship of catecholamines and (a) prolactin and (b, c) LH secretion. See text for explanation.

noradrenaline acts either by stimulating dopamine release or by presynaptic inhibition of release of prolactin-releasing factor. The actions of dopamine and noradrenaline are the same in each season (Deaver & Dailey, 1982). Therefore, seasonal changes in concentrations of prolactin (Jackson & Davis, 1979) must be mediated by whether or not prolactin-releasing factor is being secreted.

LH

Male rats with lesions of the median eminence (Zeballos & McCann, 1977), female rhesus monkeys after pituitary stalk section (Frawley *et al.*, 1981), and ewes after stalk section (Mallory *et al.*, 1986) showed increased responsiveness to exogenous LHRH. These results suggest that inhibitory factors from the median eminence, as well as LHRH, regulate LH release. After stalk-sectioning, LH continued to be secreted in a pulsatile fashion with reduced amplitude of peaks (Gust *et al.*, 1987), which was restored by exogenous LHRH (Gust, 1985). Using stalk-sectioned ewes, Mallory *et al.* (1986) concluded that progesterone exerts inhibitory effects on LH at the hypothalamus. Based on results of treatment of luteal-phase ewes with dopamine or a dopamine receptor antagonist, progesterone might exert this inhibitory effect through a dopaminergic system (Deaver & Dailey, 1983; Deaver *et al.*, 1987) as proposed initially by McNeilly (1980). Using stalk-sectioned ewes, Gust (1985) concluded that the effects of oestrogen were at the hypothalamus, in contrast to the conclusion of Clarke & Cummins (1984) that effects of oestrogen and progesterone were at the pituitary. Thiery *et al.* (1978), using hypothalamic deafferentated ewes, concluded that the positive effect of oestrogen was in the hypothalamus.

We postulate (Fig. 2b) that, as with prolactin, dopamine directly inhibits gonadotrophin secretion and noradrenaline acts presynaptically or by modulating receptor binding to reduce effects of LHRH, as can be deduced from the results of Swartz & Moberg (1986). Progesterone would stimulate dopamine turnover while low oestradiol (anoestrus) concentrations would act through noradrenaline. As proposed by Domanski *et al.* (1980) inhibitory effects would be transmitted via serotonin to reduce the production of LHRH, which might account for seasonal effects. As an alternative (Fig. 2c), noradrenaline might stimulate the dopaminergic system and reduce secretion of LHRH.

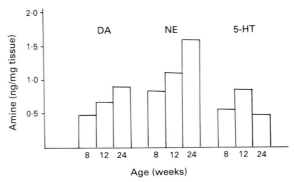

Fig. 3. Age-related changes in concentrations of dopamine (DA), noradrenaline (NE) and sero-tonin (5-HT) in the stalk-median eminence of Holstein bulls before puberty. The effect of age was significant for DA ($P < 0.01$; pooled s.e.m. = 0.022), NE ($P < 0.05$; pooled s.e.m. = 0.148) and 5-HT ($P < 0.05$; pooled s.e.m. = 0.035). Also, concentrations of DOPAC increased significantly with age ($P < 0.01$; data not shown).

Concluding remarks

One critical area in which few experiments have been conducted in ruminants is the determination of changes in activity of various neurotransmitter systems associated with the secretion of pituitary hormones. These types of studies are needed to support the hypotheses developed based on pharmacological data. Kizer *et al.* (1976) described the distribution of noradrenaline, dopamine, serotonin and enzymes necessary for the production of those and other neurotransmitters within the median eminence of steers. They concluded that only dopamine and acetylcholine were local-ized in the same regions as LHRH and might be involved with the regulation of LHRH release. Wheaton *et al.* (1975) attempted to determine effects of administration of oestradiol and L-DOPA on secretion of LH and concentrations of hypothalamic monoamines. Administration of L-DOPA blocked the occurrence of an oestradiol-induced LH surge in 3 of 4 ewes treated on Day 3 of the oestrous cycle. However, these investigators failed to detect differences in concentrations of hypothalamic dopamine, noradrenaline or serotonin after administration of oestradiol to another group of ewes on Day 3 of the cycle. This finding is not surprising considering the widely accepted view that steady-state conditions of production are often maintained even though activity of neurones is altered.

Techniques have become available, including high-performance liquid chromatography with electrochemical detectors (Mefford *et al.*, 1983) and mass spectrometry/gas chromatography (Smythe *et al.*, 1982), to determine tissue concentrations of monoamines and their metabolites. Results of these procedures agree well with the more widely accepted procedures for determining turnover rates and may provide a powerful tool for determining effects of a variety of pharmaco-logical agents known to alter the secretion of gonadotrophins and the function of various hypo-thalamic neurotransmitter systems. For example, using the liquid chromatography technique. Deaver *et al.* (1985) have reported that: (1) concentration of monoamines and their metabolites in the hypothalamus changed with age, (2) plasma concentrations of testosterone were positively correlated with concentrations of dopamine, DOPAC and noradrenaline within the stalk median eminence, and (3) when testicular development was inhibited by oestradiol, age-related changes in noradrenaline did not occur (Deaver & Peters, 1986). Based on these observations, we suggest that, in male and female ruminants, catecholamines probably mediate inhibitory effects of gonadal products on secretion of gonadotrophins.

References

Anderson, L.L., Hard, D.L., Carpenter, L.S., Awotwi, E.K. & Diekman, M.A. (1980) Prolactin secretion in hypophysial stalk-transected beef heifers during pregnancy and lactation. *J. Anim. Sci.* **51** (Suppl. 1), 255, Abstr. 364.

Anderson, L.L., Berardinelli, J.G., Malven, P.V. & Ford, J.J. (1982) Prolactin secretion after hypophysial stalk transection in pigs. *Endocrinology* **111**, 380–384.

Baird, D.T. (1978) Pulsatile secretion of LH and ovarian estradiol during the follicular phase of the sheep estrous cycle. *Biol. Reprod.* **18**, 359–364.

Barraclough, C.A. & Wise, P.M. (1982) The role of catecholamines in the regulation of pituitary luteinizing hormone and follicle stimulating hormone secretion. *Endocrine Rev.* **3**, 91–119.

Bryant, G.D., Greenwood, F.C., Kann, G., Martinet, J. & Denamur, R. (1971) Plasma prolactin in the oestrous cycle of the ewe: effect of pituitary stalk section. *J. Endocr.* **51**, 405–406.

Clarke, I.J. & Cummins, J.T. (1982) The temporal relationship between gonadotropin releasing hormone (GnRH) and luteinizing hormone (LH) secretion in ovariectomized ewes. *Endocrinology* **111**, 1737–1739.

Clarke, I.J. & Cummins, J.T. (1984) Direct pituitary effects of estrogen and progesterone on gonadotropin secretion in the ovariectomized ewe. *Neuroendocrinology* **39**, 267–274.

Clarke, I.J., Cummins, J.T. & de Kretser, D.M. (1983) Pituitary gland function after disconnection from direct hypothalamic influences in the sheep. *Neuroendocrinology* **36**, 376–384.

Dailey, R.A., Tsou, R.C., Tindall, G.T. & Neill, J.D. (1978) Direct hypophysial inhibition of luteinizing hormone release by dopamine in the rabbit. *Life Sci.* **22**, 1491–1498.

Davis, S.L. & Borger, M.L. (1973) Hypothalamic catecholamine effects on plasma levels of prolactin and growth hormone in sheep. *Endocrinology* **92**, 303–309.

Deaver, D.R. & Dailey, R.A. (1982) Effects of dopamine, norepinephrine and serotonin on plasma concentrations of luteinizing hormone and prolactin in ovariectomized and anestrous ewes. *Biol. Reprod.* **27**, 624–632.

Deaver, D.R. & Dailey, R.A. (1983) Effects of dopamine and serotonin on concentrations of luteinizing hormone and estradiol-17β in plasma of cycling ewes. *Biol. Reprod.* **28**, 870–877.

Deaver, D.R. & Peters, J.L. (1986) Effects of estradiol on development of the hypothalamic-pituitary-testicular axis in bulls. *Biol. Reprod.* **34** (Suppl. 1), 64, Abstr.

Deaver, D.R., Peters, J.L. & Rexroad, C.E., Jr (1985) Age-related changes in secretion of luteinizing hormone and metabolism of hypothalamic monoamines in prepubertal Holstein bull calves. *Biol. Reprod.* **32** (Suppl. 1), 210, Abstr. 245.

Deaver, D.R., Keisler, D.H. & Dailey, R.A. (1987) Effects of domperidone and thyrotropin-releasing hormone on secretion of luteinizing hormone and prolactin during the luteal phase and following induction of luteal regression in sheep. *Dom. Anim. Endocr.* (in press).

Domanski, E., Przekop, F., Skubiszewski, B. & Wolinska, E. (1975) The effect and site of action of indolamines on the hypothalamic centres involved in the control of LH-release and ovulation in sheep. *Neuroendocrinology* **17**, 265–273.

Domanski, E., Przekop, F. & Polkowska, J. (1980) Hypothalamic centres involved in the control of gonadotrophin secretion. *J. Reprod. Fert.* **58**, 493–499.

Everett, J.W., Sawyer, C.H. & Markee, J.E. (1949) A neurogenic timing factor in control of the ovulatory discharge of luteinizing hormone in the cyclic rat. *Endocrinology* **44**, 234–250.

Frawley, L.S., Dailey, R.A., Tindall, G.T. & Neill, J.D. (1981) Increased LH secretory response to LHRH after hypophyseal stalk-transection of monkeys. *Neuroendocrinology* **32**, 14–18.

Goodman, R.L. (1985) A role for dopamine in suppressing LH-pulse frequency during the luteal phase of the ovine estrous cycle? *Biol. Reprod.* **32** (Suppl. 1), 203, Abstr.

Goodman, R.L. & Karsch, F.J. (1980) Pulsatile secretion of luteinizing hormone: differential suppression by ovarian steroids. *Endocrinology* **107**, 1286–1290.

Goodman, R.L., Bittman, E.L., Foster, D.L. & Karsch, F.J. (1982) Alterations in the control of luteinizing hormone pulse frequency underlie the seasonal variation in estradiol negative feedback. *Biol. Reprod.* **27**, 580–589.

Gust, C.M. (1985) *Secretion of pituitary hormones in stalk-transected, ovariectomized ewes.* Ph.D. dissertation, West Virginia University, Morgantown.

Gust, C.M., Inskeep, E.K. & Dailey, R.A. (1987) Concentrations of luteinizing hormone, follicle stimulating hormone and prolactin following transection of the pituitary stalk in ovariectomized ewes. *Dom. Anim. Endocr.* (in press).

Hardin, D.R. & Randel, R.D. (1983) Effect of epinephrine, norepinephrine and (or) GnRH on serum LH in prepuberal beef heifers. *J. Anim. Sci.* **57**, 706–709.

Harris, G.W. (1955) *Neural Control of the Pituitary Gland.* Edward Arnold Ltd, London.

Jackson, G.L. (1977) Effect of adrenergic blocking drugs on secretion of luteinizing hormone in the ovariectomized ewe. *Biol. Reprod.* **16**, 543–548.

Jackson, L. & Davis, S.L. (1979) Comparison of luteinizing hormone and prolactin levels in cycling and anestrous ewes. *Neuroendocrinology* **28**, 256–263.

Johns, M.A., Azmita, E.C. & Kreiger, D.T. (1982) Specific in vitro uptake of serotonin by cells in the anterior pituitary of the rat. *Endocrinology* **110**, 754–760.

Kebabian, J.W. & Calne, D.B. (1979) Multiple receptors for dopamine. *Nature, Lond.* **277**, 93–96.

Kizer, J.S., Palkovits, M., Tappaz, M., Kebabian, J. & Brownstein, M.J. (1976) Distribution of releasing factors, biogenic amines and related enzymes in the bovine median eminence. *Endocrinology* **98**, 685–695.

Levine, J.E., Pau, K.Y., Ramirez, V.D. & Jackson, G.L. (1982) Simultaneous measurement of luteinizing hormone-releasing hormone and luteinizing hormone release in unanaesthetized, ovariectomized sheep. *Endocrinology* **111**, 1449–1455.

Louw, B.P., Lishman, A.W., Botha, W.A. & Baumgartner, J.P. (1974) Failure to demonstrate a role for the acute release of prolactin at oestrus in the ewe. *J. Reprod. Fert.* **40**, 455–458.

Mallory, D.S., Gust, C.M. & Dailey, R.A. (1986) Effects of pituitary stalk-transection and type of barrier on pituitary and luteal function during the estrous cycle of the ewe. *Dom. Anim. Endocr.* **3**, 253–259.

Malven, P.V. (1975) Immediate release of prolactin and biphasic effects of growth hormone release following electrical stimulation of the median eminence. *Endocrinology* **97**, 808–815.

McCann, S.M. (1983) Present status of LHRH: its physiology and pharmacology. In *Role of Peptides and Proteins in Control of Reproduction*, pp. 3–26. Eds S. M. McCann & D. S. Dhindsa. Elsevier Scientific, New York.

McNeilly, A.S. (1980) Prolactin and the control of gonadotrophin secretion in the female. *J. Reprod. Fert.* **58**, 537–549.

Mefford, I.N., Baker, T.L., Boehmen, R., Foutz, A.S., Ciaranello, R.D., Barchas, J.D. & Dement, W.C. (1983) Narcolepsy: biogenic deficits in an animal model. *Science, N.Y.* **220**, 629–632.

Meyer, S.L. & Goodman, R.L. (1985) Neurotransmitters involved in mediating the steroid dependent suppression of pulsatile luteinizing hormone secretion in anestrous ewes: effects of receptor antagonists. *Endocrinology* **116**, 2054–2061.

Meyer, S.L. & Goodman, R.L. (1986) Separate neural systems mediate the steroid-dependent and steroid-independent suppression of tonic luteinizing hormone secretion in the anestrous ewe. *Biol. Reprod.* **35**, 562–571.

Mondragon, A.M., Garverick, H.A., Mollett, T.A. & Elmore, R.G. (1986) Effect of serotonin on LH secretion patterns in ovariectomized cows. *Dom. Anim. Endocr.* **3**, 145–152.

Niswender, G.D. (1974) Influence of 2-Br-α-ergocryptine on serum levels of prolactin and the estrous cycle in sheep. *Endocrinology* **94**, 612–615.

Nunez, E.A., Gershon, M.D. & Silverman, A. (1981) Uptake of 5-hydroxytryptamine by gonadotrophs of the bat's pituitary: A combined immunocytochemical radioautographic analysis. *J. Histochem. Cytochem.* **29**, 1336–1346.

Oldendorff, W.H. (1971) Brain uptake of radiolabelled amino acids, amines and hexoses after arterial injection. *Am. J. Physiol.* **221**, 1629–1639.

Piezzi, R.S., Larin, F. & Wurtman, R.J. (1970) Serotonin, 5-hydroxyindoleacetic acid (5-HIAA), and monoamine oxidase in the bovine median eminence and pituitary gland. *Endocrinology* **86**, 1460–1462.

Przekop, F., Skubiszewski, B., Wolinska, E. & Domanski, E. (1975) The role of catecholamines in stimulating the release of pituitary ovulating hormone(s) in sheep. *Acta physiol. pol.* **26**, 433–437.

Riggs, B.L. & Malven, P.V. (1974) Effects of intraventricular infusion of serotonin, norepinephrine, and dopamine on spontaneous LH release in castrate male sheep. *Biol. Reprod.* **11**, 587–592.

Sawyer, C.H., Markee, J.E. & Hollingshead, W.H. (1947) Inhibition of ovulation in the rabbit by the adrenergic-blocking agent dibenamine. *Endocrinology* **41**, 395–402.

Smythe, G.A., Duncan, M.W., Bradshaw, J.E. & Cai, W.Y. (1982) Serotoninergic control of growth hormone secretion: Hypothalamic dopamine, norepinephrine, and serotonin levels and metabolism in three hyposomatotropic rat models and in normal rats. *Endocrinology* **110**, 376–383.

Steger, R.W. & Johns, A. (1985) *Handbook of Pharmacologic Methodologies for the Study of the Neuroendocrine System.* CRC Press, Inc., Boca Raton.

Swartz, S.R. & Moberg, G.P. (1986) Effects of epinephrine, norepinephrine, and dopamine on gonadotropin-releasing hormone-induced secretion of luteinizing hormone *in vitro*. *Endocrinology* **118**, 2425–2431.

Thiery, J.C., Pelletier, J. & Signoret, J.P. (1978) Effect of hypothalamic deafferentation on LH and sexual behavior in ovariectomized ewe under hormonally induced oestrous cycle. *Annls Biol. anim. Biochim. Biophys.* **18**, 1413–1426.

Weesner, G.D., McArthur, N.H. & Harms, P.G. (1986) An inhibitory effect of the medial basal hypothalamus on luteinizing hormone-releasing hormone release from the bovine median eminence *in vitro*. *J. Anim. Sci.* **63** (Suppl. 1), 56, Abstr.

Weiner, N. (1980) Drugs that inhibit adrenergic nerves and block adrenergic receptors. In *The Pharmacologic Basis of Therapeutics*, p. 176. Eds A. G. Gilman, L. S. Goodman & A. Gilman. Macmillan, New York.

Weiner, R.I. & Ganong, W.F. (1978) Role of brain monoamines and histamine in regulation of anterior pituitary secretion. *Physiol. Rev.* **58**, 905–976.

Wheaton, J.E., Martin, S.K., Swanson, L.V. & Stormshak, F. (1972) Changes in hypothalamic biogenic amines and serum LH in the ewe during the estrous cycle. *J. Anim. Sci.* **35**, 801–804.

Wheaton, J.E., Martin, S.K. & Stormshak, F. (1975) Estrogen and L-dihydroxyphenylalanine induced changes in hypothalamic biogenic amine levels and serum LH in the ewe. *J. Anim. Sci.* **40**, 1185–1191.

Zalesky, D.D., Procknor, M., McArthur, N.H. & Harms, P.G. (1986) Norepinephrine *in vitro* suppresses release of luteinizing hormone-releasing hormone (LHRH) from the bovine median eminence. *J. Anim. Sci.* **63** (Supp. 1), 55, Abstr.

Zeballos, G. & McCann, S.M. (1977) Increased responsiveness to LH-releasing hormone (LHRH) in rats with median eminence lesions. *Proc. Soc. exp. Biol. Med.* **154**, 242–245.

J. Reprod. Fert., Suppl. **34** (1987), 27–37

Regulation of the secretion of FSH in domestic ruminants

J. K. Findlay and I. J. Clarke

Medical Research Centre, Prince Henry's Hospital, St Kilda Road, Melbourne, Victoria 3004, Australia

Introduction

In the past 2–3 years there have been considerable advances in our understanding of the neuro-endocrine processes controlling pituitary function (Clarke, 1987) and the nature of steroidal and non-steroidal feedback regulation, particularly by the gonadal peptide, inhibin (Findlay, 1986). These advances offer new insights into the regulation of gonadotrophin secretion and gonadal function, and could have a considerable commercial impact in animal industry. In this review, we summarize these advances with respect to the control of FSH secretion in sheep and cattle.

Patterns of FSH secretion

The main characteristics of the patterns of FSH concentration in peripheral plasma of sheep (L'Hermite *et al.*, 1972; Salamonsen *et al.*, 1973) and cows (Walters & Schallenberger, 1984; Quirk & Fortune, 1986) are (a) relatively low levels, compared to the luteal phase, of FSH during the 1–2 days before the preovulatory surge of gonadotrophins, i.e. when LH and oestradiol secretion are increasing (Baird & McNeilly, 1981), (b) a peak of FSH secretion coincident with the preovulatory surge of LH, both of which are induced by oestradiol (Jonas *et al.*, 1973; Kesner & Convey, 1982), (c) a second surge of FSH commencing 18–24 h after onset of the preovulatory release, reaching peak concentrations just after ovulation when LH and oestradiol concentrations are both low (Baird & McNeilly, 1981) and (d) fluctuating concentrations of FSH during the remainder of the cycle. It has been estimated that the major sources of variation in peripheral concentrations of FSH during the oestrous cycle of the ewe are attributable to animals (75%) and time of sampling (16% between days and 9% between hours) (Findlay & Cumming, 1976). Unlike LH, FSH secretion in sheep is not pulsatile (see Clarke *et al.*, 1986), whereas there is evidence of pulsatility in cattle (Walters & Schallenberger, 1984).

Measurements of metabolic clearance (MCR) of FSH (Akbar *et al.*, 1974; Findlay & Cumming, 1976) support the notion that the changes in peripheral concentrations of FSH are due to changes in secretion by the pituitary gland and not removal of hormone from blood.

In addition to the cyclic changes there are other factors known to influence FSH secretion. For example, in sheep the concentrations of FSH have been related to breed (Lahlou-Kassi *et al.*, 1984; Bindon *et al.*, 1985a, b), sex (Findlay *et al.*, 1985), nutrition (Scaramuzzi & Radford, 1983), season (Findlay & Cumming, 1976; Findlay *et al.*, 1985) and age (Lee *et al.*, 1976; Bindon *et al.*, 1985b). In all of these instances, and especially during the ovarian cycle, it is apparent that the secretion of FSH can change independently of LH secretion, indicating that the control systems regulating the two pituitary gonadotrophins are not necessarily the same. There is evidence that the control processes regulating FSH, which are independent of those controlling LH, reside at the level of the hypothalamus and the pituitary gland and in the nature of the feedback influences of the gonads.

Hypothalamic control of FSH secretion

Disconnection of the direct hypothalamic influence on the pituitary gland of sheep (Clarke *et al.*, 1983) and active immunization against gonadotrophin-releasing hormone (GnRH) (McNeilly *et al.*, 1986) leads to a cessation of the secretion of FSH and LH. In these models, FSH secretion continued but at a decreased rate for days whereas LH secretion had stopped within 48–72 h. Anaesthesia of ewes at the time of luteolysis delayed the fall in FSH and the rise in LH in plasma and delayed the onset of the preovulatory surges of both gonadotrophins (Radford *et al.*, 1978). The hypothalamic influence on pituitary FSH secretion involves GnRH, which will stimulate secretion of both LH and FSH *in vivo* in sheep (Jonas *et al.*, 1973) and cattle (Kesner & Convey, 1982). The existence of a specific FSH-releasing hormone has been postulated (Bowers *et al.*, 1973) to account for the differential secretion rates and patterns of FSH and LH, but there has been no firm evidence to support the hypothesis. Recently, however, Ying *et al.* (1986a) reported that one of the cell growth factors, transforming growth factor-β (TGF-β), caused a dose-dependent stimulation of FSH secretion by rat pituitary cells *in vitro* when added at only attomolar concentrations. Subsequently, 2 polypeptides related to inhibin have been isolated from pig follicular fluid which are potent and selective stimulators of FSH release *in vitro*. One has been called follicle-stimulating hormone regulatory protein and is a homodimer of M_r 28 000 composed of two β_A subunits of inhibin (Vale *et al.*, 1986). The other heterodimer is called activin and consists of a β_A and a β_B subunit of inhibin (Ling *et al.*, 1986). The inhibin subunits share considerable sequence homology with TGF-β, itself a homodimer (Derynck *et al.*, 1985). It will be necessary to establish whether or not the β-dimers and/or TGF-β are secreted by the gonads or are produced locally in the hypothalamic–pituitary unit. A lack of glycosylation sites on the β-subunits of inhibin (see below) which would render the β-dimers very susceptible to peripheral metabolism, and the potential activation of TGF-β sites in many different tissues by circulating β-dimers argues in favour of local production of β-dimers by the hypothalamus and/or pituitary rather than secretion by the gonads.

Control of FSH secretion by higher brain centres can be shown in gonadectomized animals. Peripheral FSH concentrations are higher in sheep in the breeding season than in seasonal anoestrus and both FSH and LH are higher in males than in females (Findlay *et al.*, 1985). While these influences are likely to be mediated by the hypothalamus, the neuroendocrine processes involved are not understood. They are likely to be functional in intact animals as well because circulating FSH concentrations are reported to be higher in breeding than in anoestrous sheep (Findlay & Cumming, 1976; Lincoln & Short, 1980), an observation not shared by all investigators, however (Walton *et al.*, 1977). The sex difference in peripheral gonadotrophin concentrations could reflect greater GnRH pulse frequency and amplitude, greater sensitivity of the pituitary to releasing hormone stimulation or lower clearance of FSH in the males. The pituitary content of FSH and LH in intact rams is only 10–20% of that in cyclic ewes (Robertson *et al.*, 1984), supporting the idea of higher release rates of FSH in males.

There is evidence for a difference between the secretion of LH and FSH due to an inherent property of pituitary cells. Removal of the pituitary cells from the influence of the hypothalamus (Clarke *et al.*, 1983) or GnRH *in vivo* (McNeilly *et al.*, 1986) or *in vitro* (Miller *et al.*, 1977) results in a slower decrease in the secretion rate of FSH compared to LH that cannot be accounted for by a difference in the MCR of the two glycoproteins. It appears that FSH, unlike LH, does not require the presence of GnRH, either intermittently or continuously, to be released by the gonadotrophs. It is not known whether this is a property of FSH or the type of gonadotroph, since there are cells in sheep and cattle pituitary glands which contain only FSH immunoreactivity as well as those with immunoreactivity for LH and FSH (P. Somogyi, I. W. Chubb, I. J. Clarke, J. K. Findlay & R. Fischer-Colbie, unpublished observations). The availability of complementary DNA (cDNA) probes coding for messenger RNA for the gonadotrophin subunits and methods to study the synthesis and secretions of hormones by individual pituitary cells should contribute to answering those questions.

Gonadal control of FSH secretion

It is well established that the gonads can influence the secretion of both FSH and LH due to the inhibitory effects of gonadal steroids (Goodman & Karsch, 1981). In the female, there is also a positive feedback effect of oestradiol on FSH and LH secretion (Jonas *et al.*, 1973; Kesner & Convey, 1982) which is responsible for the preovulatory surge of gonadotrophins during the ovarian cycle.

The feedback control of gonadotrophin secretion by steroids can be manifested at both the hypothalamic and pituitary level. Clarke (1987) has defined two types of negative feedback effect of oestrogen in sheep, in addition to the positive effect mentioned above. The short-term negative feedback effect is seen when ovariectomized ewes are treated with oestradiol (Clarke *et al.*, 1982). Both LH and FSH concentrations fall in peripheral plasma. The effects of LH are apparent within 2 h, but shortly thereafter the secretion is increased to give a surge in LH. In contrast, the decrease in FSH is not observed until 4–5 h after oestrogen injection and continues for up to 24 h. The evidence available (Clarke, 1987) supports the view that this short-term negative effect is the result of a pituitary action of oestrogen and does not involve effects on GnRH secretion.

The long-term negative feedback effects are seen when oestradiol alone holds LH concentrations below those of untreated ovariectomized ewes (Diekman & Malven, 1973). Oestradiol in combination with progesterone in the correct sequence and doses can result in changes in LH concentration in ovariectomized ewes which mimic those of intact, cyclic animals (Goodman *et al.*, 1981), and the effects of these steroids are exerted via GnRH secretion (Clarke, 1987).

The situation with FSH is different. Although oestradiol has a long-term negative influence on FSH secretion, none of the steroid replacements maintained FSH concentrations in the physiological range, although some suppression was evident (Goodman *et al.*, 1981). Furthermore, the rise in oestradiol in the preovulatory period could not account for the low basal FSH concentrations observed in that period (Goodman *et al.*, 1981), and an effect of steroids could not account for the selective increase in FSH on Day 1 of the cycle. These results supported the hypothesis that an ovarian hormone other than oestradiol and progesterone inhibits FSH secretion, and that this putative hormone is selective for FSH because the secretion of LH can be accounted for by the combination of oestradiol and progesterone in the female. The remainder of this review will consider the evidence that the hormone is the gonadal peptide, inhibin.

Control of FSH secretion by inhibin

Inhibin is defined as a gonadal peptide which selectively suppresses the secretion of FSH. Although inhibin was discovered as a testicular hormone (see Baker *et al.*, 1982), it is also present in the ovary (Baker *et al.*, 1982) particularly in follicular fluid (cow: de Jong & Sharpe, 1976; sheep: Tsonis *et al.*, 1983) which has proved a useful source of inhibin activity for experimentation and purification of the hormone. In males, the activity is found in testicular extracts, rete testis fluid, testicular lymph, seminal plasma and in media from cultures of seminiferous tubules and Sertoli cells (Baker *et al.*, 1982).

The gonadal fluids and extracts containing inhibin activity have been shown to suppress FSH secretion selectively or preferentially in in-vivo and in-vitro test systems (Baker *et al.*, 1982). FSH secretion is suppressed by follicular fluid in ovariectomized (Cummins *et al.*, 1983; Findlay *et al.*, 1985) and intact (Miller *et al.*, 1982) ewes and castrated males (Findlay *et al.*, 1985), and by rete testis fluid in castrated rams (Hudson *et al.*, 1979). Bovine follicular fluid will also selectively suppress the plasma concentrations of FSH in cows (Quirk & Fortune, 1986). In many of these experiments the effects of inhibin activity *in vivo* have been dose dependent. Unlike oestradiol negative feedback (Goodman & Karsch, 1981), there does not appear to be an influence of season on the ability of inhibin to suppress FSH secretion (Findlay *et al.*, 1985). Both ovine follicular fluid

(Tsonis *et al.*, 1983, 1986) and rete testis fluid (Hudson *et al.*, 1979; Tsonis *et al.*, 1986) selectively suppress FSH in rat and sheep pituitary cell assays *in vitro*. The in-vivo post-castration suppression of FSH and the in-vitro pituitary cell culture assays are accepted as being the most specific and reliable assays at present available for inhibin (Hudson *et al.*, 1979; Baker *et al.*, 1982).

Isolation and characterization of inhibin

Inhibin has been a difficult molecule to isolate and purify and so there has been considerable confusion about its physico-chemical characteristics (see de Jong & Robertson, 1985; Findlay, 1986). This is due to a combination of factors including the source of inhibin, ill-defined bioassay systems, the lack of standard reference preparations, the nature of the molecule itself and the inability of standard purification procedures to separate inhibin. Recent reports of the purification of inhibin over 3000-fold from cow (Robertson *et al.*, 1985, 1986; Fukuda *et al.*, 1986), sheep (Leversha *et al.*, 1986) and pig (Miyamoto *et al.*, 1985; Ling *et al.*, 1985; Rivier *et al.*, 1985) follicular fluid have resulted in a convergence of opinion about the physico-chemical properties of inhibin in the ovary. Inhibin has an apparent molecular weight ranging from 31 000 to 100 000 of which the forms of M_r 31–32 000 and 55 000–65 000 have been more extensively purified. Two forms of bovine inhibin have been purified: one is a protein of M_r 58 000 which dissociated into subunits of 43 000 (α) and 15 000 (β) (Robertson *et al.*, 1985), and the other is a protein of M_r 31 000 consisting of α (M_r 20 000) and β (M_r 13–15 000) subunits (Robertson *et al.*, 1986; Fukuda *et al.*, 1986). It was shown that the form of M_r 31 000 could be generated by including an acid pH precipitation step in the purification procedure for cow follicular fluid (Robertson *et al.*, 1986), suggesting that cleavage of the α-subunit resulted in the smaller form. This was confirmed by cloning and sequence analysis of cDNA species coding for the 2 subunits of bovine inhibin (Forage *et al.*, 1986); the α subunit contained a cleavage site at residues 165, 166 which would produce the form of M_r 20 000 (αc fragment) from that of M_r 43 000 (Fig. 1).

Ovine inhibin has been isolated as forms of M_r 67 000 and 32 000, the latter being purified to homogeneity and consisting of α (M_r 20–21 000) and β (M_r 16 000) subunits (Leversha *et al.*, 1986). Similarly, porcine inhibin of M_r 32 000 has been purified and consists of α (M_r 20 000) and β (M_r 13 000) subunits (Miyamoto *et al.*, 1985; Ling *et al.*, 1985; Rivier *et al.*, 1985). Ling *et al.* (1985) reported 2 forms of porcine inhibin (A and B) based on heterogeneity of the amino acid sequence in the β subunit, which was subsequently confirmed by cloning and sequence analysis of cDNA (Mason *et al.*, 1985). This heterogeneity has not been observed with bovine or ovine inhibin. In all species examined, potential glycosylation sites appear to be confined to the α-subunit.

The results suggest that at least 3 genes are present in the ovary which code for the subunits of

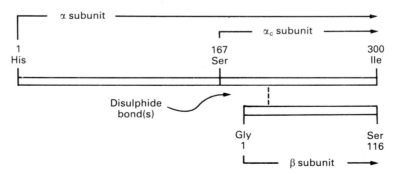

Fig. 1. Subunit structure of bovine inhibin based on sequence analysis of complementary DNA coding for the two subunits. DNA base sequence numbers are given at the beginning and end of each line and at base number 167 in the α subunit, which is a potential cleavage site. (After Forage *et al.*, 1986.)

inhibin, each of which share considerable sequence homology, indicative of a common ancestral gene (Mason *et al.*, 1985). Cloning and sequencing of cDNA coding for human inhibin (Mason *et al.*, 1986) have confirmed the similarity of the genes coding for inhibin amongst different species and underlined the extensive homology of the subunits between species. Interestingly, it is the β-subunit of inhibin which shares sequence homology with transforming growth factor-β (Derynck *et al.*, 1985) and Müllerian inhibiting substance (Cate *et al.*, 1986), raising the possibility that these are members of a family of proteins which influence cell function in similar manner. It is the β-subunits of inhibin which form homo- (Vale *et al.*, 1986) and hetero- (Ling *et al.*, 1986) dimers and are found in pig follicular fluid which have potent FSH stimulating activity *in vitro*. Although these β-dimers may be secreted by the gonad and act on the pituitary (see above), they may also have a local, intragonadal action. The fact that TGF-β enhanced and porcine inhibin of M_r 32 000 antagonized the ability of FSH to stimulate secretion of oestradiol by rat granulosa cells *in vitro* (Ying *et al.*, 1986b) underlines the possibility of a paracrine autocrine action of these inhibin-related polypeptides.

Published data on purification of inhibin from the male is not as advanced as that from ovarian follicular sources (see de Jong & Robertson, 1985; Findlay, 1986). Sairam *et al.* (1981) reported an inhibin of M_r 18 000 in bovine seminal plasma to be purified 500-fold. Activity of an M_r 20 000 protein has been isolated from ovine testicular extracts and claimed to be purified 770-fold (Moudgal *et al.*, 1985), but the assay used to monitor activity has been questioned (Hudson *et al.*, 1979).

The availability of pure bovine inhibin has allowed development of a radioimmunoassay which is applicable to cattle and humans (McLachlan *et al.*, 1986) and should ensure more rapid progress in our knowledge of the physico-chemical characteristics and physiology of inhibin. However, in view of the existence of dimers based on the β-subunits of inhibin (Vale *et al.*, 1986; Ling *et al.*, 1986) and their homology with TGF-β and Müllerian inhibiting substance (see above), it will be important to establish the specificity of the antibodies raised against inhibin. The antiserum to bovine inhibin (M_r 58 000) cross-reacts with the form of M_r 32 000 but not with the subunits (McLachlan *et al.*, 1986), suggesting that it is directed against conformational sites on the intact molecule and is therefore specific for inhibin.

Production and transport of inhibin

Inhibin activity is produced by the granulosa cells of the ovary (cow: Henderson & Franchimont, 1981) and the Sertoli cells of the testis (see Baker *et al.*, 1982). The inhibin content, and to a lesser extent the concentration, increases with follicular size and decreases with atresia of large follicles in sheep (Tsonis *et al.*, 1983) and cows (Henderson *et al.*, 1984). Treatment of hypo-physectomized ewes with PMSG increased the inhibin content of the ovary (Cahill *et al.*, 1985). However, treatment of bovine granulosa with gonadotrophin *in vitro* did not significantly alter inhibin production, although micromolar doses of androgen did cause a stimulation (Henderson & Franchimont, 1983).

The route by which inhibin leaves the testis to reach the peripheral circulation is unclear (see Baker *et al.*, 1982). It might pass through the seminiferous tubules and rete testis and be absorbed into the circulation at the head of the epididymis. It is also likely to enter the testicular capillaries from the lymph spaces surrounding the seminiferous tubules. Baker *et al.* (1982) have estimated that the daily output of inhibin in testicular lymph is at least equal to that in rete testis fluid of the ram and sufficient to maintain a tonic inhibition of FSH. They were unable to measure inhibin activity in testicular vein blood of rams, using the rat pituitary cell bioassay.

Using a similar bioassay, Findlay *et al.* (1986) measured inhibin in ovarian lymph of PMSG-treated ewes, but could not detect it in ovarian or jugular venous plasma. Destruction of visible follicles by electrocautery was followed by a rapid decline in secretion of inhibin in ovarian lymph. Findlay *et al.* (1986) concluded that inhibin could reach the peripheral circulation by the ovarian lymphatic system, particularly in the luteal phase when lymph flow rates are high. However, because venous outflow from the gonads is many times greater than lymph flow, relatively small concentrations of inhibin in venous plasma can make a major contribution to peripheral concentrations. Tsonis *et al.* (1986) have developed a more sensitive in-vitro assay using sheep pituitary cells which detected inhibin activity in ovarian venous plasma of a control and an FSH-treated ewe, and in peripheral plasma of the FSH-treated ewe. They showed higher concentrations of inhibin in the ovarian vein than in peripheral blood, particularly after treatment with FSH.

The half-life of ovine inhibin activity in peripheral blood of ovariectomized ewes was calculated to be around 45–50 min (Findlay *et al.*, 1986). In intact ewes, Tsonis *et al.* (1986) observed 2 half-life components of 18–24 and 50–60 min after injection of ovine follicular fluid. In both studies, the inhibin concentrations in plasma remained elevated after injection for up to 30 min before exponentially decaying, and in the intact ewes duration of the plateau was dose-dependent. This may reflect equilibration of inhibin to a carrier protein, dissociation of bound or polymeric forms, or saturation of a cleavage enzyme necessary for clearance or release from a polymeric form. It could also reflect the presence of the FSH-stimulating β-dimers (Vale *et al.*, 1986; Ling *et al.*, 1986) in ovine follicular fluid which would reduce the 'apparent' inhibitory effect of inhibin in the in-vitro bioassay until such time as they are cleared from the circulation. We suggest that the β-dimers would have a more rapid clearance than inhibin because of the lack of glycosylation sites on the β-subunits (see above).

Action of inhibin

That inhibin has a pituitary site of action is amply demonstrated by the effect of extracts and fluids containing inhibin activity (Baker *et al.*, 1982) and pure inhibins from cows, pigs and sheep (Robertson *et al.*, 1985, 1986; Miyamoto *et al.*, 1985; Rivier *et al.*, 1985; Ling *et al.*, 1985; Fukuda *et al.*, 1986; Leversha *et al.*, 1986) on FSH release by pituitary cell cultures *in vitro*. The effect of pure inhibin (forms of $M_r \sim 58\,000$ and $\sim 32\,000$) at lower doses is specific for FSH release in that there were no significant changes in LH or prolactin release *in vitro*. At higher doses, pure inhibin will suppress LH *in vitro* (Farnworth *et al.*, 1986). A direct pituitary site of action was recently demonstrated *in vivo* (Clarke *et al.*, 1986) when it was shown that treatment with ovine follicular fluid caused a 95% reduction in plasma FSH concentrations in ovariectomized ewes that had been subjected to hypothalamo–pituitary disconnection and in which gonadotrophin secretion had been reinstated with 250 ng pulses of GnRH every 2 h.

Whether or not inhibin has a hypothalamic action to limit secretion of GnRH is less certain. Intrahypothalamic and intraventricular injections of inhibin activity suppressed plasma FSH concentrations in laboratory rodents (see Clarke *et al.*, 1986, for references) but the mechanisms for these effects were not clear.

High doses of inhibin activity are known to inhibit both FSH and LH release induced by endogenous GnRH in ovariectomized ewes (Findlay *et al.*, 1985). Since LH pulses reflect GnRH secretion (Clarke & Cummins, 1982) it is possible to assess, indirectly, the effects of high doses of inhibin on hypothalamic function by examining LH pulse frequency in ovariectomized ewes. Similar studies in intact ewes (McNeilly, 1984; Wallace & McNeilly, 1985) are difficult to interpret because of confounding effects of oestrogens on GnRH pulse frequency (Clarke, 1987). Large doses of ovine follicular fluid given to ovariectomized ewes causes a significant reduction in plasma FSH and LH concentration, the latter probably due to a reduction in LH pulse amplitude in 3/4 animals, but no change in LH pulse frequency (Clarke *et al.*, 1986). The ability of follicular fluid containing inhibin activity to reduce LH pulse amplitude by a direct action on the pituitary was confirmed in experiments using ovariectomized ewes with hypothalamo–pituitary disconnection (Clarke *et al.*, 1986). Overall, these experiments suggest that inhibin does not act on hypothalamic or extra-hypothalamic centres that determine GnRH pulse frequency. The major effect of inhibin is to

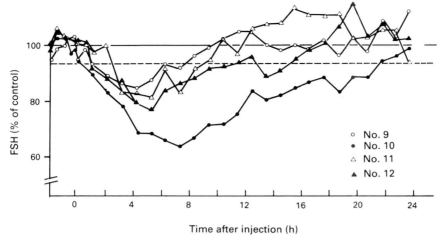

Fig. 2. Time course of suppression of plasma FSH in 4 ovariectomized ewes after intracarotid injection of 470 units of pure bovine inhibin (M_r 31 000). The broken line represents the range of FSH in plasma of 4 control ewes which received 2 ml serum from an hypophysectomized ewe. (Unpublished observations of J. K. Findlay, I. J. Clarke & D. M. Robertson.)

reduce FSH secretion by an action on the pituitary gland and there are components in follicular fluid, including inhibin, which can affect LH pulse amplitude, probably by a direct pituitary action.

It is possible to attribute the FSH-suppressing activity in follicular fluid to inhibin in the experiments described above because of the in-vitro actions of the purified inhibins (see above) and the recent demonstration that pure bovine inhibin (M_r 31 000) is also active *in vivo* (J. K. Findlay, I. J. Clarke & D. M. Robertson, unpublished observations). Intracarotid injection of pure inhibin into ovariectomized ewes resulted in a suppression of plasma FSH concentrations (Fig. 2) with a time course similar to that produced by bovine follicular fluid (J. K. Findlay & I. J. Clarke, unpublished), and with no significant effect on plasma LH. Unlike the responses of intact ewes (Miller *et al.*, 1982; Wallace & McNeilly, 1985) and cows (Quirk & Fortune, 1986), there was no rebound effect on plasma FSH in ovariectomized ewes after injection with bovine follicular fluid (J. K. Findlay & I. J. Clarke, unpublished) or pure bovine inhibin. This indicates that the rebound effect is associated with the gonads, presumably via a reduction in the endogenous secretion of oestradiol and inhibin as a result of decreasing gonadotrophin stimulation. There could also have been a direct effect of follicular fluid (Cahill *et al.*, 1985) and inhibin (Ying *et al.*, 1986b) on ovarian function in the intact ewes, to reduce gonadal feedback.

Repeated treatment of intact sheep with bovine follicular fluid resulted in a gradual onset of refractoriness of the FSH-suppressing activity (Wallace & McNeilly, 1985; Henderson *et al.*, 1986). This phenomenon is unlikely to be due to an immune response to bovine follicular fluid, given the relatively short duration of treatment (Wallace & McNeilly, 1985). It may be due to a long-term reduction in ovarian production of inhibin and oestradiol allowing a partial increase in pituitary FSH secretion (Henderson *et al.*, 1986). Or there could be a refractory (down-regulation?) effect at the pituitary level.

Inhibin and fecundity

FSH controls gamete production in the gonads and therefore manipulation of FSH secretion through an influence on inhibin production and action offers a means of improving fecundity in domestic animals.

One way to achieve this would be to select animals with low inhibin production. The highly fecund Booroola Merino ewe has a lower ovarian content of inhibin than control Merinos (Cummins *et al.*, 1983). This would lead to higher FSH secretion and the consequent effects on the gonads and suggests variation in the inhibin gene may be the basis of the increased fecundity in the Booroolas (Bindon *et al.*, 1985b). D'man sheep are also highly fecund and have high circulatory FSH concentrations (Lahlou-Kassi *et al.*, 1984) but their inhibin levels have not been reported.

A second method to increase fecundity would be to neutralize circulating inhibin by active or passive immunization. Cummins *et al.* (1986) observed an increase in ovulation rate after immunization of ewes with a fraction of bovine follicular fluid containing inhibin activity. The plasma FSH concentrations in the immunized animals were increased compared with controls (Al-Obaidi *et al.*, 1987), and plasma from immunized ewes inhibited the FSH-suppressing activity of bovine follicular fluid administered to ovariectomized ewes (Al-Obaidi *et al.*, 1986), suggesting that antibodies had been raised against inhibin. Now that inhibin has been purified and the cDNA cloned (see above) it should be possible to repeat these experiments with pure inhibin, its subunits and fragments to assess whether this is a practical method of improving fecundity.

Finally, it has been observed (Miller *et al.*, 1982; Wallace & McNeilly, 1985; Henderson *et al.*, 1986; Quirk & Fortune, 1986) that, 24–36 h after cessation of treatment of intact ewes and cows with follicular fluid, there is a rebound effect on plasma FSH such that concentrations exceed those in control animals. This treatment increased the proportion of multiple ovulating ewes (Wallace & McNeilly, 1985; Henderson *et al.*, 1986), an effect attributed to the increase in FSH. If similar increases in ovulation rate can be achieved using pure or synthesized inhibin during the late luteal phase of the cycle, this method may also improve fecundity of animals.

Conclusions

The secretion of FSH and LH has been shown to vary independently in physiological and experimental situations, indicating independent regulatory mechanisms for each gonadotrophin. The regulation of FSH can occur at three levels, i.e. hypothalamus and higher brain centres, the pituitary gland and the gonad. At the hypothalamus, specific FSH-releasing hormones have been postulated but never proven conclusively. The demonstration that homo- and heterodimers of the β-chain of inhibin have potent FSH-releasing activities offers new scope in that area. At the pituitary, there is evidence that the gonadotrophs which synthesize and secrete FSH have a much greater capacity to continue that function than those which produce LH. This indicates intra- and intercellular differences in the regulation of LH and FSH and offers an experimental challenge in the area of control of synthesis, processing and packaging of the gonadotrophin subunits. Finally, regulation of FSH occurs at the level of the gonad through production of agents such as inhibin which selectively influence FSH secretion.

Inhibin has now been isolated from follicular fluid of cows, sheep and pigs and purified to homogeneity. The cDNA coding for bovine, porcine and human inhibin messenger RNA has been cloned and sequenced. Inhibin is a heterodimer in which a larger, glycosylated α-chain (M_r 20 000–43 000) is joined to a smaller β-chain (M_r 13–15 000) by disulphide bridges. Pure inhibin is biologically active *in vivo* and *in vitro* and radioimmunoassays for measuring inhibin have been developed. Collectively, this should provide an impetus to describe the physiology of inhibin in the next few years.

There are already some surprises. Inhibin shares amino acid sequence homology with the cell regulator, TGF-β, and with Müllerian inhibiting substance, suggesting that these peptides belong to a gene family whose products have cell regulatory properties at endocrine, paracrine and/or autocrine levels. In the case of inhibin, reassembly of its subunits to form β-dimers can reverse the activity from inhibitory to stimulatory, indicating that the point of assembly of the subunits is a key control point for subsequent biological activity. Manipulation of inhibin and its β-chain dimers by analogue treatment, immunization or genetic selection and engineering will provide new insights into the control of reproduction and may offer a means of regulating fecundity of domestic animals.

We thank Jenny Judd and Robyn Redman for their help in producing the manuscript. The authors were supported by grants from the National Health and Medical Research Council of Australia and the Australian Meat Research Committee.

References

Akbar, A.M., Nett, T.M. & Niswender, G.D. (1974) Metabolic clearance and secretion rates of gonadotropins at different stages of the estrous cycle in ewes. *Endocrinology* **94**, 1318–1324.

Al-Obaidi, S.A.R., Bindon, B.M., Hillard, M.A., O'Shea, T. & Piper, L.R. (1986) Suppression of ovine plasma FSH by bovine follicular fluid: neutralization by plasma from ewes immunized against an inhibin-enriched preparation from bovine follicular fluid. *J. Endocr.* **111**, 1–5.

Al-Obaidi, S.A.R., Bindon, B.M., Findlay, J.K., Hillard, M.A. & O'Shea, T. (1987) Plasma follicle stimulating hormone in Merino ewes immunized with an inhibin-enriched fraction from bovine follicular fluid. *Anim. Reprod. Sci.* (in press).

Baird, D.T. & McNeilly, A.S. (1981) Gonadotrophic control of follicular development and function during the oestrous cycle of the ewe. *J. Reprod. Fert., Suppl.* **30**, 119–133.

Baker, H.W.G., Eddie, L.W., Higginson, R.E., Hudson, B. & Niall, H.D. (1982) Clinical context, neuroendocrine relationships, and nature of inhibin in males and females. In *Clinical Neuroendocrinology*, Vol. II, Chap. 7, pp. 283–330. Eds G. M. Besser & L. Martini. Academic Press, New York.

Bindon, B.M., Findlay, J.K. & Piper, L.R. (1985a) Plasma FSH and LH in prepubertal Booroola ewe lambs. *Aust. J. biol. Sci.* **38**, 215–220.

Bindon, B.M., Piper, L.R., Cummins, L.J., O'Shea, T., Hillard, M.A., Findlay, J.K. & Robertson, D.M. (1985b) Reproductive endocrinology of prolific sheep: studies of the Booroola merino. In *Genetics of Reproduction in Sheep*, Ch. 23, pp. 217–235. Eds R. B. Land & D. W. Robinson. Butterworths, London.

Bowers, C.Y., Currie, B.L., Johansson, K.M.G. & Folkers, K. (1973) Biological evidence that separate hypothalamic hormones release the follicle stimulating and luteinizing hormones. *Biochem. Biophys. Res. Commun.* **50**, 20–26.

Cahill, L.P., Clarke, I.J., Cummins, J.T., Driancourt, M.A., Carson, R.S. & Findlay, J.K. (1985) An inhibitory effect at the ovarian level of ovine follicular fluid on PMSG-induced folliculogenesis in hypophysectomized ewes. In *Proc. 5th Ovarian Workshop*, pp. 35–38. Eds D. Toft & R. J. Ryan. Plenum Press, New York.

Cate, R.L., Mattaliano, R.J., Hession, C., Tizard, R., Farber, N.M., Cheung, A., Ninfa, E.G., Frey, A.Z., Gash, D.J., Chow, E.P., Fisher, R.A., Bertonis, J.M., Torres, G., Wallner, B.P., Ramachandran, K.L.,

Ragin, R.C., Manganaro, T.F., McLaughlin, D.T. & Donahoe, P.K. (1986) Isolation of the bovine and human genes for Mullerian inhibiting substance and expression of the human gene in animal cells. *Cell* **45**, 685–698.

Clarke, I.J. (1987) Control of GnRH secretion. *J. Reprod. Fert., Suppl.* **34**, 1–8.

Clarke, I.J. & Cummins, J.T. (1982) The temporal relationship between gonadotrophin-releasing hormone (GnRH) and luteinizing hormone (LH) secretion in ovariectomized ewes. *Endocrinology* **116**, 2376–2383.

Clarke, I.J., Funder, J.W. & Findlay, J.K. (1982) Relationship between pituitary nuclear oestrogen receptors and the release of LH, FSH and prolactin in the ewes. *J. Reprod. Fert.* **64**, 355–362.

Clarke, I.J., Cummins, J.T. & de Kretser, D.M. (1983) Pituitary gland function after disconnection from direct hypothalamic influences in the sheep. *Neuroendocrinology* **36**, 376–384.

Clarke, I.J., Findlay, J.K., Cummins, J.T. & Ewens, W.J. (1986) Effects of ovine follicular fluid on plasma LH and FSH secretion in ovariectomized ewes to indicate the site of action of inhibin. *J. Reprod. Fert.* **77**, 575–585.

Cummins, L.J., O'Shea, T., Bindon, B.M., Lee, V.W.K. & Findlay, J.K. (1983) Ovarian inhibin content and sensitivity to inhibin in Booroola and control strain Merino ewes. *J. Reprod. Fert.* **67**, 1–7.

Cummins, L.J., O'Shea, T., Al-Obaidi, S.A.R., Bindon, B.M. & Findlay, J.K. (1986) Increase in ovulation rate after immunization of Merino ewes with a fraction of bovine follicular fluid containing inhibin activity. *J. Reprod. Fert.* **77**, 365–372.

de Jong, F.H. & Robertson, D.M. (1985) Inhibin: 1985 update on action and purification. *Molec. cell. Endocr.* **42**, 95–103.

de Jong, F.H. & Sharpe, R.M. (1976) Evidence for inhibin-like activity in bovine follicular fluid. *Nature, Lond.* **263**, 71–72.

Derynck, R., Jarrett, J.A., Chen, E.Y., Eaton, D.W., Bell, J.R., Assoian, R.K., Roberts, A.H., Sporn, M.B. & Goeddel, D.V. (1985) Human transforming growth factor-β complementary DNA sequence and expression in normal and transformed cells. *Nature, Lond.* **316**, 701–705.

Diekman, M.A. & Malven, P.V. (1973) Effect of ovariectomy and estradiol on LH patterns in ewes. *J. Anim. Sci.* **37**, 562–567.

Farnworth, P.G., Robertson, D.M., de Kretser, D.M. &

Burger, H.G. (1986) Studies on the mechanism of action of purified bovine inhibin on pituitary gonadotrophs. *Proc. Endocr. Soc. Aust., Brisbane*, Abstr.

Findlay, J.K. (1986) The nature of inhibin and its use in the regulation of fertility and diagnosis of infertility. *Fert. Steril.* **46,** 770–783.

Findlay, J.K. & Cumming, I.A. (1976) FSH in the ewe: effects of season, liveweight and plane of nutrition on plasma FSH and ovulation rate. *Biol. Reprod.* **15,** 335–342.

Findlay, J.K., Gill, T.W. & Doughton, B.W. (1985) Influence of season and sex on the inhibitory effect of ovine follicular fluid on plasma gonadotrophins in gonadectomized sheep. *J. Reprod. Fert.* **73,** 329–335.

Findlay, J.K., Tsonis, C.G., Staples, L.D. & Cahill, R.N.P. (1986) Inhibin secretion by the sheep ovary. *J. Reprod. Fert.* **76,** 751–761.

Forage, R.G., Ring, J.M., Brown, R.W., McInerney, B.V., Cobon, G.S., Gregson, R.P., Robertson, D.M., Morgan, F.J., Hearn, M.T.W., Findlay, J.K., Wettenhall, R.E.H., Burger, H.G. & de Kretser, D.M. (1986) Cloning and sequence analysis of cDNA species coding for the two subunits of inhibin from bovine follicular fluid. *Proc. natn. Acad. Sci. U.S.A.* **83,** 3091–3095.

Fukuda, M., Miyamoto, K., Hasegawa, Y., Nomura, M., Igarashi, M., Kangawa, K. & Matsuo, H. (1986) Isolation of bovine follicular fluid inhibin of about 32 kDa. *Molec. cell. Endocr.* **44,** 55–60.

Goodman, R.L. & Karsch, R.J. (1981) A critique of the evidence on the importance of steroid feedback to seasonal changes in gonadotrophin secretion. *J. Reprod. Fert., Suppl.* **30,** 1–13.

Goodman, R.L., Pickover, S.M. & Karsch, F.J. (1981) Ovarian feedback control of follicle-stimulatory hormone in the ewe: evidence for selective suppression. *Endocrinology* **108,** 772–777.

Henderson, K.M. & Franchimont, P. (1981) Regulation of inhibin production by bovine ovarian cells *in vitro. J. Reprod. Fert.* **63,** 431–442.

Henderson, K.M. & Franchimont, P. (1983) Inhibin production by bovine ovarian tissue *in vitro* and its regulation by androgens. *J. Reprod. Fert.* **67,** 291–298.

Henderson, K.M., Franchimont, P., Charlet-Renard, Ch. & McNatty, K.P. (1984) Effect of follicular atresia on inhibin production by bovine granulosa cells *in vitro* and inhibin concentrations in follicular fluid. *J. Reprod. Fert.* **72,** 1–8.

Henderson, K.M., Prisk, M.D., Hudson, N., Ball, K., McNatty, K.P., Lun, S., Heath, D., Kieboom, L. & McDiarmid, J. (1986) Use of bovine follicular fluid to increase ovulation rate or prevent ovulation in sheep. *J. Reprod. Fert.* **76,** 623–635.

Hudson, B., Baker, H.W.G., Eddie, L.W., Higginson, R.E., Burger, H.G., de Kretser, D.M., Dobos, M. & Lee, V.W.K. (1979) Bioassays of inhibin: a critical review. *J. Reprod. Fert., Suppl.* **26,** 17–29.

Jonas, H.A., Salamonsen, L.A., Burger, H.G., Chamley, W.A., Cumming, I.A., Findlay, J.K. & Goding, J.R. (1973) Release of FSH after administration of gonadotrophin-releasing hormones or estradiol to the anestrous ewe. *Endocrinology* **92,** 862–865.

Kesner, J.S. & Convey, E.M. (1982) Interaction of estradiol and luteinizing hormone releasing hormone

on follicle stimulating hormone release in cattle. *J. Anim. Sci.* **54,** 817–821.

Lahlou-Kassi, A., Schams, D. & Glatzel, P. (1984) Plasma gonadotrophin concentrations during the oestrous cycle and after ovariectomy in two breeds of sheep with low and high fecundity. *J. Reprod. Fert.* **70,** 165–173.

Lee, V.W.K., Cumming, I.A., de Kretser, D.M., Findlay, J.K., Hudson, B. & Keogh, E.J. (1976) Regulation of gonadotrophin secretion in rams from birth to sexual maturity. I. Plasma, LH, FSH and testosterone levels. *J. Reprod. Fert.* **46,** 1–6.

Leversha, L.J., Robertson, D.M., de Vos, F.L., Findlay, J.K., Burger, H.G., Morgan, F.J., Hearn, M.T.W., Wettenhall, R.E.H. & de Kretser, D.M. (1986) Isolation and purification of ovine inhibin. *Proc. Endocr. Soc. Aust., Brisbane,* Abstr.

L'Hermite, M., Niswender, G.D., Reichert, L.E., Jr & Midgely, A.R., Jr (1972) Serum follicle-stimulating hormone in sheep as measured by radioimmunoassay. *Biol. Reprod.* **6,** 325–332.

Lincoln, G.A. & Short, R.V. (1980) Seasonal breeding: Nature's contraceptive. *Recent Prog. Horm. Res.* **36,** 1–52.

Ling, N., Ying, S-Y., Ueno, N., Esch, F., Denoroy, L. & Guillemin, R. (1985) Isolation and partial characterisation of a Mr 32 000 protein with inhibin activity from porcine follicular fluid. *Proc. natn. Acad. Sci. U.S.A.* **82,** 7217–7221.

Ling, N., Ying, S-Y., Ueno, N., Shimasaki, S., Esch, F., Hotta, M. & Guillemin, R. (1986) Pituitary FSH is released by a heterodimer of the β-subunits from the two forms of inhibin. *Nature, Lond.* **321,** 779–782.

Mason, A.J., Hayflick, J.S., Ling, N., Esch, F., Ueno, N., Ying, S-Y., Guillemin, R., Niall, H.D. & Seeburg, P.H. (1985) Complementary DNA sequences of ovarian follicular fluid inhibin show precursor structure and homology with transforming growth factor-β. *Nature, Lond.* **318,** 659–663.

Mason, A.J., Niall, H.D. & Seeburg, P.H. (1986) Structure of two human ovarian inhibins. *Biochem. Biophys. Res. Commun.* **135,** 957–964.

McLachlan, R.I., Robertson, D.M., Burger, H.G. & de Kretser, D.M. (1986) The radioimmunoassay of bovine and human follicular fluid and serum inhibin. *Molec. cell. Endocr.* **46,** 175–185.

McNeilly, A.S. (1984) Changes in FSH and the pulsatile secretion of LH during the delay in oestrus induced by treatment of ewes with bovine follicular fluid. *J. Reprod. Fert.* **72,** 165–172.

McNeilly, A.S., Jonassen, J.A. & Fraser, H.M. (1986) Suppression of follicle development after chronic LHRH immunoneutralization in the ewe. *J. Reprod. Fert.* **76,** 481–490.

Miller, K.F., Critser, J.K. & Ginther, O.J. (1982) Inhibition and subsequent rebound of FSH secretion following treatment with bovine follicular fluid in the ewe. *Theriogenology* **18,** 45–53.

Miller, W.L., Knight, M.M., Grimek, H.J. & Gorski, J. (1977) Estrogen regulation of follicle-stimulating hormone in cell cultures of sheep pituitaries. *Endocrinology* **100,** 1306–1316.

Miyamoto, M., Hasegawa, Y., Fukuda, M., Nomura, M., Igarashi, M., Kangawa, K. & Matsuo, M. (1985) Isolation of porcine follicular fluid inhibin of 32k daltons. *Biochem. Biophys. Res. Commun.* **129**, 396–403.

Moudgal, N.R., Murthy, H.M.S., Murthy, G.S. & Rao, A.J. (1985) Regulation of FSH secretion in the primate by inhibin: studies in the bonnet monkey (*M. radiata*). In *Gonadal Proteins and Peptides and their Biological Significance*, pp. 21–38. Eds M. R. Sairam & L. E. Atkinson. World Scientific Publishing, Singapore.

Quirk, S.M. & Fortune, J.E. (1986) Plasma concentrations of gonadotrophins, preovulatory follicular development and luteal function associated with bovine follicular fluid-induced delay of oestrus in heifers. *J. Reprod. Fert.* **76**, 609–621.

Radford, H.M., Nancarrow, C.D. & Findlay, J.K. (1978) Effect of anaesthesia on ovarian follicular development and ovulation in the sheep subsequent to prostaglandin-induced luteolysis. *J. Endocr.* **78**, 321–327.

Rivier, J., Spiess, J., McClintock, V.J. & Vale, W. (1985) Purification and partial characterisation of inhibin from porcine follicular fluid. *Biochem. Biophys. Res. Commun.* **133**, 120–127.

Robertson, D.M., Ellis, S., Foulds, L.M., Findlay, J.K. & Bindon, B.M. (1984) Pituitary gonadotrophins in Booroola and control Merino sheep. *J. Reprod. Fert.* **71**, 189–197.

Robertson, D.M., Foulds, L.M., Leversha, L., Morgan, F.J., Hearn, M.T.W., Burger, H.G., Wettenhall, R.E.H. & de Kretser, D.M. (1985) Isolation of inhibin from bovine follicular fluid. *Biochem. Biophys. Res. Commun.* **126**, 220–226.

Robertson, D.M., de Vos, F.L., Foulds, L.M., McLachlan, R.I., Burger, H.G., Morgan, F.J., Hearn, M.T.W. & de Kretser, D.M. (1986) Isolation of a 31 kDa form of inhibin from bovine follicular fluid. *Molec. cell. Endocr.* **44**, 271–277.

Sairam, M.R., Ranganathan, M.R., Ramasharma, K. & Lamothe, P. (1981) Isolation and characterization of a bovine seminal plasma protein inhibiting pituitary FSH secretion. *Molec. cell. Endocr.* **22**, 231–250.

Salamonsen, L.A., Jonas, H.A., Burger, H.G., Buckmaster, J.M., Chamley, W.A., Cumming, I.A., Findlay, J.K. & Goding, J.R. (1973) A heterologous radioimmunoassay for follicle-stimulating hormone: application to measurement of FSH in the ovine estrous cycle and in several other species including man. *Endocrinology* **93**, 610–618.

Scaramuzzi, R.J. & Radford, H.M. (1983) Factors regulating ovulation rate in ewes. *J. Reprod. Fert.* **69**, 353–367.

Tsonis, C.G., Quigg, H., Lee, V.W.K., Leversha, L., Trounson, A.O. & Findlay, J.K. (1983) Inhibin in individual follicles in relation to diameter and atresia. *J. Reprod. Fert.* **67**, 83–90.

Tsonis, C.G., McNeilly, A.S. & Baird, D.T. (1986) Measurement of exogenous and endogenous inhibin in sheep serum using a new and extremely sensitive bioassay for inhibin based on inhibition of ovine pituitary FSH secretion *in vitro*. *J. Endocr.* **110**, 341–352.

Vale, W., Rivier, J., Vaughan, J., McClintock, R., Corrigan, A., Woo, W., Carr, D. & Spiess, J. (1986) Purification and characterization of an FSH releasing protein from porcine ovarian follicular fluid. *Nature, Lond.* **321**, 776–779.

Wallace, J.M. & McNeilly, A.S. (1985) Increase in ovulation rate after treatment of ewes with bovine follicular fluid in the luteal phase of the oestrous cycle. *J. Reprod. Fert.* **73**, 505–515.

Walters, D.L. & Schallenberger, E. (1984) Pulsatile secretion of gonadotrophins, ovarian steroids and ovarian oxytocin during the periovulatory phase of the oestrous cycle of the cow. *J. Reprod. Fert.* **71**, 503–512.

Walton, J.S., McNeilly, J.R., McNeilly, A.S. & Cunningham, F.J. (1977) Changes in concentrations of follicle-stimulating hormone, luteinizing hormone, prolactin and progesterone in the plasma of ewes during the transition from anoestrus to breeding season. *J. Endocr.* **75**, 127–136.

Ying, S-Y., Becker, A., Baird, A., Ling, N., Ueno, N., Esch, F. & Guillemin, R. (1986a) Type beta transforming growth factor (TGF-β) is a potent stimulator of the basal secretion of follicle stimulating hormone (FSH) in a pituitary monolayer system. *Biochem. Biophys. Res. Commun.* **135**, 950–956.

Ying, S-Y., Becker, A., Ling, N., Ueno, N. & Guillemin, R. (1986b) Inhibin and beta type transforming growth factor (TGF-β) have opposite modulating effects on the follicle stimulating hormone (FSH)-induced aromatase activity of cultured rat granulosa cells. *Biochem. Biophys. Res. Commun.* **136**, 969–975.

J. Reprod. Fert., Suppl. **34** (1987), 39–54

Printed in Great Britain
© 1987 Journals of Reproduction & Fertility Ltd

Control of follicular growth and development

J. J. Ireland*

Michigan State University, Department of Animal Science and Physiology, East Lansing, MI 48823, U.S.A.

Summary. During folliculogenesis a group of growing preantral follicles becomes responsive and dependent upon gonadotrophins, especially FSH, for their continued growth and differentiation. However, most of these follicles undergo atresia. The mechanisms that result in survival of a specific number of ovulatory (dominant) follicles appear to depend upon: (a) responsiveness of preantral follicles to gonadotrophins, (b) inhibitory and stimulatory factors from a dominant follicle, and (c) an exquisitely sensitive long-loop feedback system between the dominant follicle and pituitary gland.

Introduction

Folliculogenesis is spectacular not only because a primordial follicle may increase in diameter 400- to 600-fold before ovulation but because the growth of 500 to 1000 primordial follicles each oestrous cycle usually results in development of only a few ovulatory follicles. The fact that during the lifespan of an animal 99·9% of the primordial follicles fail to ovulate illustrates not only that development of an ovulatory follicle is an extremely rare biological event but also that the process of folliculogenesis is complex.

Goodman & Hodgen (1983) suggest the use of the following terms for describing folliculogenesis: *recruitment*—a gonadotrophin-dependent event during which a group of follicles gain the ability to respond to gonadotrophins and require gonadotrophins for continued growth; *selection*—a process whereby only a few of the 'recruited' follicles are 'selected' to escape atresia and survive to ovulate; and *dominance*—the mechanism that an ovulatory (or dominant) follicle(s) uses to escape atresia. The phenomenon of dominance is central to understanding folliculogenesis since it suggests that some follicles survive in a milieu suppressive to growth of other follicles, or that some follicles prevent growth of other follicles (Goodman & Hodgen, 1983).

This review will focus on the characteristics of development of dominant follicles during oestrous cycles, the mechanism of action of gonadotrophins during folliculogenesis, the potential intragonadal regulators of follicular growth and function, and the possible role of intragonadal regulators on recruitment and selection of the dominant follicle.

Endocrine and receptor changes that characterize the development of dominant follicles

The change in concentration of oestradiol in each ovarian vein during an oestrous cycle is the best endocrine marker for depicting the selection and dominance processes. These processes are associated with a unique follicular hierarchy, symmetrical and asymmetrical ovarian production of oestradiol, and marked changes in binding of human chorionic gonadotrophin (hCG) to follicles,

---

*Present address: Yale University School of Medicine, Department of Obstetrics and Gynecology, 333 Cedar Street, New Haven, CT 06510, U.S.A.

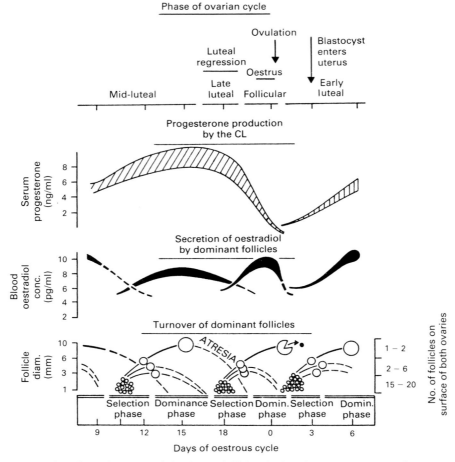

Fig. 1. Cycles of development of dominant follicles during the oestrous cycle of a cow. (After Ireland & Roche, 1987.)

especially in primates and cattle (Goodman & Hodgen, 1983; Ireland & Roche, 1987). In single-ovulating species, symmetrical ovarian production of oestradiol indicates that the selection process for a dominant follicle is on-going, whereas asymmetrical production of oestradiol indicates that selection is complete and dominance is underway. Endocrine markers for recruitment are ill-defined, although considerable evidence (discussed later) indicates that changes in FSH concentrations in blood may be the stimulus for recruitment (Ireland & Roche, 1987).

Cattle have at least 3 cycles of development of dominant follicles during an oestrous cycle (Fig. 1; Ireland & Roche, 1987). Unlike primates (Goodman & Hodgen, 1983), cattle have dominant non-ovulatory follicles that develop during the early- and mid-luteal phase (Matton *et al.*, 1981; Ireland & Roche, 1983a). These follicles are responsible for the increases in concentration of oestradiol in blood that occur several days after ovulation (Glencross *et al.*, 1973) and again after mid-cycle (Hansel & Echternkamp, 1972). Although these changes in concentrations of oestradiol are difficult to detect in the peripheral circulation, analysis of utero-ovarian venous blood samples established that symmetrical and asymmetrical production of oestradiol from each ovary occur not only during the follicular phase but also during the luteal phase of the oestrous cycle of cattle (Ireland *et al.*, 1985; Fogwell *et al.*, 1985). Coinciding with each cycle of development of a dominant follicle is a marked increase in the ability of the dominant follicle to bind hCG specifically (Ireland & Roche, 1982, 1983a, b). Each cycle of development of the dominant follicle occurs in remarkably different hormonal environments (Fig. 1; Rahe *et al.*, 1980), suggesting that peri-ovulatory changes in gonadotrophins are not necessary for development of a dominant follicle.

Sheep, like cattle, have several periods during an oestrous cycle when concentrations of oestradiol increase in blood (Baird *et al.*, 1976). This supports the idea that cycles of development of dominant follicles occur in sheep (Smeaton & Robertson, 1971).

Physiological events such as luteolysis, movement of the blastocyst into the uterus, and oestrus and the preovulatory gonadotrophin surge, correspond with a different cycle of development of a dominant follicle in cattle (Fig. 1). Since oestradiol, produced primarily from a dominant follicle rather than other follicles, can influence each of these physiological events (Ireland & Roche, 1987), dominant follicles not only function to provide oocytes for fertilization but may have an important role in regulating a cascade of physiological events necessary for successful reproduction.

Gonadotrophins are required for development of dominant follicles, synthesis and secretion of oestradiol from these follicles and ovulation. However, the mechanisms that result in the cyclic appearance of ovulatory and non-ovulatory dominant follicles and secretion of oestradiol from these follicles in remarkably different hormonal environments during an oestrous cycle (Fig. 1) are unknown. In addition, the mechanisms that result in selection of only a few dominant follicles out of the hundreds of follicles that are recruited to grow each oestrous cycle are unknown.

Mechanism of action of gonadotrophins during folliculogenesis

Follicle-stimulating hormone (FSH) and luteinizing hormone (LH) are the primary protein hormones involved in folliculogenesis (Hisaw, 1947). Gonadotrophins enhance steroidogenic enzyme activity in granulosa and theca cells after interaction of the gonadotrophins with their receptor sites and activation of cAMP-dependent processes. This results in an increased synthesis and accumulation of steroids especially oestradiol in the general circulation and in the follicular fluid of antral follicles. Since oestradiol is the key hormone for promoting folliculogenesis (Richards, 1980) and for triggering physiological events necessary for reproduction (Fig. 1), the dominant follicle(s) must possess an enhanced capacity over other follicles to synthesize and release this steroid. This enhanced capacity to produce oestradiol involves the action of both FSH and LH on theca and granulosa cells. The mechanism of action of FSH and LH on granulosa and theca cells has been exceptionally well-reviewed by Richards (1980), Hsueh *et al.* (1984) and Erickson *et al.* (1985). Much of our current understanding of the mechanisms of action of FSH and LH on folliculogenesis is derived from studies using immature and/or hypophysectomized rats and primary cultures of granulosa and theca-interstitial cells from rats and domestic species. The following is a brief summary of the results of these studies as they relate to folliculogenesis.

Receptors

Granulosa and theca cells contain a plethora of receptor sites for numerous hormonal and non-hormonal factors (Fig. 2), indicating that many factors other than gonadotrophins influence folliculogenesis. Many of these receptor sites are up-regulated or down-regulated ($\pm$) by FSH and LH, which supports the concept that gonadotrophins are the primary stimulators of folliculo-genesis (Hisaw, 1947). Granulosa cells in preantral follicles possess FSH receptors and do not gain LH receptors until a follicle forms an antrum. Beginning at the preantral stage of development, theca cells contain LH receptors, but never gain FSH receptors (Richards, 1980). Follicles are, therefore, capable of responding to both FSH and LH at a very early stage of development. During folliculogenesis, FSH induces appearance of its own receptor and then LH receptor in granulosa cells. Regulation of the thecal LH receptor is unclear. FSH is also required for maintenance of gonadotrophin receptors throughout folliculogenesis (Richards, 1980).

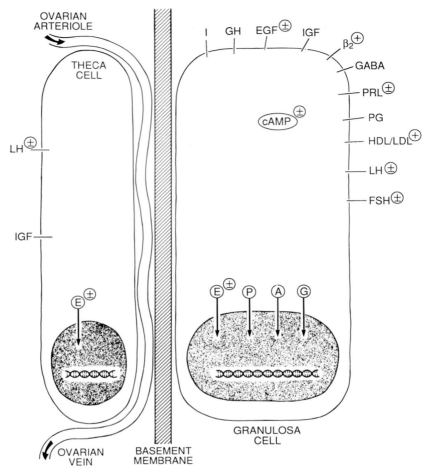

Fig. 2. Receptor sites that are present on granulosa or theca cells. I = insulin, GH = growth hormone, EGF = epidermal growth factor, IGF = insulin-like growth factor, β_2 = β_2-adrenergic hormones, GABA = gamma amino butyric acid, PRL = prolactin, PG = prostaglandin, HDL/LDL = high and low density lipoproteins, LH = luteinizing hormone, FSH = follicle-stimulating hormone, cAMP = cyclic adenosine monophosphate, P = progesterone, A = androstenedione, G = glucocorticoids. $\pm$ indicates that gonadotrophins can up- or down-regulate the receptor site. (Modified from Hsueh *et al.*, 1984.)

Response systems

     Gonadotrophins bind to their receptor sites and activate the adenylate cyclase–cAMP response system. This results in activation of protein kinases and phosphorylation of various proteins that may be involved in folliculogenesis (Fig. 3). Adenylate cyclase is comprised of a regulatory subunit which interacts with the hormone–receptor complex. This subunit, called the G or N protein, is activated by guanine nucleotides and magnesium. Adenylate cyclase also contains a catalytic subunit which metabolizes ATP into cAMP when it is bound to the N subunit (Birnbaumer & Kirchick, 1983). Sustained high concentrations of gonadotrophins and slow rates of dissociation of gonadotrophins from their receptors result in abrogation of the receptor–adenylate cyclase–cAMP response system. This results in a temporary or permanent desensitization of the adenylate cyclase in follicular cells to further gonadotrophin stimulation (Birnbaumer & Kirchick, 1983). During folliculogenesis, changes in episodic patterns of secretion of gonadotrophins, such as during a preovulatory gonadotrophin surge, may not only promote luteinization of granulosa cells and ovulation but also result in the demise of some populations of follicles (permanent desensitization) and enhance growth of others.

FSH increases the proposed regulatory subunit for protein kinase (R_{11}, Richards & Rolfes, 1979), the mRNA for this protein (Hedin *et al.*, 1986) and phosphorylation of several granulosa cell proteins (Richards *et al.*, 1983). The specificity of action of gonadotrophins therefore resides in the specific proteins which are phosphorylated in the granulosa and theca cells. It is not known whether proteins that are phosphorylated in response to gonadotrophin action are involved in induction of receptors, activation of enzymes for steroidogenesis, control of the luteinization process or other facets of folliculogenesis.

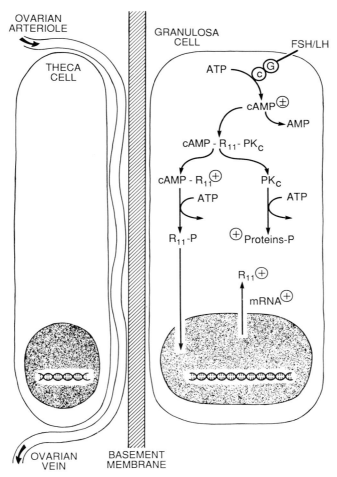

Fig. 3. Diagram representing the proposed mechanism of action for gonadotrophins. G = guanine and magnesium activated regulatory subunit of adenylate cyclase, C = catalytic subunit of adenylate cyclase, cAMP = 3'5' cyclic adenosine monophosphate, AMP = 5'-adenosine monophosphate, R_{11} = regulatory subunit of protein kinase, PK_c = catalytic subunit of protein kinase, cAMP-R_{11} = phosphorylated R_{11}, Proteins-P = phosphorylated proteins. ± indicates regulation by LH or FSH.

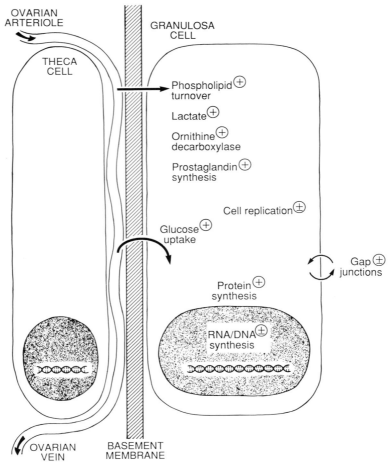

Fig. 4. Some of the major functions of the granulosa cell that are positively (+) or negatively (−) regulated by gonadotrophins. (Modified from Hsueh *et al.*, 1984.)

Several neurotransmitters, peptides and growth factors that have a role in folliculogenesis not only activate the cAMP response system but also activate phosphoinositide metabolism after binding to membrane receptors. Phosphoinositide metabolites, at least in other cell types, are involved in release of arachidonic acid for prostaglandin synthesis, activation of protein kinase C, calcium mobilization, and activation of guanylate cyclase for cGMP production. Oncogenes which are proposed to regulate all aspects of cellular growth are also linked to the phosphoinositide-response system (Berridge, 1984). Although it is unknown whether activation of the phosphoinositide-response system explains many of the well-known actions of gonadotrophins during folliculogenesis (Fig. 4), incubation of pig granulosa cells with LH, but not FSH or cAMP, results in accumulation of inositol phosphates in the medium (Dimino & Snitzer, 1986).

Steroidogenesis

The undifferentiated thecal cell is incapable of synthesizing androgens although it contains LH receptor (Erickson *et al.*, 1985). During folliculogenesis, the theca cell responds to LH primarily by activation of the side-chain cleavage enzymes and the 17α-hydroxyprogesterone and 17–20-desmolase enzyme systems (Erickson *et al.*, 1985), allowing the theca cell to synthesize androgens. During folliculogenesis a granulosa cell, which does not produce androgens, is incapable of producing oestradiol until the theca cell differentiates into an androgen-producing cell (Fig. 5).

FSH and LH enhance the uptake of lipoproteins, liberation of cholesterol from lipoproteins, mobilization of cholesterol, conversion of cholesterol into pregnenolone through activation of the mitochondrial side-chain cleavage enzyme system and conversion of pregnenolone to progesterone through activation of the 3β-hydroxysteroid dehydrogenase enzyme system in granulosa cells.

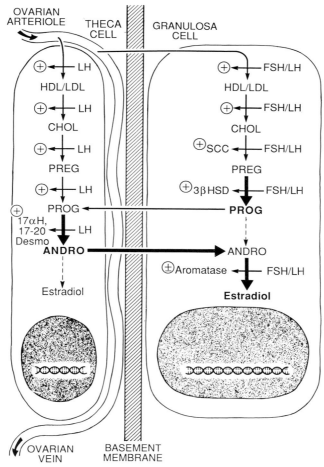

Fig. 5. Proposed steroidogenic pathways and enzymes for the theca and granulosa cell (modified from Erickson *et al.*, 1985; Hsueh *et al.*, 1984; Leung & Armstrong, 1980). Degree of darkness of arrows between steroids represent primary steroid synthesized. Arrows pointing to enzymes indicate whether enzyme activity is stimulated (+) or inhibited (−) by FSH or LH. Arrows from one cell to the other represent diffusion of steroid either from cell to cell or through the capillary network surrounding a follicle. Arrows from arteriole represent uptake. HDL/LDL = high or low density lipoproteins, CHOL = cholesterol, PREG = pregnenolone, PROG = progesterone, ANDRO = androgens, SCC = mitochondrial side-chain cleavage enzymes, 3βHSD = 3β-hydroxysteroid dehydrogenase, 17αH, 17–20 DESMO = 17α-hydroxylase and 17–20-desmolase.

Although a granulosa cell responds to LH by increasing the production of progesterone and oestradiol, FSH priming is required for this steroidogenic effect (Hsueh *et al.*, 1984).

In some species, such as pigs, the theca cell possesses an active aromatase enzyme system (Evans *et al.*, 1981) which diminishes the role of FSH in oestradiol production, at least during some stages of folliculogenesis. However, in most species, granulosa rather than theca cells possess the major aromatase activity (Leung & Armstrong, 1980).

Ability of the theca cell to produce androgens precedes acquisition of the aromatase system by granulosa cells in cattle (McNatty *et al.*, 1984). Moreover, on each day of the cycle in cows, 20–60 antral follicles may produce androgens, whereas only 1–3 antral follicles are able to metabolize androgens into oestrogens. This clearly illustrates that acquisition of the aromatase system, which is regulated by FSH (Leung & Armstrong, 1980), is a key maturation step during folliculogenesis for initiation of the dominance process.

Many factors, which could inhibit synthesis of oestradiol, such as FSH and LH binding inhibitors and follicular regulatory protein, are present in follicular fluid (Table 1). It is unknown whether dominant follicles use these factors to inhibit development of other follicles or are spared from the inhibitory effects of such factors. The activity of the steroidogenic enzymes and, in turn, oestradiol synthesis by a dominant follicle may not only be controlled by inhibitory proteins but by a complex interaction of ovarian steroids with steroidogenic enzymes (Leung & Armstrong, 1980). For example, androgens enhance aromatase activity whereas oestrogens inhibit androgen production. It is possible that ovarian steroids mediate their effects on steroidogenesis through synthesis of various intragonadal regulators of steroidogenic enzyme activity.

Modulators of gonadotrophin action

Various steroids and proteins enhance FSH action. For example, oestradiol, androgens, progesterone, LH, insulin, platelet-derived growth factor, insulin-like growth factors (Hsueh *et al.*, 1984) and transforming growth factor beta (Dodson & Schomberg, 1986) all enhance various aspects of FSH action, including induction of FSH or LH receptors and progesterone or oestradiol synthesis. In contrast, pharmacological doses of LH or FSH, glucocorticoids, fibroblastic growth factor, epidermal growth factor, calcium inhibitors (Hsueh *et al.*, 1984), transforming growth factor beta (Knecht & Feng, 1986), and follicular regulatory protein (diZerega *et al.*, 1984) all inhibit FSH action. The mechanisms of action of these modulators of FSH action are unknown.

Oestradiol is the best studied amplifier of FSH action. For example, in laboratory species, oestradiol markedly enhances the ability of FSH to increase its receptor in granulosa cells (Richards, 1980). Indeed, synthesis of oestradiol is required for FSH to promote folliculogenesis (Tonetta & Ireland, 1984; Tonetta *et al.*, 1985). Since oestradiol does not increase the FSH receptor (Richards *et al.*, 1976), the catalytic subunit of protein kinase (Richards *et al.*, 1984), the regulatory subunit of protein kinase (Ratoosh & Richards, 1985), and only slightly increases adenylate cyclase activity (Jonassen *et al.*, 1982), its action remains a mystery. Nevertheless, oestradiol has a well-known role in protein synthesis in many tissues, and so oestradiol-induced granulosa or theca cell proteins may be involved in enhancement of FSH action. It awaits to be determined whether these oestradiol-induced proteins include some of the various factors known to enhance FSH action.

Potential intragonadal regulators of follicular growth and function

The role of most of the various intragonadal hormonal and non-hormonal and putative regulatory factors (Table 1) in folliculogenesis is unknown. The intriguing possibility exists, however, that many of these factors operate in an endocrine, paracrine and autocrine fashion to regulate development of dominant follicles.

Follicular fluid contains numerous factors that are potential intragonadal modulators of folliculogenesis. In addition to ovarian steroids, follicular fluid contains non-gonadal protein hormones such as LH, FSH (McNatty *et al.*, 1975), insulin (Rein & Schomberg, 1982), prolactin (Meloni *et al.*, 1986) and prorenin (Glorioso *et al.*, 1986). Granulosa cells synthesize many hormonal and non-hormonal products other than steroids, including relaxin (Bryant-Greenwood *et al.*, 1980), oxytocin (Sheldrick & Flint, 1984), inhibin (Ling *et al.*, 1985), activin (Ling *et al.*, 1986) or follicle-stimulating hormone releasing protein (Vale *et al.*, 1986), proteoglycans (Yanagishita *et al.*, 1981), prostacyclins (Ranta *et al.*, 1986), plasminogen activator (Beers, 1975), anti-Müllerian hormone (Josso, 1986) and fibronectin (Skinner *et al.*, 1985). Inhibin and activin or FSH releasing protein, which directly regulate FSH secretion and modulate FSH-induced oestradiol synthesis, stand alone as the only protein hormones in follicular fluid to have their amino acid sequences determined following purification of preparations of follicular fluid (Ling *et al.*, 1985, 1986; Vale *et al.*, 1986; Ying *et al.*, 1986b). Hypoxanthine and adenosine have been isolated from follicular fluid (Downs *et al.*, 1985) and shown to possess the ability to inhibit oocyte maturation *in vitro*. Whether these purines represent the elusive oocyte maturation inhibitor (OMI) or a mediator of OMI activity remains to be determined.

An abundance of factors of unknown molecular identities also exists in follicular fluid and may be involved in regulating granulosa and theca cell function (Hammond, 1981; Sairam & Atkinson, 1984; see Table 1). The most recent potential intragonadal regulator of follicular function was discovered by Aten *et al.* (1986) in rat, human, sheep and cow ovaries. This factor, which is absent from follicular fluid, inhibits binding of radioactive GnRH to rat ovarian membranes. GnRH modulates gonadotrophin action in rats but not in other species, and high-affinity receptors for GnRH are found in rat ovaries but not ovaries of other species (Hsueh & Jones, 1981). The newly discovered ovarian factor is not GnRH since, unlike GnRH, it is heat-sensitive, its activity elutes differently than GnRH during HPLC, and it does not possess GnRH immuno-reactivity. Whether this ovarian factor has anti-gonadal properties in humans and domestic species awaits determination.

Recruitment and selection of the dominant follicle

In theory, the cyclic appearance of the ovulatory quota of dominant follicles throughout an oestrous cycle (Fig. 1) may involve the interaction of gonadotrophins with various intragonadal factors. Intragonadal factors would modulate the action of gonadotrophins such that only a few follicles are selected to become dominant.

Recruitment

Hypophysectomy, or short-term suppressions of FSH in blood, clearly inhibits development of preantral and antral follicles (Nakano *et al.*, 1975; Carter *et al.*, 1961). FSH alone restores antral follicle development (Richards, 1980). Since preantral follicles possess FSH receptors (Hsueh *et al.*, 1984), gonadotrophin responsiveness (recruitment) may be developed during this stage of folliculogenesis. Once this occurs, the preantral follicle is thereafter dependent on FSH support for continued development into a dominant follicle (Richards, 1980). In laboratory species, the pro-oestrous and oestrous rises in FSH during one cycle regulate the recruitment of follicles for ovulation in the next cycle (Hirshfield & Midgley, 1978; Chappel & Selker, 1979). These data illustrate that FSH has a major role in the recruitment process. Similar studies were attempted in sheep but failed to alter the ovulation quota. However, unlike the study of Chappel & Selker (1979) in laboratory species, the post-ovulatory increase in FSH was only partly obliterated in sheep (Bindon & Piper, 1984).

Table 1. Putative ovarian factors and their biological actions

Factor	Tissue source	Species	Physical characteristics	Biological action	Reference
Angiogenic factor	Non-luteal ovaries, theca	Pig	Heat-labile, protease-sensitive	Blood vessel growth	Makris et al. (1984)
Renin-like activity	Follicular fluid	Human	Samples liberate radioimmunoassayable angiotensin	Blood vessel formation	Fernandez et al. (1985)
Luteinization inhibitor	Follicular fluid	Pig, cattle	M_r 100 000	Blocks LH-induced cAMP and progesterone production by granulosa cells in culture	Hammond (1981)
Gonadotrophin surge inhibiting factor	Follicular fluid	Pig	Heat-labile	Blocks preovulatory gonadotrophin surge	Littman & Hodgen (1984)
FSH binding inhibitor	Follicular fluid, serum	Cattle, pig, human	M_r < 5000, 10–11 amino acid peptide, protease-sensitive, putrescine (?)	Blocks binding of [125I]-labelled FSH to bovine granulosa cells or testes	Sluss & Reichert (1984); Darga & Reichert (1979); Sato et al. (1982); Sanzo & Reichert (1982)
LH binding inhibitors	Follicular fluid, serum	Human, pig, sheep	M_r 20 000	Blocks binding of [125I]-labelled LH to testes or luteal cells	Sanzo & Reichert (1982); Kumari et al. (1984)

Factor	Source	Species	Properties	Actions	References
Oocyte maturation inhibitor	Follicular fluid, granulosa cells	Human, pig, cattle	$M_r < 2000$, heat stable, may be hypoxanthine or adenosine	Inhibits in-vitro maturation of cumulus-enclosed oocytes of various species	Tsafriri et al. (1976) Hillensjo et al. (1980) Downs et al. (1985) Miller & Behrman (1986)
Follicular regulatory protein	Follicular fluid, serum	Pig, cattle, human	M_r 12 500–16 000, heat and trypsin sensitive	Inhibits aromatase activity by rat and porcine granulosa cells in culture, also inhibits progesterone synthesis	diZerega et al. (1984)
Transforming growth factor, beta-like activity	Follicular fluid	Pig	Similar to inhibin (?)	Actions unknown, TGF_B alters FSH-induced increases in LH receptor and epidermal growth factor in granulosa cells and promotes FSH release from pituitary cells in culture	Dodson & Schomberg (1986) Knecht & Feng (1986) Ying et al. (1986a,b)
Gonadocrinin	Follicular fluid, ovary	Pig, cattle	M_r 1000–10 000, heat and trypsin sensitive	Stimulates release of LH and FSH from rat pituitary cells in culture	Ying et al. (1981)
GnRH-like ovarian hormone	Ovary	Human, cattle, sheep	M_r 1000–10 000, heat and protease sensitive, not GnRH	Inhibits binding of ^{125}I-labelled GnRH to rat ovarian luteal homogenates	Aten et al. (1986)
Insulin-like growth factors	Follicular fluid, granulosa cells	Pig	Immunoassay and radioreceptor assay activity	Stimulate granulosa cell replication, enhance FSH action	Adashi et al. (1985) Hammond et al. (1985) Hsu & Hammond (1986)

Selection and dominance

   Selection may be a three-part process involving the ability of preantral follicles to respond to gonadotrophins, elaboration of inhibitory factors from a dominant follicle and feedback between dominant follicles and the pituitary gland.

   Atresia is observed in preantral follicles of all species. Selection of the ovulatory quota of follicles may therefore begin at the preantral stage of folliculogenesis. Initiation of follicular growth from the non-growing pool of avascular follicles is probably asynchronous (Lintern-Moore & Moore, 1979). Therefore, all growing follicles may not be at the same stage of development during recruitment. One reason that growing preantral follicles do not continue to grow, although each in theory receives a similar gonadotrophin stimulus during recruitment, may be the result of a gradient of subtle differences in number of granulosa or theca cells and/or amount of gonadotrophin receptor per follicle. The number of follicles responding to gonadotrophins would then depend upon the strength of the recruitment stimulus (i.e. amount of gonadotrophins) and the inherent ability of each follicle to respond. Direct evidence for these assumptions does not exist for any species.

   A popular theory is also held that dominant follicles elaborate various hormonal and non-hormonal factors (Table 1) that control selection by modulating the gonadotrophin-induced recruitment process. For example, gonadotrophin-induced follicular growth is more effective in promoting multiple follicle growth before establishment of follicular dominance in primates (diZerega & Hodgen, 1980). Furthermore, factors from a dominant follicle or corresponding ovarian vein inhibit steroidogenesis (Table 1) by granulosa cells, supporting the idea that a dominant follicle may inhibit growth of other follicles but not itself. A strong possibility also exists that during folliculogenesis antral follicles receive unequal amounts of gonadotrophin because of differences in vasculature (Zeleznik *et al.*, 1981). This could also result in the selective growth of a dominant follicle.

   A feedback system between a dominant follicle and the pituitary gland and the ability of a dominant follicle to survive in an hormonal milieu suppressive to growth of other follicles may explain, in part, the selection process. In cattle, a dominant follicle possesses markedly greater inhibin bioactivity and oestradiol content than do other follicles (Padmanabhan *et al.*, 1984). Follicular fluid, which contains inhibin, and oestradiol both depress secretion of FSH (Zeleznik, 1981; Kesner & Convey, 1982; Ireland *et al.*, 1983). FSH enhances oestradiol and inhibin production in cultures of granulosa cells (Hsueh *et al.*, 1984). Thus, a classical long-loop negative feedback system may exist between inhibin and oestradiol from the dominant follicle and FSH from the pituitary gland. Results of several studies support this concept. After removal of a single ovary in most species, the remaining ovary compensates for the loss of the other and maintains the original ovulatory quota. In sheep, removal of one ovary results in only a 20–30% increase in FSH lasting 6–7 h (Findlay & Cumming, 1977). This transient increase in FSH is sufficient to enhance growth of more follicles and/or prevent atresia of existing follicles resulting in a normal ovulation quota. Compensatory follicular growth is blocked if the rise in FSH is suppressed after removal of an ovary (Welschen *et al.*, 1979). Further evidence for the sensitivity of the dominant follicle–pituitary feedback system is best shown in the studies of Zeleznik (1981), Zeleznik *et al.* (1985), Dierschke *et al.* (1985), Pathiraja *et al.* (1984) and Webb & Gauld (1987). Physiological doses of oestradiol given during the follicular phase result in short-term suppression in concentrations of FSH in blood and atresia of the existing dominant follicle. Moreover, infusion of oestradiol antibodies enhances concentrations of LH and FSH and results in development of multiple dominant follicles in primates and sheep. Finally, several studies in sheep suggest that an association exists between FSH and inhibin which may affect ovulation rate. For example, prolific breeds of sheep, especially the Booroola Merino, have higher concentrations of FSH in blood and lower ovarian values of inhibin bioactivity than do less prolific breeds of sheep. Immunization of sheep against follicular fluid, which contains inhibin and other potential regulatory factors, enhances ovulation rates (Bindon &

Piper, 1984). In summary, modulation of many factors involved in the selection process, especially in sheep, results in enhanced rates of ovulation (Driancourt, 1987).

If inhibin and oestradiol interact to regulate the number of dominant follicles, how then would a dominant follicle(s) survive in an hormonal milieu (especially low FSH) that is suppressive to growth of other follicles? Oestradiol, which markedly enhances FSH and LH action (Hisaw, 1947; Richards, 1980), is in remarkably greater concentrations in dominant follicles than in other co-existing follicles (Ireland & Roche, 1982, 1983a, b). Because of its greater oestradiol levels, a dominant follicle may survive in an oestradiol- or inhibin-induced FSH-deficient milieu. Other follicles with diminished levels of oestradiol undergo atresia.

Because the interaction of oestradiol and FSH alone could provide a model which explains selection and dominance, what role, if any, do intragonadal factors have on regulation of development of dominant follicles? Since most of these factors (Table 1) are inhibitory to follicular function, they may represent metabolic products secreted from each follicle (or a dominant follicle alone) that are necessary to cause atresia. These factors could also be degradative by-products of atresia. Clearly, the predominant ovarian event is atresia, and follicles may synthesize and secrete factors for stimulation of this process.

In summary, gonadotrophins, especially FSH, regulate the recruitment process. FSH, oestradiol, and perhaps inhibin, control, in part, the selection process. It is unknown whether a dominant follicle secretes factors, other than oestradiol, that participate in the dominance process. In conclusion, the concepts of recruitment and selection seem indisputable, whereas evidence for the dominance process remains equivocal.

I thank Dr H. R. Behrman, Department of OB/GYN, Yale University, for the time permitted to prepare this manuscript; and Janet Ireland and Judith Luborsky for helpful suggestions during preparation of this text. The author is the recipient of a Senior Investigator Fellowship from NICHHD.

References

Adashi, E.Y., Resnick, C.E., D'ercole, A.J., Svoboda, M.E. & YanWyk, J.J. (1985) Insulin-like growth factors as intraovarian regulators of granulosa cell growth and function. *Endocr. Reviews* **6**, 400–420.

Aten, R.F., Ireland, J.J., Polan, M.L., Weems, C.W. & Behrman, H.R. (1986) GnRH-like ovarian hormone (GLOH) is present in several species which do not have GnRH receptors. *Endocrinology* **118** (Suppl.), 221, Abstr. 763.

Baird, D.T., Land, R.B., Scaramuzzi, R.J. & Wheeler, A.G. (1976) Endocrine changes associated with luteal regression in the ewe: The secretion of ovarian oestradiol, progesterone and androstenedione and uterine prostaglandin $F_{2\alpha}$ throughout the oestrous cycle. *J. Endocr.* **69**, 275–286.

Beers, W. (1975) Follicular plasminogen and plasminogen activator and effect of plasmin on ovarian follicle wall. *Cell* **6**, 379–386.

Berridge, M.J. (1984) Inositol trisphosphate and diacylglycerol as second messengers. *Biochem. J.* **220**, 345–360.

Bindon, B.M. & Piper, L.R. (1984) Endocrine basis of genetic differences in ovine prolificacy. *Proc. 10th Int. Congr. Anim. Reprod. & A.I., Urbana-Champaign* **IV**, pp. 17–26.

Birnbaumer, L. & Kirchick, H.J. (1983) Regulation of gonadotropic action: The molecular mechanisms of gonadotropin-induced activation of ovarian adenylyl cyclases. In *Factors Regulating Ovarian Function*, pp. 287–310. Eds G. S. Greenwald & P. F. Terranova. Raven Press, New York.

Bryant-Greenwood, G.D., Jeffrey, R., Ralph, M.M. & Seamark, R.F. (1980) Relaxin production by the porcine ovarian graafian follicle in vitro. *Biol. Reprod.* **23**, 792–800.

Carter, F., Woods, M.C. & Simpson, M.E. (1961) The role of the pituitary gonadotropins in induction of ovulation in the hypophysectomized rat. In *Control of Ovulation*, pp. 1–23. Ed C. A. Villee. Pergamon Press, New York.

Chappel, S.C. & Selker, F. (1979) Relation between the secretion of FSH during the periovulatory period and ovulation during the next cycle. *Biol. Reprod.* **21**, 347–352.

Darga, N.C. & Reichert, L.E., Jr (1979) Evidence for the presence of a low molecular weight follitropin binding inhibitor in bovine follicular fluid. *Adv. exp. Med. Biol.* **112**, 383–388.

Dierschke, D.J., Hutz, R.J. & Wolf, R.C. (1985) Induced follicular atresia in rhesus monkeys: strength-duration relationships of the estrogen stimulus. *Endocrinology* **117**, 1397–1403.

Dimino, M.J. & Snitzer, J. (1986) Luteinizing hormone (LH) causes rapid increases in accumulation of inositol phosphates in granulosa from medium follicles of porcine ovaries. *Biol. Reprod.* **34**, Suppl. 1, 168, Abstr. 237.

diZerega, G.S. & Hodgen, G.D. (1980) The primate ovarian cycle: Suppression of human menopausal gonadotropin-induced follicular growth in the presence of the dominant follicle. *J. clin. Endocr. Metab.* **50**, 819–825.

diZerega, G.S., Campeau, J.D., Ujita, E.L., Schreiber, J.R., Battin, D.A., Montz, F.J. & Nakamura, R.M. (1984) Follicular regulatory proteins: Paracrine regulators of follicular steroidogenesis. In *Gonadal Proteins and Peptides and their Biological Significance*, pp. 215–228. Eds M. R. Sairam & L. E. Atkinson. World Scientific Publ. Co., Singapore.

Dodson, W.C. & Schomberg, D.W. (1986) Transforming growth factor beta enhances FSH-stimulated LH receptor induction in granulosa cells. *Endocrinology* **118** (Suppl.), 290, Abstr. 1037.

Downs, S.M., Coleman, D.L., Ward-Bailey, P.F. & Eppig, J.J. (1985) Hypoxanthine is the principal inhibitor of murine oocyte maturation in a low molecular weight fraction of porcine follicular fluid. *Proc. natn. Acad. Sci., U.S.A.* **82**, 454–458.

Driancourt, M.A. (1987) Follicular dynamics and intra-ovarian control of follicular development in the ewe. In *Follicular Growth and Ovulation Rate in Farm Animals* (in press). Eds J. Roche & D. O'Callaghan. Martinus Nijhoff, The Hague.

Erickson, G.F., Magoffin, D.A., Dyer, C.A. & Hofeditz, C. (1985) The ovarian androgen producing cells: A review of structure/function relationships. *Endocr. Reviews* **6**, 371–399.

Evans, G., Dobias, M., King, G.J. & Armstrong, D.T. (1981) Estrogen, androgen and progesterone biosynthesis by theca and granulosa of preovulatory follicles in the pig. *Biol. Reprod.* **25**, 673–682.

Fernandez, L.A., Tarlatzis, B.C., Rzasa, P.J., Caride, J., Laufer, N., Negro-Vilar, A.F., DeCherney, A.H. & Naftolin, F. (1985) Renin-like activity in ovarian follicular fluid. *Fert. Steril.* **44**, 219–223.

Findlay, J.K. & Cumming, I.A. (1977) The effect of unilateral ovariectomy on plasma gonadotropin levels, estrus and ovulation rate in sheep. *Biol. Reprod.* **17**, 178–183.

Fogwell, R.L., Cowley, J.L., Wortman, J.A., Ames, N.K. & Ireland, J.J. (1985) Luteal function in cows following destruction of ovarian follicles at midcycle. *Theriogenology* **23**, 389–398.

Glencross, R.G., Munro, I.B., Senior, B.D. & Pope, G.S. (1973) Concentrations of oestradiol-17β, oestrone and progesterone in jugular venous plasma of cows during the estrous cycle and in early pregnancy. *Acta endocr., Copenh.* **73**, 374–384.

Glorioso, N., Atlas, S.A., Laragh, J.H., Jewelewicz, R. & Sealey, J.E. (1986) High concentrations of prorenin in human ovarian follicular fluid. *Endocrinology* **118** (Suppl.), 222, Abstr. 765.

Goodman, A.L. & Hodgen, G.D. (1983) The ovarian triad of the primate menstrual cycle. *Recent Prog. Horm. Res.* **39**, 1–73.

Hammond, J.M. (1981) Peptide regulators in the ovarian follicle. *Aust. J. biol. Sci.* **34**, 491–504.

Hammond, J.M., Lino, J., Baranao, S., Skaleris, D.,

Knight, A.B., Romanus, J.A. & Rechler, M. (1985) Production of insulin-like growth factors by ovarian granulosa cells. *Endocrinology* **117**, 2553–2555.

Hansel, W. & Echternkamp, S.E. (1972) Control of ovarian function in domestic animals. *Am. Zool.* **12**, 225–243.

Hedin, L., Ratoosh, S., Lifka, J., Durica, J., Jahnsen, T. & Richards, J.S. (1986) Hormonal regulation of mRNA for the regulatory subunit of type II cAMP-dependent protein kinase in ovarian granulosa cells. *Endocrinology* **118** (Suppl.), 221, Abstr. 761.

Hillensjo, T., Pomerantz, S.H., Schwartz-Kripner, A., Anderson, L.D. & Channing, C.P. (1980) Inhibition of cumulus cell progesterone secretion by low molecular weight fractions of porcine follicular fluid which also inhibit oocyte maturation. *Endocrinology* **106**, 584–591.

Hirshfield, A.N. & Midgley, A.R., Jr (1978) The role of FSH in the selection of large ovarian follicles in the rat. *Biol. Reprod.* **19**, 606–611.

Hisaw, F.L. (1947) Development of the graafian follicle and ovulation. *Physiol. Rev.* **27**, 95–119.

Hsu, C.J. & Hammond, J.M. (1986) Gonadotropins and estradiol regulate the production of insulin-like growth factor-I/Somatomedin C by porcine granulosa cells in culture. *Endocrinology* **118** (Suppl.), 290, Abstr. 761.

Hsueh, A.J. & Jones, P.B.C. (1981) Extrapituitary actions of gonadotropin-releasing hormone. *Endocr. Reviews* **2**, 437–461.

Hsueh, A.J., Adashi, E.Y., Jones, P.B.C. & Welsh, T.H., Jr (1984) Hormonal regulation of the differentiation of cultured ovarian granulosa cells. *Endocr. Reviews* **5**, 76–127.

Ireland, J.J. & Roche, J.F. (1982) Development of antral follicles in cattle after prostaglandin-induced luteolysis: changes in serum hormones, steroids in follicular fluid, and gonadotropin receptors. *Endocrinology* **111**, 2077–2086.

Ireland, J.J. & Roche, J.F. (1983a) Development of non-ovulatory antral follicles in heifers: changes in steroids in follicular fluid and receptors for gonadotropins. *Endocrinology* **112**, 150–156.

Ireland, J.J. & Roche, J.F. (1983b) Growth and differentiation of large antral follicles after spontaneous luteolysis in heifers: changes in concentration of hormones in follicular fluid and specific binding of gonadotropins to follicles. *J. Anim. Sci.* **57**, 157–167.

Ireland, J.J. & Roche, J.F. (1987) Hypothesis regarding development of dominant follicles during a bovine estrous cycle. In *Follicular Growth and Ovulation Rate in Farm Animals* (in press). Eds J. Roche & D. O'Callaghan. Martinus Nijhoff, The Hague.

Ireland, J.J., Curato, A. & Wilson, J. (1983) Effect of charcoal-treated bovine follicular fluid on secretion of LH and FSH in ovariectomized heifers. *J. Anim. Sci.* **57**, 1512–1515.

Ireland, J.J., Fogwell, R.L., Oxender, W.D., Ames, K. & Cowley, J.L. (1985) Production of estradiol by each ovary during the estrous cycle of cows. *J. Anim. Sci.* **59**, 764–771.

Jonassen, J.A., Bose, K. & Richards, J.S. (1982) Enhancement and desensitization of hormone responsive adenylate cyclase in granulosa cells of preantral and antral ovarian follicles: effects of estradiol and follicle-stimulating hormone. *Endocrinology* **111**, 74–79.

Josso, N. (1986) Anti Müllerian hormone: new perspective for a sexist molecule. *Endocr. Revs* **7**, 421–433.

Kesner, J.S. & Convey, E.M. (1982) Interaction of estradiol and luteinizing hormone releasing hormone on follicle stimulating hormone release in cattle. *J. Anim. Sci.* **54**, 817–825.

Knecht, M. & Feng, P. (1986) Transforming growth factor-beta inhibits LH receptor formation and enhances EGF receptor formation induced by FSH in rat granulosa cells. *Endocrinology* **118** (Suppl.), 32, Abstr. 17.

Kumari, G.L., Kumar, N., Duraiswami, S. & Roy, S. (1984) Ovarian lutropin receptor binding inhibitor. In *Gonadal Proteins and Peptides and their Biological Significance*, pp. 161–175. Eds M. R. Sairam & L. E. Atkinson. World Scientific Publ. Co., Singapore.

Leung, P.C.K. & Armstrong, D.T. (1980) Interactions of steroids and gonadotropins in the control of steroidogenesis in the ovarian follicle. *Ann. Rev. Physiol.* **42**, 71–82.

Ling, N., Ying, S.Y., Ueno, N., Esch, F., Denoroy, L. & Guillemin, R. (1985) Isolation and partial characterization of a Mr 32 000 protein with inhibin activity from porcine follicular fluid. *Proc. natn. Acad. Sci. U.S.A.* **82**, 7217–7221.

Ling, N., Ying, S.Y., Ueno, N., Shimasaki, S., Esch, F., Hotta, M. & Guillemin, R. (1986) Pituitary FSH is released by a heterodimer of the B-subunits from the two forms of inhibin. *Nature, Lond.* **19**, 779–782.

Lintern-Moore, S. & Moore, G.P.M. (1979) The initiation of follicle and oocyte growth in the mouse ovary. *Biol. Reprod.* **20**, 773–778.

Littman, B.A. & Hodgen, G.D. (1984) Pharmacologic ovarian stimulation in monkeys: Blockade of the luteinizing hormone surge by a highly transient ovarian factor. In *Gonadal Proteins and Peptides and their Biological Significance*, pp. 111–125. Eds M. R. Sairam & L. E. Atkinson. World Scientific Publ. Co., Singapore.

Makris, A., Ryan, K.J., Yasumizu, T., Hill, C.L. & Zetter, B.R. (1984) The nonluteal porcine ovary as a source of angiogenic activity. *Endocrinology* **15**, 1672–1677.

Matton, P., Adelakoun, V., Couture, Y. & Dufour, J.J. (1981) Growth and replacement of the bovine ovarian follicles during the estrous cycle. *J. Anim. Sci.* **52**, 813–818.

McNatty, K.P., Hunter, W.M., McNeilly, A.S. & Sawers, P.S. (1975) Changes in the concentration of pituitary and steroid hormones in the follicular fluid of human graafian follicle throughout the menstrual cycle. *J. Endocr.* **64**, 555–571.

McNatty, K.P., Heath, D.A., Henderson, K.M., Lun, S., Hurst, P.R., Ellis, L.M., Montgomery, G.W., Morrison, L. & Thurley, D.C. (1984) Some aspects of thecal and granulosa cell function during follicular development in the bovine ovary. *J. Reprod. Fert.* **72**, 39–53.

Meloni, F., Ben-Rafael, Z., Fateh, M. & Flickinger, G.L. (1986) Molecular weight variants of prolactin in human follicular fluid. *Biol. Reprod.* **34** (Suppl. 1), 104, Abstr. 110.

Miller, J.G.O. & Behrman, H.R. (1986) Oocyte maturation is inhibited by adenosine in the presence of FSH. *Biol. Reprod.* **35**, 833–837.

Nakano, R., Mizuno, T., Katayama, K. & Toja, S. (1975) Growth of ovarian follicles in rats in the absence of gonadotrophins. *J. Reprod. Fert.* **45**, 545–548.

Padmanabhan, V., Convey, E.M., Roche, J.F. & Ireland, J.J. (1984) Changes in inhibin-like bioactivity in ovulatory and atretic follicles and utero-ovarian venous blood after prostaglandin-induced luteolysis in heifers. *Endocrinology* **115**, 1332–1340.

Pathiraja, N., Carr, W.R., Fordyce, M., Forster, J., Land, R.B. & Morris, B.A. (1984) Concentration of gonadotrophins in the plasma of sheep given gonadal steroid antisera to raise ovulation rate. *J. Reprod. Fert.* **72**, 93–100.

Rahe, C.H., Owens, R.E., Fleeger, J.L., Newton, J.H. & Harms, P.G. (1980) Pattern of plasma luteinizing hormone in the cyclic cow: dependence upon period of the cycle. *Endocrinology* **107**, 498–503.

Ranta, T., Huhtaniemi, I., Jalkanen, J., Ritvos, O. & Ylikorkala, O. (1986) Activation of protein kinase C stimulates human granulosa-luteal cell prostacyclin production. *Endocrinology* **118** (Suppl.), 67, Abstr. 147.

Ratoosh, S.L. & Richards, J.S. (1985) Regulation of the content and phosphorylation of R_{11} by adenosine 3′,5′-monophosphate, follicle-stimulating hormone, and estradiol in cultured granulosa cells. *Endocrinology* **117**, 917–927.

Rein, M.S. & Schomberg, D.W. (1982) Characterization of insulin receptors on porcine granulosa cells. *Biol. Reprod.* **26** (Suppl. 1), 113, Abstr.

Richards, J.S. (1980) Maturation of ovarian follicles: Actions and interactions of pituitary and ovarian hormones on follicular cell differentiation. *Physiol. Rev.* **60**, 51–89.

Richards, J.S. & Rolfes, A.I. (1979) Hormonal regulation of cyclic AMP binding to specific receptor proteins in rat ovarian follicles. *J. biol. Chem.* **255**, 5481–5489.

Richards, J.S., Haddox, M., Tash, J.S., Walter, U. & Lohmann, S. (1984) Adenosine 3′5′-monophosphate-dependent protein kinase and granulosa cell responsiveness to gonadotropins. *Endocrinology* **114**, 2190–2198.

Richards, J.S., Ireland, J.J., Rao, M.C., Bernath, G.A. & Midgley, A.R., Jr (1976) Ovarian follicular development in the rat: Hormone receptor regulation by estradiol, follicle stimulating hormone and luteinizing hormone. *Endocrinology* **99**, 1562–1570.

Richards, J.S., Sehgal, N. & Tash, J.S. (1983) Changes in content and cAMP-dependent phosphorylation of specific proteins in granulosa cells of preantral and preovulatory ovarian follicles and in corpora lutea. *J. biol. Chem.* **258**, 5227–5232.

Sairam, M.R. & Atkinson, L.E. (1984) The trail of gonadal proteins and peptides. In *Gonadal Proteins and Peptides and their Biological Significance*, pp. 3–6. Eds M. R. Sairam & L. E. Atkinson. World Scientific Publishing Co., Singapore.

Sanzo, M.A. & Reichert, L.E., Jr (1982) Gonadotropin receptor binding regulators in serum. *J. biol. Chem.* **257**, 6033–6040.

Sato, E., Ishibashi, T. & Iritani, A. (1982) Purification and action sites of a follicle stimulating hormone inhibitor from bovine follicular fluid. *J. Anim. Sci.* **55**, 873–877.

Sheldrick, E.L. & Flint, A.P.F. (1984) Ovarian oxytocin. In *Gonadal Proteins and Peptides and their Biological Significance*, pp. 257–272. Eds M. R. Sairam & L. E. Atkinson. World Scientific Publ. Co., Singapore.

Skinner, M.K., McKeracher, H.L. & Dorrington, J.H. (1985) Fibronectin as a marker of granulosa cell cytodifferentiation. *Endocrinology* **117**, 886–892.

Sluss, P.M. & Reichert, L.E., Jr (1984) Porcine follicular fluid contains several low molecular weight inhibitors of follicle-stimulating hormone binding to receptor. *Biol. Reprod.* **30**, 1091–1104.

Smeaton, T.C. & Robertson, H.A. (1971) Studies on the growth and atresia of Graafian follicle in the ovary of the sheep. *J. Reprod. Fert.* **25**, 243–252.

Tonetta, S.A. & Ireland, J.J. (1984) Effect of cyanoketone on follicle-stimulating hormone (FSH) induction of receptors for FSH in granulosa cells of the rat. *Biol. Reprod.* **31**, 487–493.

Tonetta, S.A., Spicer, L.J. & Ireland, J.J. (1985) CI-628 inhibits follicle-stimulation hormone (FSH)-induced increases in FSH receptors of the rat ovary: requirement of estradiol for FSH action. *Endocrinology* **116**, 715–722.

Tsafriri, A., Pomerantz, S.H. & Channing, C.P. (1976) Inhibition of oocyte maturation by porcine follicular fluid: partial characterization of the inhibitor. *Biol. Reprod.* **14**, 511–516.

Vale, W., Rivier, J., Vaughan, J., McClintock, R., Corrigan, A., Woo, W., Karr, D. & Spiess, J. (1986) Purification and characterization of an FSH releasing protein from porcine follicular fluid. *Nature, Lond.* **19**, 776–779.

Webb, R. & Gauld, I.K. (1987) Endocrine control of follicular growth in the ewe. In *Follicular Growth and Ovulation Rate in Farm Animals* (in press). Eds J. Roche & D. O'Callaghan. Martinus Nijhoff, The Hague.

Welschen, R., Dullaart, J. & DeJong, F.H. (1979) Interrelationships between circulating levels of estradiol-17β, progesterone, FSH, and LH immediately after unilateral ovariectomy in the cycling rat. *Biol. Reprod.* **18**, 421–427.

Yanagishita, M., Hascall, V.C. & Rodbard, D. (1981) Biosynthesis of proteoglycans by rat granulosa cells cultured in vitro: modulation by gonadotropins, steroid hormones, prostaglandins, and a cyclic nucleotide. *Endocrinology* **109**, 1641–1649.

Ying, S.Y., Ling, N., Bohlen, P. & Guillemin, R. (1981) Gonadocrinins: peptides in ovarian follicular fluid stimulating the secretion of pituitary gonadotropins. *Endocrinology* **108**, 1206–1215.

Ying, S.Y., Becker, A., Baird, A., Ling, N., Ueno, N., Esch, F. & Guillemin, R. (1986a) Type beta transforming growth factor (TGF-B) is a potent stimulator of the basal secretion of follicle stimulating hormone (FSH) in a pituitary monolayer system. *Biochem. Biophys. Res. Commun.* **135**, 950–956.

Ying, S.Y., Becker, A., Ling, N., Ueno, N. & Guillemin, R. (1986b) Inhibin and beta type transforming growth factor (TGFβ) have opposite modulating effects on the follicle stimulating hormone (FSH)-induced aromatase activity of cultured rat granulosa cells. *Biochem. Biophys. Res. Commun.* **136**, 969–974.

Zeleznik, A.J. (1981) Premature elevation of systemic estradiol reduces serum levels of follicle-stimulating hormone and lengthens the follicular phase of the menstrual cycle in Rhesus monkeys. *Endocrinology* **109**, 352–355.

Zeleznik, A.J., Schuler, H.M. & Reichert, L.E., Jr (1981) Gonadotropin-binding sites in the Rhesus monkey ovary: role of the vasculature in the selective distribution of human chorionic gonadotropin to the preovulatory follicle. *Endocrinology* **109**, 356–361.

Zeleznik, A.J., Hutchison, J.S. & Schuler, H.M. (1985) Interference with the gonadotropin-suppressing actions of estradiol in Macaques overrides the selection of a single preovulatory follicle. *Endocrinology* **117**, 991–999.

J. Reprod. Fert., Suppl. **34** (1987), 55–69

Molecular and cellular changes associated with maturation and early development of sheep eggs

R. M. Moor* and F. Gandolfi*

AFRC Institute of Animal Physiology, 307 Huntingdon Road, Cambridge CB3 0QJ, U.K.

Introduction

It has long been postulated in non-mammalian species that successful embryogenesis depends directly on an ordered sequence of events in oogenesis (Wilson, 1925). Oogenesis in these species is recognized as the phase of synthesis and storage of intracellular components whilst early embryogenesis is the period of distribution and utilization of stored product. A clear example of the interrelationship between oogenesis and embryogenesis is provided by the eggs of the toad, *Xenopus laevis*. In this species the single-celled egg at fertilization contains all the components and information required for the development of the swimming tadpole which consists of approximately 50 000 cells (Gurdon, 1974). Although this extreme degree of independence from extracellular support is unlikely to apply directly to mammals, it is nevertheless critical to identify the extent to which mammalian embryogenesis is regulated by the products of oogenesis. It is our purpose to answer this question by describing the intracellular events during oogenesis in sheep and relating these to the control of fertilization and early development.

Over what developmental time-scale does oogenesis occur in mammals? The process is initiated when the primordial germ cells invade the genital ridge of the embryo. After colonization of the early gonad the germ cells undergo a period of mitotic activity before entering meiosis and progressing to the dictyate stage of meiotic prophase. At this point the cell cycle is interrupted and the oocyte, containing a large nucleus referred to as a germinal vesicle (GV), remains in meiotic arrest for all but the last few hours of oogenesis. In addition to nuclear arrest the oocyte, surrounded by a single layer of flattened cells, constitutes part of the non-growing or resting pool of primordial follicles for much of its postnatal existence. The limited amount of biochemical evidence available about oocytes in primordial follicles suggests that they are merely synthesizing 'housekeeping' proteins during the resting period. However, a small number of primordial follicles enter the growing pool each day. It is with the developmental events initiated in the oocyte at this time and terminating during embryogenesis when maternal regulation ceases that this paper deals. Although the sheep oocyte and embryo will serve as a model for the paper, information from other mammals is used to compensate for deficiencies in our knowledge of oogenesis in this species.

At least three distinct developmental programmes direct the molecular changes which occur during oogenesis and early embryogenesis. A growth programme regulates differentiation in the immature oocyte (Canipari *et al.*, 1984) while a separate maturation programme regulates the reprogramming of the oocyte before ovulation. The entry of the spermatozoon initiates an early embryonic programme which persists until maternal regulation is terminated and development becomes directed by the embryonic genome (Howlett & Bolton, 1985). A later embryonic programme, which will not be discussed in this communication, begins with the expression of the

---

*Present address: AFRC Institute of Animal Physiology and Genetics Research, Babraham, Cambridge CB2 4AT, U.K.

embryonic genes and is an important regulator of events culminating in the formation of the blastocyst (Johnson *et al.*, 1984).

If the growth, maturation and fertilization programme provide the framework for early development what molecular and cellular events occur during each of these phases? Furthermore, how is each programme regulated and what effects do early events exert on later development?

Oocyte growth phase

Morphology and biology of oocyte growth

Two features characterize the oocyte at the start of its growth phase. The first is its small size and the second is the total inability of the early oocyte to progress from the G2 to the M phase of the meiotic cycle (Szybek, 1972). Both of these characteristics are radically altered during the growth phase. Firstly, the oocyte, after exit from the resting state, grows rapidly and its volume increases 350-fold within the first 2 weeks (Wassarman *et al.*, 1981). About half the amount of protein required for this rapid growth is synthesized by the oocyte (Schultz *et al.*, 1979). The remainder is almost certainly synthesized by somatic cells and taken up by the oocyte in a non-degraded form (Glass & Cons, 1968).

The highly active metabolic state of the growing oocyte imposes demands on the oolemma which are far in excess of the limited uptake characteristics of this membrane (reviewed by Moor, 1983; Schultz, 1985). Thus, oocytes isolated from direct contact with surrounding follicle cells fail to grow, while those maintaining normal oocyte–follicle cell contacts grown normally *in vitro* (Eppig, 1979). Investigations of the precise nature of these heterologous contacts reveal that they are formed by cellular processes which extend across the zona pellucida and terminate on the oocyte membrane where gap junctions are formed (Anderson & Albertini, 1976). The intercellular passage of molecules through these junctions provides the high substrate uptake required by oocytes during the growth phase (Moor, 1983; Schultz, 1985).

The second feature of the resting oocyte, its inability to progress beyond diplotene, changes during the final part of the growth phase. Studies of mice, pigs and sheep indicate that oocytes acquire the ability to proceed to metaphase I when they reach about 80% of their full size, but are unable to progress beyond this to metaphase II until their growth is complete (Sorensen & Wassarman, 1976; Motlik *et al.*, 1984; R. M. Moor, unpublished observations).

The growth of the oocyte, the dependence on intact oocyte–follicle cell contact and the acquisition of meiotic competence provide valuable markers of cell differentiation and development. What molecular changes occur during this phase of differentiation and what subsequent biological actions do these changes regulate?

Biochemistry of oocyte growth

RNA synthesis. Autoradiographic analyses (Crozet *et al.*, 1981) demonstrate that growing pig oocytes actively synthesize both ribosomal (r)RNA and heterogeneous (hn)RNA. A significant reduction in the synthesis of rRNA is observed as pig oocytes approach their full size. More rigorous quantitative analyses of RNA synthesis have regrettably not yet been undertaken for this or other domestic animals. By contrast the detailed analyses of RNA synthesis and stability undertaken in mouse oocytes provide a model for domestic animals. The major classes of RNA have been studied during mouse oogenesis by pulse-labelling procedures (see Kaplan *et al.*, 1982, for references). Total RNA, rRNA, tRNA, poly(A)$^+$RNA and ribosomes accumulate steadily throughout growth. The bulk of the newly synthesized RNA is remarkably stable during the growth phase but degradation increases, once maturation is initiated. About 20% of maternal RNA is degraded during maturation, a further 40% by the 2-cell stage and 30% on Day 3. Despite this dramatic post-fertilization loss of RNA, Bachvarova & De Leon (1980) calculated that the amount of mRNA stored in the mouse egg is sufficient to direct protein synthesis up to the 8-cell stage.

In summary, the accumulated information indicates that mammalian oocytes, like those of non-mammalian species (Davidson, 1976), sequester large amounts of RNA during the growth phase. The sequestered ribosomes and RNA are packaged into lattices or ribonuclear particles and remain inactive in the oocyte for extended periods of time. To what extent are proteins like RNA, also synthesized and stored during the growth phase for utilization during early development?

Protein synthesis. Oocyte growth involves not only a substantial enlargement of the cell but also a major reorganization of its metabolism and gene expression. The increase in oocyte volume is accompanied by a proportional increase in the amino acid pool and a linear increase in protein synthesis (Wassarman *et al.*, 1981). Detailed electrophoretic analyses of proteins at different stages of growth reveal that the synthesis of structural and other housekeeping proteins is accompanied by the synthesis of a changing pattern of stage-specific proteins. In a series of investigations on these proteins Wassarman *et al.* (1981) have identified some early proteins whose function becomes critical in later embryonic development. For example, histone-H4, some of the ribosomal proteins and a so-called germinal vesicle associated protein (GVAP) are all sequestered in the nucleus and used during maturation and development. An even more dramatic example of the early synthesis of proteins which will then be sequestered and used specifically for later developmental events relates to the production of three glycoproteins. These products, representing at least 10% of the total synthesis of growing oocytes, are secreted and form the zona pellucida (Wassarman *et al.*, 1981). The smallest of these glycoproteins in the mouse zona will subsequently function as the sperm receptor; modifications to the other zona glycoproteins fulfil the further role of forming the block of polyspermy after fertilization has occurred.

Regulators of oocyte growth. Indirect evidence suggests that the synthesis of at least some of the developmental regulator proteins is controlled by a programme which is unrelated to the growth of the oocyte. For example, growing oocytes freed of follicle cells and cultured thereafter on fibroblast monolayers survive but do not grow (Canipari *et al.*, 1984). Nevertheless, even these non-growing oocytes acquire meiotic competence at precisely the same temporal stage as do their growing counterparts. The results of Canipari *et al.* (1984) suggest that a programme is initiated in the newly activated primordial follicle which then progresses independently through a series of molecular sequences to its conclusion when meiotic competence is attained.

At the completion of the growth phase the oocyte has a store of components required for subsequent development. It has also acquired the ability to progress from the dictyate to the metaphase II stage of the cell cycle. The oocyte is not, however, able to become fertilized or support early embryonic development at this stage. The acquisition of developmental competence depends on a further phase of differentiation, initiated *in vivo* by the release of LH before ovulation. What intracellular changes confer developmental competence on the ovine oocyte?

Oocyte maturation phase

Biology of maturation

After completion of growth the germinal vesicle oocyte is characterized by a relatively impermeable membrane, intimate association with adjacent follicle cells through gap-junctions, a peripheral location of many cellular organelles, a low rate of transcription and a stable pattern of protein synthesis (Fig. 1). The role of the somatic compartment at this time is crucial both for the provision of metabolic support and for the regulation of nuclear and synthetic activity in the oocyte (Moor & Osborn, 1983). The increased amounts of gonadotrophins released at oestrus bind to the follicle cells and alter both the signals (Fig. 2) and their means of transmission within the follicle

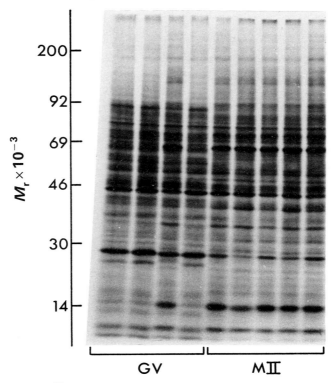

Fig. 1. Fluorographs of [^{35}S]methionine-labelled polypeptides from individual oocytes before the initiation of maturation (GV) and after its completion (MII). Cumulus-enclosed oocytes were removed from the ovaries of sheep 40 h after FSH stimulation (GV stage) or from super-ovulated sheep slaughtered 18 h after the onset of oestrus (MII). Note that the protein pattern in one of the GV oocytes (track 3) has been prematurely activated by the FSH treatment.

(Moor, 1983). The resultant changes in the oocyte and the consequences of these maturational changes on fertilization and early embryogenesis are the subject of the remainder of this paper.

The maturation timetable

The acquisition of developmental competence in the sheep oocyte occurs over a 24-h period before ovulation and involves two phases. In the first 6-h phase the oocyte undergoes few structural or synthetic changes but appears instead to be undergoing programming by the somatic elements (Moor & Warnes, 1977). In the longer (18 h) second phase most components of the oocyte undergo reorganization; it is unlikely that the follicle cells have an important influence during this period.

Membrane-related changes during maturation

Immediately before the LH surge the oocyte is coupled to the surrounding follicle cells through a highly developed system of gap-junctional contacts. These junctions, which occupy approxi-mately 1% of the total membrane area, mediate the entry of nucleosides, sugars, phospholipid precursors, amino acids and signal molecules (Moor, 1983). The intercellular passage of these substrates into the oocyte remains at a consistently high level for the first 12 h of maturation and then declines sharply in the following 3 h. This reduction in junctional transmission is compensated

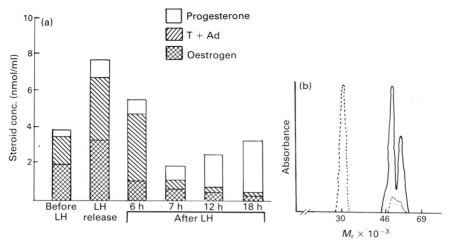

Fig. 2. Changes in intrafollicular signals generated during the 24 h period of maturation in the sheep. (a) The time-dependent fluctuations in both total steroid output and in the proportions of unconjugated oestradiol-17β, androgen (testosterone plus androstenedione) and progesterone in the follicular fluid of preovulatory follicles (Osborn & Moor, 1983b). (b) A densitometric trace of the changes in the profile of polypeptides secreted by the granulosa cells during maturation; a polypeptide complex (M_r 46 000–60 000) secreted before the LH surge (solid trace) disappears and is replaced during maturation by a polypeptide with a relative molecular mass of 30 000 (R. M. Moor & I. M. Crosby, unpublished observations).

for by a significant increase in the carrier-mediated uptake of substrates across the membrane itself (Moor & Smith, 1979).

Degenerative changes in the fine processes and junctions, induced primarily by FSH rather than LH, account for the reduced transmission of substrates into maturing oocytes (Moor & Cran, 1980). Some workers assert that the disruption of junctional contact also prevents the entry of an inhibitor of meiosis into the oocyte and thereby initiates nuclear maturation (Dekel & Beers, 1978). This explanation appears unlikely to apply directly to meiotic regulation in sheep oocytes because nuclear change precedes junctional disruption in this species (Moor et al., 1981). The possibility of an alternative form of junctional regulation has, however, recently been highlighted by Larsen et al. (1986). These workers report on an early and selective uncoupling of adjacent cumulus cells in rats injected with hCG, at a stage when the heterologous coupling between the oocyte and the corona is still fully maintained. The corona–oocyte unit may therefore be isolated from the granulosa compartment (and its inhibiting signals) at an early stage in maturation by a band of non-communicating cumulus cells.

It has further been postulated that the disruption of cumulus–oocyte coupling initiates cytoplasmic remodelling in the maturing oocytes of sheep and other species (Szollosi et al., 1978). To what extent and for what purpose does this postulated relocation of organelles take place in the oocyte during maturation?

Structural remodelling during oocyte maturation

Ultrastructural studies by Zamboni (1970) were the first to show that maturation is accompanied by a major reorganization of organelles such as mitochondria, lysosomes and cortical granules within the oocyte. The development of perinuclear microtubule organizing centres (MTOCs) during maturation, and the establishment of a temporal relationship between LH release and intracellular remodelling have extended these initial observations (Szollosi et al., 1972; Kruip

R. M. Moor and F. Gandolfi

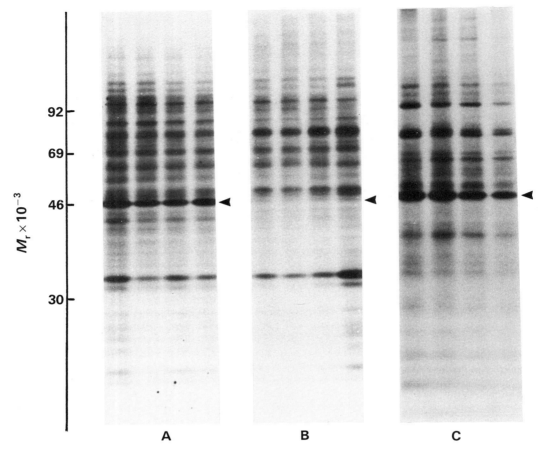

Fig. 3. Effect of somatic cells on the synthesis of β-actin (arrowed) by sheep oocytes incubated in [^{35}S]methionine. Fluorographs on gel A show that individual cumulus-enclosed oocytes synthesize substantial amounts of actin. Actin synthesis is immediately terminated by the removal of the cumulus cells although the synthesis of all the other polypeptides continues (gel B). Actin synthesis is re-initiated in the denuded oocyte when the zona pellucida is removed to enable the gamete to reform junctional contact with a wide range of somatic cells (gel C).

et al., 1983; Van Blerkom & Runner, 1984). The mitochondria are located in a peripheral position whilst rough endoplasmic reticulum (RER) and membrane-bound vesicles including cortical granules occupy a more central position in germinal vesicle oocytes of domestic animals. After the breakdown of the germinal vesicle the RER disappears and clusters of mitochondria and lipid droplets become apparent (Kruip *et al.*, 1983). During the final phase of maturation the clusters of organelles disperse and become centrally located leaving the cortical granules in the relatively organelle-free region immediately adjacent to the plasma membrane.

It is reported by Szollosi *et al.* (1978) that the migration and alignment of cortical granules beneath the plasma membrane depend on the disruption of the junctional contact between the oocyte and somatic cells. Equally, Van Blerkom (1985) argues that the relocation of organelles is mediated by microtubules radiating from perinuclear MTOCs that develop during the GV to metaphase I transition. Although the inductive signals and molecular mechanisms underlying intracellular modelling are still uncertain, no doubt exists about the biological importance of remodelling

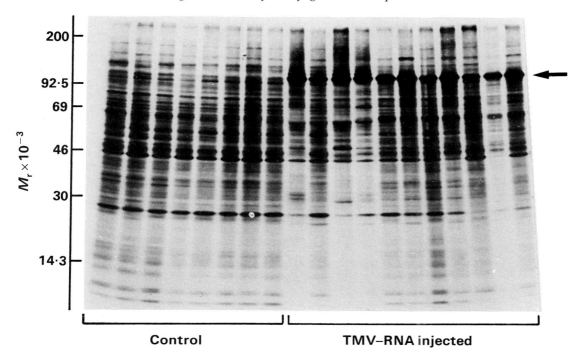

Fig. 4. Fluorograph of ^{35}S-labelled polypeptides from individual oocytes injected with buffer (controls) or with TMV-RNA (10 pg TMV-RNA per oocyte). The dominant early TMV poly-peptide (M_r 110 000) is arrowed. Production of TMV-encoded proteins is accompanied by a corresponding decrease in the synthesis of endogenous polypeptides (R. M. Moor, unpublished observations).

during maturation. At fertilization exocytosis of cortical granules occurs, and the membrane and zona pellucida are modified and form a block to the entry of additional spermatozoa. Any failure in the alignment of cortical granules beneath the membrane prevents exocytosis and results invariably in polyspermy.

Reprogramming of protein synthesis

The final 24 h of differentiation in sheep oocytes is accompanied by major changes in protein synthesis (Fig. 1); these changes are obligatory events in the preparation of female gametes for fertilization (Warnes *et al.*, 1977; Thibault, 1977). This reprogramming process involves a slight decline in the overall rate of synthesis and changes in the relative rates of synthesis, degradation and post-translational modification of specific proteins (Wassarman *et al.*, 1981; Crosby *et al.*, 1984; Johnson *et al.*, 1984). Whilst the majority of these changes take place after the breakdown of the nucleus some important changes precede and regulate GVBD in domestic animals (Moor & Crosby, 1986).

Two major questions arise from the reprogramming of protein synthesis during maturation. Firstly, what mechanisms regulate the reprogramming process and secondly, what role do the new proteins play in development?

Regulation of protein reprogramming. Various regulatory mechanisms influence protein synthesis during the maturation process. Indirect evidence from RNA inhibitor studies suggests that new transcription and the resultant limited change in mRNA populations regulates a small number

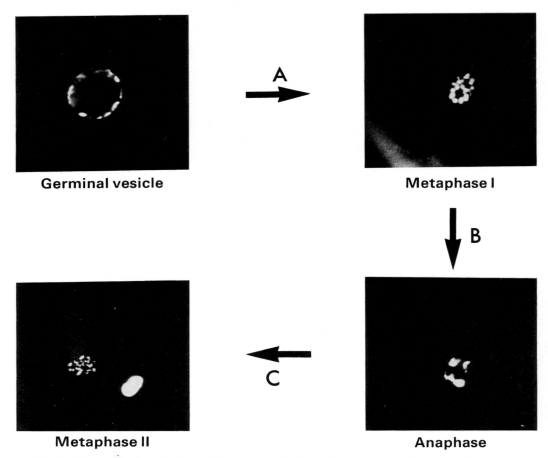

Fig. 5. Changes in the meiotic cycle from the initiation of maturation to its completion. New protein synthesis is required to drive the meiotic cycle into metaphase I (phase A), for the transition from metaphase I to anaphase (phase B) and for the inhibition of the cycle at metaphase II (phase C).

of early protein changes in sheep oocytes (Osborn & Moor, 1983a; Moor & Crosby, 1986). The great majority of changes in the oocytes of sheep are, however, not dependent on new transcription but are induced primarily by alterations to existing polypeptides by phosphorylation or glycosylation (Crosby *et al.*, 1984).

Changes in the rates of utilization of specific subsets of mRNA represent a further major regulator of protein change during maturation. The means by which stored message is selectively activated is unclear but cannot be explained by general changes in the 5'-capped or 3'-tailed status of the mRNA or by variations in the ability to process primary transcripts (see Johnson *et al.*, 1984). On the other hand, translation of at least one species of mRNA, that coding for β-actin, is regulated by the degree of junctional coupling that exists between the oocyte and the adjacent somatic cells (Fig. 3a). Disruption of junctional contact immediately terminates actin mRNA translation in sheep oocytes despite the continual presence of the transcriptions in the cytoplasm (Fig. 3b). Re-establishment of junctional contact with follicle cells or other somatic cells (Fig. 3c) rapidly re-initiates actin mRNA translation (Moor & Osborn, 1983). Experimental evidence from somatic cells suggests that changes in cell cytoarchitecture, induced by alterations to cell

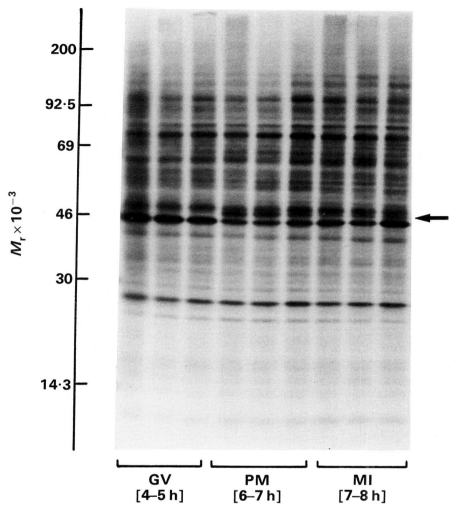

Fig. 6. Fluorographs of polypeptide profiles synthesized by oocytes labelled with [^{35}S]methionine for short periods during the 4 h before germinal vesicle breakdown. The appearance of a polypeptide (arrowed) with a relative molecular mass of 47 000 correlates closely with the breakdown of the germinal vesicle (Moor & Crosby, 1986).

shape and contact, affect actin mRNA translation by changing the association between mRNA, initiation factor and ribosomes (Farmer *et al.*, 1983; Howe & Hershey, 1984). A similar alteration to the cytoarchitecture of the oocyte might occur when junctional contact is disrupted and this could account for the cessation of actin mRNA translation in these cells. An alternative explanation has arisen from a careful examination of the actin mRNA molecule itself (Bachvarova *et al.*, 1987). Their work has indicated that actin mRNA becomes deadenylated in mouse oocytes after junctional disruption; it is postulated that the process of deadenylation may itself regulate translation of actin mRNA. However, junctional coupling, whilst important for the translation of some messages, does not appear to regulate translation of many other mRNA subsets. Although of central importance to maturation, the mechanisms involved in the translation of these other mRNA species are still uncertain. Similar uncertainty surrounds the mechanisms that selectively

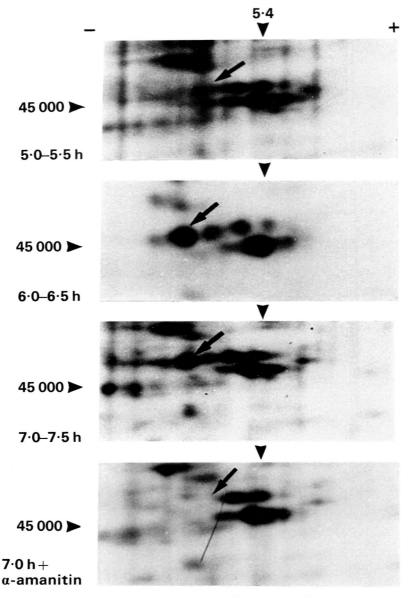

Fig. 7. Two-dimensional separations of the polypeptide m (arrowed) which is closely temporally related to the breakdown of the germinal vesicle. The inhibition of transcription with α-amanitin during the first 2 h after the induction of maturation prevents the synthesis of polypeptide m and blocks the breakdown of the germinal vesicle (Moor & Crosby, 1986).

alter the rate of synthesis and stability of certain polypeptides during maturation. Our approach to the study of these questions has firstly been to determine the efficiency of the translation apparatus in sheep oocytes (R. M. Moor, unpublished observation). Maturing oocytes have been micro-injected with Tobacco Mosaic Virus–RNA (TMV–RNA), then labelled with [^{35}S]methionine and the resultant protein profiles quantitated by densitometry. The results clearly indicate that TMV-polypeptides are synthesized by injected oocytes (Fig. 4). In addition, it is clear from the quantitative analyses that the proportion of endogenous polypeptides synthesized is reduced in direct proportion to the amount of TMV-polypeptide produced (data not shown).

The results demonstrate that exogenous mRNA competes for a strictly limited translational capacity within the oocyte. This is consistent with the hypothesis of Laskey *et al.* (1977) that the translational capacity of the oocyte is fully saturated during maturation. It follows that the translation of new mRNA must be accompanied by a reduction in the translation of existing messages.

Role of proteins synthesized during maturation. The occurrence of abnormalities in oocytes during fertilization is the most obvious consequence of incomplete maturation (Thibault, 1977). Thus, both the block to polyspermy and the cytoplasmic factors required to decondense sperm chromatin are absent in ovine oocytes fertilized during the GV or metaphase I stage of meiosis. Similar effects on fertilization can be induced by altering the steroid signals during the early phase of maturation (Moor *et al.*, 1980). Quantitative analyses of polypeptide patterns indicate that the ability to decondense sperm chromatin is associated with the synthesis of a small group of polypeptides between 12 and 18 h after the induction of maturation (Osborn & Moor, 1983b).

It is, moreover, known that some proteins synthesized during maturation persist throughout early development. While we are unable to ascribe particular functions to these proteins indirect evidence suggests that they may still exert an important effect up to blastulation (Moor & Trounson, 1977). It was observed in those experiments that intrafollicular oocytes matured *in vitro* in the presence of sub-optimal levels of gonadotrophin reached metaphase II, became fertilized and cleaved but failed to blastulate. If some proteins synthesized during maturation are possibly important at the blastocyst stage, others are required immediately after synthesis as regulators of the meiotic cycle. We have used protein blocks and two-dimensional gel electrophoresis to determine the relationship between protein synthesis and progression through the meiotic cycle (Moor & Crosby, 1986).

Meiotic cycle control mechanisms

Relationship between RNA synthesis, proteins and meiosis. Figure 5 shows both the sequence of nuclear events and the stages during meiosis at which new protein synthesis is essential for the continued progression of the cycle. Our results show that an early set of new proteins is required to drive the meiotic cycle from prophase to metaphase I and a second period of synthesis is required to facilitate progression from metaphase I to anaphase (Moor & Crosby, 1986). The third set of meiotic proteins is synthesized in the last 4–6 h of maturation and exerts an inhibitory influence on meiosis. These results are interesting, not only because they identify an essential period of synthesis, but also because they underline important differences between meiosis in rodents and domestic animals. No protein synthesis or transcription is required for the breakdown of the germinal vesicle in mouse oocytes (Stern *et al.*, 1972; Wassarman *et al.*, 1981). By contrast, both transcription and new protein synthesis are essential for the breakdown of the germinal vesicle in sheep oocytes. Experiments using the polymerase II inhibitor, α-amanitin, indicate that a 2 h period of RNA synthesis at the beginning of maturation is a prerequisite for GVBD (Osborn & Moor, 1983a). Moreover, protein synthesis is required 4–6 h later; the germinal vesicle breaks down about 90 min after the synthesis of these proteins (Moor & Crosby, 1986). The results of short-term radio-labelling experiments (Fig. 6) indicate that a protein with a relative molecular mass of $\sim 47\,000$ and an isoelectric point of 5·8 is probably involved in the transition from prophase to metaphase I. Moreover, this protein is probably coded for by transcripts synthesized at the initiation of maturation since inhibition of transcription selectively inhibits the synthesis of this protein (Fig. 7).

Intrafollicular inhibitors of meiosis. From the above results it is perhaps not surprising that the regulators of meiotic maturation in mice and sheep do not appear to be similar. Exhaustive studies, initiated in amphibian oocytes and extended to the mouse, indicate that a phosphoprotein, a

R. M. Moor and F. Gandolfi

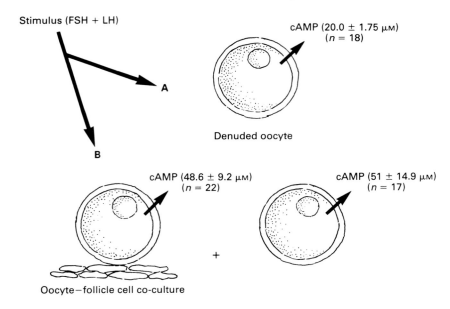

Fig. 8. The effect of gonadotrophins and follicle cell signals on the concentrations of cyclic AMP in ovine oocytes. Oocytes denuded of follicle cells and then exposed to gonadotrophins (panel A) have cyclic AMP concentrations not significantly different from those of untreated oocytes (see Moor & Heslop, 1981). By contrast gonadotrophins added to cumulus-enclosed oocytes (data not shown) or denuded oocytes co-cultured directly on follicle cells or in the same culture dish as follicle cells (panel B) significantly increase intraoocyte concentrations of cyclic AMP (Crosby *et al.*, 1985).

substrate of protein kinase, maintains meiotic arrest in mouse oocytes (Maller & Krebs, 1977; Bornslaeger *et al.*, 1986). A decrease in intraoocyte cyclic AMP and an accompanying decrease in protein kinase results in the net dephosphorylation of the inhibitory phosphoprotein and the consequent resumption of meiosis (Bornslaeger *et al.*, 1986). In sheep oocytes, cyclic AMP concentrations do not decline at the initiation of maturation but instead increase significantly at this time (Moor & Heslop, 1981). It appears further that the increase in intraoocyte cyclic AMP levels during maturation is initiated by gonadotrophins acting on the follicle cells (Crosby *et al.*, 1985). Signals from the follicle cells, acting as distance activators, stimulate the adenylate cyclase system, and consequently increase the cyclic AMP concentration, in the oocyte (Fig. 8). However, even though cyclic AMP does not appear to act as the meiotic inhibitor in sheep oocytes no other inhibitory factor has yet been identified in this species. It should be stressed that the requirement for early transcription and protein synthesis before GVBD in sheep oocytes presents novel possibilities for cell cycle regulation which warrant investigation.

The end of oogenesis is marked by the completion of intracellular reorganization and the formation of the second metaphase plate. However, it is not until the pool of stored mRNA is exhausted in early embryogenesis that the oocyte programmes cease to be a dominant directive force in development. At this time maternal regulation is replaced by the regulation exerted by the embryonic genome. The period from the end of oogenesis to the onset of genomic control by the embryo will be designated as the early embryonic phase. This phase encompasses both fertilization and the earliest materno–embryonic recognition processes. Since both these events are the subject of detailed consideration elsewhere (First & Parrish, 1987; Sasser & Ruder, 1987) attention will be focussed here solely on the utilization of maternal transcripts during early embryogenesis.

Early embryonic phase

The early embryonic phase, representing a period of distribution and utilization of stored components, is under dual molecular control in mouse eggs (Pratt *et al.*, 1983). The maturation programme, which confers on the oocyte the potential to initiate development, continues throughout early development to direct the housekeeping functions of the cell. Superimposed on this is a sequence of specialized changes in synthesis which are initiated or accelerated by sperm penetration. Howlett & Bolton (1985) postulate further that this fertilization programme is not only the direct initiator of embryogenesis but also performs the function of terminating the oocyte programme.

The results of our experiments in sheep are entirely compatible with the above dual control concept. Protein synthesis in unfertilized sheep oocytes continues unchanged for many hours after ovulation (I. M. Crosby and R. M. Moor, unpublished observations). Fertilization imposes on this oocyte-directed pattern of synthesis specific but limited changes in protein profiles. Despite similarities in control mechanisms, the duration of the resultant early embryonic phase differs markedly between species. Thus, the transition from the maternally-directed early embryonic phase to the later embryo-directed phase occurs at the mid-2-cell stage in mice, the 4–8-cell stage in rabbits and the 8–16-cell stage in sheep (Manes, 1973; Young *et al.*, 1978; I. M. Crosby & R. M. Moor, unpublished observations). It is with the molecular events that occur in the egg before the transition that our current work is directed.

Protein and RNA utilization during the early embryonic phase

Protein modulation. Fertilization is the trigger for a number of protein changes in the sheep egg. Firstly, studies on the effects of acrosin suggest that certain macromolecules associated with the zona undergo limited proteolysis during fertilization (Brown, 1986). The glycoproteins most affected by sperm enzymes are probably not native to the egg: these glycoproteins are produced by the oviduct and thereafter bind avidly to the zona pellucida (Brown, 1986). Other modifications induced by fertilization include changes to two intracellular polypeptides ($M_r = 47\,000$ and $67\,000$), but we are not able at present to ascribe specific functions to these proteins.

Once these fertilization-induced modifications are completed no other consistent changes in the polypeptide profiles are seen until the onset of genomic control by the embryo (I. M. Crosby & R. M. Moor, unpublished observations).

Although the profile of polypeptides remains constant at this time major changes occur in the rates of amino acid incorporation during the early embryonic phase (I. M. Crosby & R. M. Moor, unpublished observations). Between the 4-cell and 8-cell stage the rate of [^{35}S]methionine incorporation, after correction for changes in precursor uptake, declines precipitously and remains low until after the transition to the embryo-directed phase of development. The fall in amino acid incorporation, and thus in the synthesis of new protein, does not imply that sequestered proteins are also degraded during this period. Indeed, indirect evidence presented earlier suggests that proteins synthesized during oogenesis persist until at least the blastocyst stage.

RNA utilization. No direct evidence exists in sheep on the rate of mRNA degradation during the maternally-directed early embryonic phase. By contrast, careful measurements in mouse eggs show that virtually all the maternally derived mRNA is degraded by the late 2-cell stage in this species (Bachvarova & De Leon, 1980). Whilst most of this loss occurs at the 2-cell stage, earlier losses at the 1-cell stage have also been detected (Piko *et al.*, 1984). It is tempting to speculate on the possibility that the drop in protein synthesis in the 8-cell sheep egg reflects a similar but delayed period of degradation of maternal transcripts. However, until measurements of poly(A)$^+$RNA levels are made in cleaving sheep eggs nothing definite on RNA metabolism can be concluded from our experiments.

Concluding remarks

Oogenesis in the sheep represents a period of synthesis and sequestration of products required for the support of early embryogenesis. However, the fragmentary nature of our knowledge about these early processes in sheep is clear: corresponding information on other large mammals is even more sparse. The progressive miniaturization of molecular technology and the increased availability of oocytes and embryos from in-vitro sources provide the embryologist with unique opportunities for critical experimentation. It is imperative that work on the oocytes and embryos of domestic animals should be based on these modern embryological techniques.

References

Anderson, E. & Albertini, D.F. (1976) Gap junctions between the oocyte and companion follicle cells in mammalian ovary. *J. Cell Biol.* **71**, 680–686.

Bachvarova, R. & De Leon, V. (1980) Polyadenylated RNA of mouse ova and loss of maternal RNA in early development. *Devl Biol.* **74**, 1–8.

Bachvarova, R., De Leon, V., Johnson, A., Kaplan, G. & Paynton, V. (1987) Changes in total RNA, polyadenylated RNA and actin mRNA during meiotic maturation of mouse oocytes. *Devl Biol.* (in press).

Bornslaeger, E.A., Mattei, P. & Schultz, R.M. (1986) Involvement of cAMP-dependent protein kinase and protein phosphorylation in regulation of mouse oocyte maturation. *Devl Biol.* **114**, 453–462.

Brown, C.R. (1986) The morphological and molecular susceptibility of sheep and mouse zona pellucida to acrosin. *J. Reprod. Fert.* **77**, 411–417.

Canipari, R., Palombi, F., Riminucci, M. & Mangia, F. (1984) Early programming of maturation competence in mouse oogenesis. *Devl Biol.* **102**, 519–524.

Crosby, I.M., Osborn, J.C. & Moor, R.M. (1981) Follicle cell regulation of protein synthesis and developmental competence in sheep oocytes. *J. Reprod. Fert.* **62**, 575–582.

Crosby, I.M., Osborn, J.C. & Moor, R.M. (1984) Changes in protein phosphorylation during the maturation of mammalian oocytes *in vitro*. *J. exp. Zool.* **221**, 459–466.

Crosby, I.M., Moor, R.M., Heslop, J. & Osborn, J.C. (1985) cAMP in ovine oocytes: localization of synthesis and its action on protein synthesis, phosphorylation and meiosis. *J. exp. Zool.* **234**, 307–318.

Crozet, N., Motlik, J. & Szollosi, D. (1981) Nucleolar fine structure and RNA synthesis in porcine oocytes during the early stages of antrum formation. *Biol. Cell* **41**, 35–42.

Davidson, E.M. (1976) *Gene Activity in Early Development*, 2nd edn. Academic Press, New York.

Dekel, N. & Beers, W.H. (1978) Rat oocyte maturation *in vitro*: relief of cyclic AMP inhibition by gonadotrophins. *Proc. natn. Acad. Sci. U.S.A.* **75**, 4369–4373.

Eppig, J.J. (1979) A comparison between oocyte growth in co-culture with granulosa cells and oocytes with granulosa cell-oocyte junctional contact maintained *in vitro*. *J. exp. Zool.* **209**, 345–353.

Farmer, S.R., Wan, K.M., Ben-Ze'Ev, A. & Penman, S. (1983) Regulation of actin mRNA levels and translation responds to changes in cell configuration. *Molec. cell. Biol.* **2**, 182–189.

First, N.L. & Parrish, J.J. (1987) In-vitro fertilization of ruminants. *J. Reprod. Fert., Suppl.* **34**, 151–165.

Glass, L.E. & Cons, J.M. (1968) Stage dependent transfer of systemically injected foreign antigen and radio label into mouse ovarian follicles. *Anat. Rec.* **162**, 139–156.

Gurdon, J.B. (1974) *The Control of Gene Expression in Animal Development*. Oxford and Harvard University Presses.

Howe, J.G. & Hershey, J.W.B. (1984) Translational initiation factor and ribosome association with the cytoskeletal framework fraction from Hela cells. *Cell* **37**, 85–93.

Howlett, S.K. & Bolton, V.N. (1985) Sequence and regulation of morphological and molecular events during the first cell cycle of mouse embryogenesis. *J. Embryol. exp. Morph.* **87**, 175–206.

Johnson, M.H., McConnell, J. & Van Blerkom, J. (1984) Programmed development in the mouse embryo. *J. Embryol. exp. Morph.* **83**, 197–231.

Kaplan, G., Abreu, S.L. & Bachvarova, R. (1982) rRNA accumulation and protein synthetic patterns in growing mouse oocytes. *J. exp. Zool.* **220**, 361–370.

Kruip, T.A.M., Cran, D.G., Van Beneden, T.H. & Dieleman, S.J. (1983) Structural changes in bovine oocytes during final maturation *in vitro*. *Gamete Res.* **8**, 29–48.

Larsen, W.J., Wert, S.E. & Brunner, G.D. (1986) A dramatic loss of cumulus cell gap junctions is correlated with germinal vesicle breakdown in rat oocytes. *Devl Biol.* **113**, 517–521.

Laskey, R.A., Mills, A.D., Gurdon, J.B. & Partington, G.A. (1977) Protein synthesis in oocytes in *Xenopus Laevis* is not regulated by the supply of messenger RNA. *Cell* **11**, 345–351.

Maller, J.L. & Krebs, E.G. (1977) Progesterone stimulated meiotic cell division in Xenopus oocytes. Induction by regulatory subunit and inhibition by catalytic subunit of adenosine 3',5'-monophosphate-dependent protein kinase. *J. biol. Chem.* **252**, 1712–1718.

Manes, C. (1973) The participation of the embryonic genome during early cleavage in the rabbit. *Devl Biol.* **32**, 453–459.

Moor, R.M. (1983) Contact, signalling and cooperation between follicle cells and dictyate oocytes in mammals. In *Current Problems in Germ cell Differentiation*, pp. 307–326. Eds A. McLaren & C. C. Wylie. University Press, Cambridge.

Moor, R.M. & Cran, D.G. (1980) Intercellular coupling

in mammalian oocytes. In *Development in Mammals*, vol. 4, pp. 3–37. Ed. M. H. Johnson. Elsevier/North Holland Biomedical Press, Amsterdam.

Moor, R.M. & Crosby, I.M. (1986) Protein requirements for germinal vesicle breakdown in ovine oocytes. *J. Embryol. exp. Morph.* **94**, 207–220.

Moor, R.M. & Heslop, J.P. (1981) Cyclic AMP in mammalian follicle cells and oocytes during maturation. *J. exp. Zool.* **216**, 205–209.

Moor, R.M. & Osborn, J.C. (1983) Somatic control of protein synthesis in mammalian oocytes during maturation. In *Molecular Biology of Egg Maturation* (Ciba Fdn Symp. No. 98), pp. 178–196. Pitman, London.

Moor, R.M. & Smith, M.W. (1979) Amino acid transport in mammalian oocytes. *Expl Cell Res.* **119**, 333–341.

Moor, R.M. & Trounson, A.O. (1977) Hormonal and follicular factors affecting maturation of sheep oocytes *in vitro* and their subsequent developmental capacity. *J. Reprod. Fert.* **49**, 101–109.

Moor, R.M. & Warnes, G.M. (1977) Regulation of oocyte maturation in mammals. In *Control of Ovulation*, pp. 159–176. Eds G. E. Lamming & D. B. Crighton. Butterworths, London.

Moor, R.M., Polge, C. & Willadsen, S.M. (1980) Effect of follicular steroids on the maturation and fertilization of mammalian oocytes. *J. Embryol. exp. Morph.* **56**, 319–335.

Moor, R.M., Osborn, J.C., Cran, D.G. & Walters, D.E. (1981) Selective effect of gonadotrophins on cell coupling, nuclear maturation and protein synthesis in mammalian oocytes. *J. Embryol. exp. Morph.* **61**, 347–365.

Motlik, J., Crozet, N. & Fulka, J. (1984) Meiotic competence in vitro of pig oocytes isolated from early antral follicles. *J. Reprod. Fert.* **72**, 323–328.

Osborn, J.C. & Moor, R.M. (1983a) Time-dependent effects of α-amanitin on nuclear maturation and protein synthesis in mammalian oocytes. *J. Embryol. exp. Morph.* **73**, 317–338.

Osborn, J.C. & Moor, R.M. (1983b) The role of steroid signals in the maturation of mammalian oocytes. *J. Steroid Biochem.* **19**, 133–137.

Piko, L., Hammons, M.D. & Taylor, K.D. (1984) Amounts, synthesis and some early properties of intracisternal A particle-related RNA in early mouse embryos. *Proc. natn. Acad. Sci. U.S.A.* **81**, 488–492.

Pratt, H.P.M., Bolton, V.N. & Gudgeon, K.A. (1983) The legacy from the oocyte and its role in controlling early development of the mouse embryo. In *Molecular Biology of Egg Maturation* (Ciba Fdn Symp. No. 98), pp. 197–227. Pitman, London.

Sasser, R.G. & Ruder, C.A. (1987) Detection of early pregnancy in domestic ruminants. *J. Reprod. Fert., Suppl.* **34**, 261–271.

Schultz, R.M. (1985) Roles of cell-to-cell communication in development. *Biol. Reprod.* **32**, 27–42.

Schultz, R.M., Letourneau, G.E. & Wassarman, P.M. (1979) Programme of early development in the mammal: changes in patterns and absolute rates of tubulin and total protein synthesis during oogenesis and early embryogenesis in the mouse. *Devl Biol.* **73**, 120–133.

Sorensen, R.A. & Wassarman, P.M. (1976) Relationship between growth and meiotic maturation of the mouse oocyte. *Devl Biol.* **50**, 531–536.

Stern, S., Rayyis, A. & Kennedy, J.F. (1972) Incorporation of amino acids during maturation *in vitro* by the mouse oocyte: effect of puromycin on protein synthesis. *Biol. Reprod.* **7**, 341–346.

Szollosi, D., Calarco, P. & Donahue, R.P. (1972) Absence of centrioles in the first and second meiotic spindles of mouse oocytes. *J. Cell Sci.* **11**, 521–541.

Szollosi, D., Gerard, M., Menezo, Y. & Thibault, C. (1978) Permeability of ovarian follicles, corona cell-oocyte relationship in mammals. *Annls Biol. anim. Biochim. Biophys.* **18**, 511–521.

Szybek, K. (1972) *In vitro* maturation of oocytes from sexually immature mice. *J. Endocr.* **54**, 527–528.

Thibault, C. (1977) Are follicular maturation and oocyte maturation independent processes? *J. Reprod. Fert.* **51**, 1–15.

Van Blerkom, J. (1985) Extragenomic regulation and autonomous expression of a developmental program in the early mouse embryo. *Ann. N.Y. Acad. Sci.* **442**, 58–72.

Van Blerkom, J. & Runner, M.N. (1984) A cytoplasmic reorganization provides mitochondria needed for resumption of arrested meiosis in the mouse oocyte. *Am. J. Anat.* **171**, 335–355.

Warnes, G.M., Moor, R.M. & Johnson, M.H. (1977) Changes in protein synthesis during maturation of sheep oocytes *in vivo* and *in vitro*. *J. Reprod. Fert.* **49**, 331–335.

Wassarman, P.M., Bleil, J.D., Cascio, S.M., La Marca, M.J., Letourneau, G.E., Mrozak, S.C. & Schultz, R.M. (1981) Programming of gene expression during mammalian oogenesis. In *Bioregulators of Reproduction*, pp. 119–150. Eds G. Jagiello & G. Vogell. Academic Press, New York.

Wilson, E.B. (1925) *The Cell in Development and Heredity*, 3rd edn. Macmillan, New York.

Young, R.J., Sweeney, K. & Bedford, J.M. (1978) Uridine and guanosine incorporation by the mouse one-cell embryo. *J. Embryol. exp. Morph.* **44**, 133–148.

Zamboni, L. (1970) Ultrastructure of mammalian oocytes and ova. *Biol Reprod.* (Suppl.) **2**, 44–63.

J. Reprod. Fert., Suppl. **34** (1987), 71–85

Heterogeneous cell types in the corpus luteum of sheep, goats and cattle

J. D. O'Shea

Department of Veterinary Preclinical Sciences, The University of Melbourne, Parkville, Victoria 3052, Australia

Summary. Data on the structure, quantitation, origins and functions of the large luteal (LL) and small luteal (SL) cells of sheep, goats and cattle are reviewed. Both LL and SL cells show ultrastructural features consistent with a steroidogenic function. However, in addition to differences in size and shape, LL cells differ from SL cells primarily in possessing large numbers of secretory granules, suggesting an additional protein/polypeptide synthetic and secretory function. In sheep, morphometric estimates show that the corpus luteum (CL) contains $\simeq 10 \times 10^6$ LL cells and $\simeq 50$–60×10^6 SL cells: individual LL cells are $\simeq \times 6$ greater in volume than SL cells. During formation of the CL, granulosa and theca cells are incorporated, and evidence suggests that granulosa cells give rise to LL cells and theca cells to SL cells. However, SL cells, or cells of thecal origin, may also give rise to some LL cells. Both LL and SL cells produce progesterone *in vitro*. On a per cell basis, LL cells produce more progesterone than do SL cells, but SL cells show a much greater progesterone-secretory response to LH. Oxytocin is synthesized, and secreted in granule form, only by the LL cells, and relaxin, whose presence has been demonstrated convincingly only in cattle, also appears to be produced only by LL cells. The two types of luteal cell in ruminants therefore show major differences in function: the occurrence of any significant functional interaction remains to be established.

Introduction

The corpus luteum (CL) in ruminants, as in other mammals, contains specific hormone-producing (luteal) cells and other cells of several types. The latter include endothelial cells and pericytes, smooth muscle cells, fibrocytes, macrophages, leucocytes and occasional plasma cells (O'Shea *et al.* 1979; Rodgers *et al.*, 1984). Although there is some evidence that macrophages may serve a specific stimulatory function in relation to progesterone synthesis (Kirsch *et al.*, 1983), the non-luteal cell populations are presumed to be primarily concerned with their conventional ancillary roles.

In several groups of mammals, including perissodactyls (e.g. horse), cetaceans (e.g. whales, dolphins) and artiodactyls (e.g. pig, ruminants), cellular heterogeneity is also seen within the population of steroidogenic luteal cells. In all of these groups, two distinct populations of luteal cells have been recognized histologically (Mossman & Duke, 1973), and variously termed large and small luteal cells or granulosa and theca luteal (or lutein) cells on account of their putative origins. This review will deal only with heterogeneity within the luteal cell population, and consider data relevant to qualitative and quantitative aspects of structure, to the origins, and to the functions of the large and small luteal cells of domestic ruminants. Reasons why it may be advantageous for the CL to possess two types of luteal cell, in terms of distinctive functions and possible functional interactions, are also considered.

Structural features

Sheep

Large luteal cells. The large luteal (LL) cells of cyclic ewes attain their mature size and structure (Figs 1–3) about 6 or 7 days after ovulation, and show little change thereafter until the onset of structural luteal regression. They are large, polyhedral cells with a single, round or oval, pale vesicular nucleus situated close to the centre of the cell and containing one or more prominent nucleoli. Nuclear:cytoplasmic ratio, at $\simeq 1:27$, is very low (Rodgers *et al.*, 1984).

Details of the ultrastructure of these cells have been reported by many workers, including Deane *et al.* (1966), Gemmell *et al.* (1974), O'Shea *et al.* (1979) and Paavola & Christensen (1981), and the description below draws heavily on these reports. The cytoplasm of the LL cells shows specializations consistent with both a steroidogenic and a protein synthetic and secretory function. Features characteristic of steroidogenic cells include large quantities of smooth endoplasmic reticulum and high numbers of mitochondria whose cristae are frequently tubular in form. Lipid droplets are not abundant until the onset of luteal regression. Smooth endoplasmic reticulum occurs in these cells in the form of anastomosing networks of branching tubules, and does not form whorl-like arrays. As in many other cells which secrete proteins or polypeptides, LL cells contain multiple parallel arrays of flattened cisternae of rough endoplasmic reticulum, prominent Golgi complexes and numerous membrane-bound granules ('secretory granules'), $\simeq 0.15$–$0.3 \, \mu m$ in diameter, whose electron-dense contents can frequently be seen to be released by exocytosis (Corteel, 1973; Gemmell *et al.*, 1974). Not all granules in the LL cells are related to protein secretion, as lysosomes, multivesicular bodies and peroxisomes are also present (Paavola & Christensen, 1981).

The surfaces of these cells, which are covered by a moderately well-defined basal lamina except at sites of close contact with neighbouring cells, are extensively folded to form microvillous or flattened projections which usually extend parallel to the cell surface. These projections commonly interdigitate with others from the same or neighbouring cells, including both large and small luteal cells. At sites of interdigitation the membranes of adjacent processes are frequently separated by a rather constant space of $\simeq 20 \, nm$, and resemble 'septate-like' junctions (Friend & Gilula, 1972). Other forms of specialized junction are rarely seen, and although occasional gap junctions ('abutment nexuses') have been observed (O'Shea *et al.*, 1979), Higuchi *et al.* (1976) were unable to demonstrate electrical coupling between luteal cells of sheep.

Characteristic LL cells persist throughout pregnancy (Gemmell *et al.*, 1977; O'Shea *et al.*, 1979) and into the early post-partum period (O'Shea & Wright, 1985), but show progressive structural modifications. As pregnancy advances these cells become more rounded, and the extent of surface contacts is reduced. Cytoplasmic lipid droplets accumulate in large numbers, and many mitochondria acquire large, dense, rounded matrix granules which may be $\geqslant 1 \, \mu m$ in diameter. Secretory granules become reduced in number, with less evidence of exocytosis, and both smooth and rough endoplasmic reticulum become less prominent.

Small luteal cells. These cells (SL; Figs 4–7), described in detail by O'Shea *et al.* (1979), are smaller, more angular and often elongated or spindle-shaped. Their nuclei are less regular in outline

---

Fig. 1. Large luteal cell from a cyclic sheep, showing abundant mitochondria and many dense granules (arrows). $\times 7850$ (bar $= 1 \, \mu m$).

Fig. 2. Detail from Fig. 1, showing mitochondria with tubular cristae, smooth endoplasmic reticulum, and the release of secretory granules (arrow). $\times 23\,200$ (bar $= 0.5 \, \mu m$).

Fig. 3. Detail from Fig. 1. Membrane-bound secretory granules, showing the dense core and lighter periphery (arrow) which characterize many of these granules. $\times 23\,200$ (bar $= 0.5 \, \mu m$).

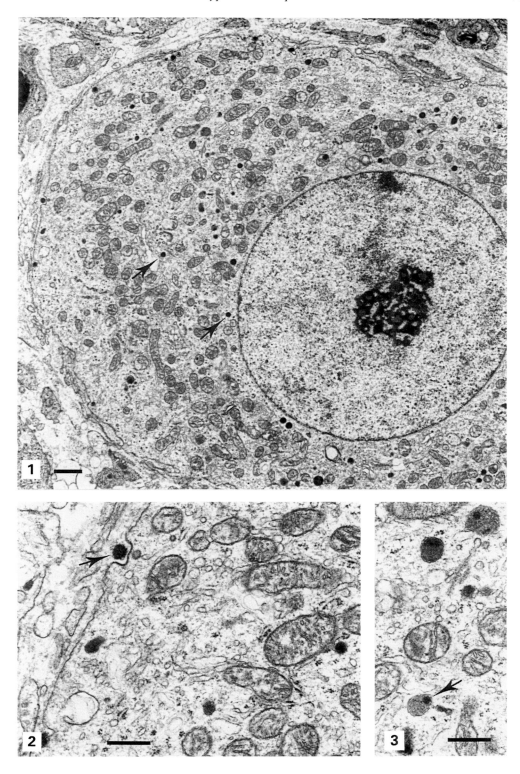

and more densely staining due to a coarser pattern of heterochromatin. One or more small nucleoli are commonly seen. Complex infoldings of the nuclear envelope are frequently present, and large nuclear inclusions containing modified cytoplasmic organelles, often apparently wholly detached from the remaining cell cytoplasm, are a striking feature of a small proportion of these cells. Nuclear:cytoplasmic ratio, at $\simeq 1{:}8$, is higher than in the LL cells (Rodgers *et al.*, 1984).

The cytoplasm shows features consistent with a steroidogenic function, including large amounts of tubular endoplasmic reticulum on which small groups of ribosomes are sparsely distributed, many mitochondria which contain both tubular and lamellar cristae, and scattered lipid droplets. Mitochondria are, however, less densely packed, and more irregular in size, than in the large luteal cells. Free ribosomes, and one or more small Golgi complexes, are also present, but these cells lack focal aggregates of rough endoplasmic reticulum and secretory granules. Small numbers of dense, membrane-bound granules, possibly lysosomes, are present, but there is no evidence of exocytosis.

Surface folding and basal lamina formation are less marked than in LL cells, and surface projections are more often microvillous in character. Changes associated with the advancement of pregnancy are also less marked, some of these cells appearing essentially unchanged even into the immediate post-partum period (O'Shea & Wright, 1985).

Goat

The limited ultrastructural data available for this species indicate a close similarity between sheep and goats in LL and SL cell structure (Gemmell *et al.*, 1977; Azmi & Bongso, 1985). The only significant difference so far reported is in relation to secretory granules in LL cells during the last part of pregnancy. In goats, granules remain abundant at least until Day 140, with continuing release by exocytosis (Gemmell *et al.*, 1977).

Cow

It is clear from the ultrastructural studies of Priedkalns & Weber (1968) and Fields *et al.* (1985), of cyclic and pregnant cows respectively, that LL and SL cells broadly comparable to those in sheep occur also in the cow. However, some apparent contradictions are found between the reports of different workers studying the bovine CL, which suggest that clearcut distinction between LL and SL cells is less easy in cattle than in sheep or goats. In particular, Singh (1975) described LL ("granulosa lutein") cells in the pregnant cow but incorporated several ultrastructural features (notably whorls of smooth endoplasmic reticulum and stacks of rough endoplasmic reticulum) which, based on the subsequent report of Fields *et al.* (1985), may in fact have been present in adjoining SL cells. Parry *et al.* (1980), describing LL cells in cyclic cows, were unable to distinguish

Fig. 4. Small luteal cell from a cyclic sheep, showing irregularity of nuclear outline and absence of secretory granules. A capillary (C) is seen at top, and part of a macrophage (M) at upper left. × 7850 (bar = 1 μm).

Fig. 5. Detail from Fig. 4, showing cytoplasm rich in smooth endoplasmic reticulum but containing relatively few mitochondria. × 23 200 (bar = 0·5 μm).

Fig. 6. Nucleus of a small luteal cell showing an inclusion containing cytoplasmic material, surrounded by an inverted nuclear envelope. × 23 200 (bar = 0·5 μm).

Fig. 7. Detail from Fig. 6, showing a narrow neck of continuity (arrow) between the cell cytoplasm and the contents of the inclusion. × 46 400 (bar = 0·2 μm).

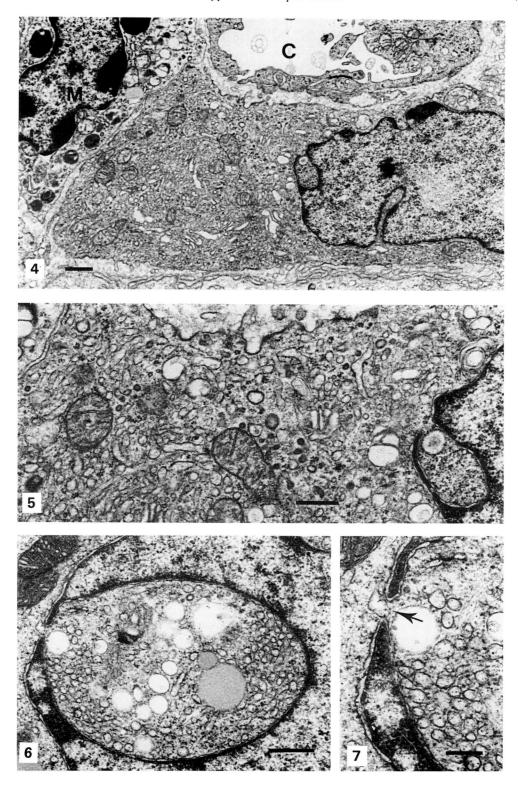

between these and SL cells, and regarded the two types as "part of the same population". Contrasts with the sheep, as outlined below, are therefore based mainly on the recent report of Fields *et al.* (1985).

The LL cells of the cow appear to differ from those of sheep in that they lack parallel stacks of cisternae of rough endoplasmic reticulum, and in that their secretory granules commonly aggregate in the central regions of the cell and subsequently migrate *en masse* towards the cell periphery where they are released by exocytosis. Secretory granules reach their maximum numbers per cell early in the last third of pregnancy, and then decline towards term (Fields *et al.*, 1985). The SL cells, on the other hand, possess stacks of cisternae of rough endoplasmic reticulum, and also whorl-like arrays of smooth endoplasmic reticulum, neither of which are characteristic of sheep SL cells. Crystalline-like inclusions in mitochondria have also been observed in SL cells of cows (Fields *et al.*, 1985), and lipid droplets are particularly abundant during both the oestrous cycle and pregnancy (Priedkalns & Weber, 1968; Fields *et al.*, 1985). Although small numbers of 'secretory granules' have been described in bovine SL cells by Fields *et al.* (1985), no evidence of their release by exocytosis has been presented.

Quantitative aspects of structure

Volume density

Estimates of the percentage composition (volume density) of mature, functional CL from naturally cyclic ewes, based on standard point- (hit-) counting methods, have been made using both light (Niswender *et al.*, 1976) and electron (Rodgers *et al.*, 1984; O'Shea *et al.*, 1986) microscopy. Findings in relation to LL and SL cells are summarized in Table 1, the remaining proportions of the tissue being composed of cells of other types, connective tissue fibres and intercellular spaces, and vascular lumina. The combined populations of LL and SL cells thus constitute up to $\simeq 50\%$ of the total luteal tissue, with LL cells occupying the greater volume. A higher ultrastructural estimate ($> 70\%$) of the volume density of combined luteal cells in cyclic CL of cows was obtained by Parry *et al.* (1980).

Cell size

Three types of estimate of the sizes of LL and SL cells in the sheep are available. These are based on measurement of diameters of cells in tissue sections and in enzymically-dissociated cell populations, and on tissue morphometry. The first method is clearly of limited value, being based on random two-dimensional images of cells whose shape is irregular, and will not be considered further.

Since dissociated luteal cells in suspension adopt a more-or-less spherical form, measured diameters of free-floating cells should be meaningful. Data derived using this approach, coupled in

Table 1. Volume density (% $\pm$ s.e.m.) of the corpus luteum of the cyclic ewe occupied by large and small luteal cells

Source of data	Day of cycle	Large luteal (granulosa lutein) cells	Small luteal (theca lutein) cells	Total
Niswender *et al.* (1976)	10	33·8 $\pm$ 2·5	16·5 $\pm$ 1·6	50·3
Rodgers *et al.* (1984)	Not known	25·4 $\pm$ 2·7	17·5 $\pm$ 0·6	42·9
O'Shea *et al.* (1986)	9	28·3 $\pm$ 2·1	22·2 $\pm$ 1·4	50·5

Table 2. Measured diameters of large and small luteal cells after enzymic dissociation

Source of data	Species	Reproductive status	Large luteal cells (μm)	Small luteal cells (μm)
Rodgers & O'Shea (1982)	Sheep	Cycle, days unknown	>19	13·5–19
Fitz *et al.* (1982)	Sheep	Cycle, Days 8–12, after superovulation	23–25	12–22
Ursely & Leymarie (1979)	Cow	Pregnant, days 60–100	30–50 (mean = 37)	15–20 (mean = 18)
Koos & Hansel (1981)	Cow	Cycle, Days 11–13	>25	10–20
Chegini *et al.* (1984)	Cow	Pregnant, days unknown	18–45	15–18

Table 3. Cell volume (mean ± s.e.m.) and derived* cell diameters of large and small luteal cells from cyclic ewes

Source of data	Day of cycle	Cell volume ($\mu m^3 \times 10^{-3}$)		Cell diameter (μm)	
		Large luteal cells	Small luteal cells	Large luteal cells	Small luteal cells
Rodgers *et al.* (1984)	Not known	13·1 ± 2·1	2·1 ± 0·2	29·2	15·8
O'Shea *et al.* (1986)	9	15·6 ± 1·3	2·1 ± 0·3	31·0	16·0

*Diameters derived from measured cell volume (volume = $4/3\pi r^3$) assuming cells were to take on a spherical form.

some cases with ultrastructural (Koos & Hansel, 1981; Rodgers & O'Shea, 1982) or enzyme histochemical (Fitz *et al.*, 1982) methods to aid identification of the cell populations in question, are summarized in Table 2. This method, as a means of estimating the sizes of LL and SL cells in the original tissue does, however, have some disadvantages. The actual cells measured cannot be identified as to type by any objective criteria other than size, and there is no certainty that the populations obtained by tissue dissociation are accurately representative of those in the original tissue. Nor, in view of the somewhat arbitrary size limits used to categorize cells as LL or SL types, is it possible to be certain that the populations defined as LL and SL cells by different workers are necessarily identical.

The third type of estimate, at present available only for cyclic sheep, is based on the morphometric measurement of mean cell volume for each type of luteal cell in whole luteal tissue. Volume thus measured can also be converted to cell diameter if it be assumed that the cells were to take on a spherical form (Table 3). These estimates indicate that individual LL cells are at least 6 times the volume of SL cells, an important factor when comparing the functional capacity of the two cell types on a per cell basis.

Cell numbers

Estimates of the total number of cells of all types per CL in the sheep have been obtained using three independent methodologies. In the first, based on counts of free cells in suspension following enzymic dissociation of functional luteal tissue from cyclic ewes, totals of $\simeq 40 \times 10^6$ (Rodgers & O'Shea, 1982) and $\simeq 60 \times 10^6$ (Niswender *et al.*, 1985) have been reported. This approach has the disadvantage that some cells are almost certainly lost during dissociation, and the extent of this loss cannot be quantitated. Hence some degree of underestimate appears inevitable.

The second method, based on measurement of cell and tissue DNA, gives a much higher estimate of 238×10^6 cells (Rodgers *et al.*, 1984). This estimate is, however, reasonably close to estimates obtained by the third method, ultrastructural morphometry, giving values of 258×10^6 (Rodgers *et al.*, 1984) and 201×10^6 (O'Shea *et al.*, 1986). The relatively close correspondence between the findings from these last two methods, in which questions of non-measurable losses do not arise, suggests that the cyclic CL of the sheep probably contains $\geqslant 200 \times 10^6$ cells, indicating the loss of very significant numbers of cells during tissue dissociation.

Morphometric estimates further place the mean numbers of LL and SL cells per CL at $10{\cdot}7 \times 10^6$ and $49{\cdot}2 \times 10^6$ (Rodgers *et al.*, 1984) and $9{\cdot}6 \times 10^6$ and $57{\cdot}5 \times 10^6$ (O'Shea *et al.*, 1986), respectively. Thus the combined total of both types of luteal cells, at $60–70 \times 10^6$, represents a minority of the total cells per CL, and the data indicate a ratio of 5 or 6 SL cells for each LL cell.

Origins of luteal cells

There is now clear evidence that both the granulosa and theca layers contribute cells to the developing CL in ruminants. Many studies in sheep (McClellan *et al.*, 1975; O'Shea *et al.*, 1980) and cattle (Priedkalns *et al.*, 1968; Alila & Hansel, 1984) provide support for this conclusion. However, the precise nature of the contributions of granulosa and theca cells to developing and established CL remains more controversial.

Genesis of the CL

Fate of the granulosa cells. All of the studies cited above are consistent in reaching the conclusion that following ovulation the granulosa cells differentiate into LL cells. This conclusion is based on sequential observations of light and electron microscopic structure, together with the use of 'markers' in the form of glycogen bodies and monoclonal antibodies. Glycogen bodies (Le Beux, 1969), involving whorled arrangements of smooth membranes with or without the presence of associated glycogen particles, are a characteristic feature of granulosa cells in sheep preovulatory follicles (Corteel, 1973) and persist for 24–48 h after ovulation. Since these structures are not seen in thecal cells they form a convenient ultrastructural marker of granulosa-derived cells in the early postovulatory period (O'Shea *et al.*, 1980). Monoclonal antibodies raised against granulosa cell surface antigens have been used as markers of granulosa-derived cells in the cow CL (Alila & Hansel, 1984). Finally, numbers of granulosa cells in preovulatory sheep follicles (O'Shea *et al.*, 1985) are similar to numbers of LL cells per CL (Rodgers *et al.*, 1984; O'Shea *et al.*, 1986), which would be expected in view of evidence that granulosa-derived cells undergo little if any post-ovulatory mitosis (McClellan *et al.*, 1975).

Fate of the theca cells. Many morphological studies have indicated that SL cells may be of thecal origin. This possibility has been supported in cattle by Donaldson & Hansel (1965) and Priedkalns *et al.* (1968), and in sheep by Deane *et al.* (1966) and McClellan *et al.* (1975). Markers, including the enzyme alkaline phosphatase and monoclonal antibodies, have also been used in tracing the fate of theca-derived cells after ovulation. In sheep, alkaline phosphatase is confined to the theca, and particularly the theca interna, in preovulatory follicles. Use of this marker has clearly demonstrated the incorporation of theca cells into the developing CL, and supported the concept of a thecal origin for the SL cells (O'Shea *et al.*, 1980). However, the value of alkaline phosphatase as a thecal marker is limited to the first 48 h after ovulation, as this enzyme subsequently becomes widely distributed throughout the CL. Monoclonal antibodies against bovine theca cell surface antigens, used to identify cells of thecal origin after ovulation, have similarly pointed to the theca as a source of at least a large proportion of the SL cells in the cow CL (Alila & Hansel, 1984).

Cellular transformation in the CL

The question as to whether LL and SL cells persist as discrete and closed populations throughout the lifespan of the CL is more controversial. Several lines of evidence suggest that SL cells may be able to differentiate into LL cells, and it has further been suggested that there may be a population of stem cells in the CL which differentiate into small steroidogenic cells which in turn differentiate into LL cells (Niswender *et al.*, 1985).

Morphological evidence. If SL cells do differentiate into LL cells then it would be expected that some cells of intermediate structure would be demonstrable by electron microscopy. In an early study of the ultrastructure of luteinization in the cow Priedkalns *et al.* (1968) did appear to recognize the presence of such cells, and Parry *et al.* (1980) were subsequently unable to distinguish bovine LL and SL cells as separate cell types. On the other hand, O'Shea *et al.* (1979) and Fields *et al.* (1985) were able to identify two distinctive cell populations, in sheep and cattle, respectively, and failed to observe cells of intermediate structure. However, the subjective nature of all of these observations needs to be recognized, and the question cannot be regarded as definitively resolved.

Evidence based on cell numbers. In a study of numbers of large and small steroidogenic cells obtained by enzymic tissue dissociation, Fitz *et al.* (1981) observed a rise in numbers of large cells and a fall in numbers of small cells during the later stages of the oestrous cycle in the sheep. While these findings are suggestive of the possibility of transformation of some SL cells to LL cells, cells were categorized solely on the basis of diameter in suspension. Hence simple enlargement of some or all of the SL cells, without true cytodifferentiation to LL cells, could equally validly explain these observations. Furthermore there is a risk that cell populations obtained by tissue dissociation may not be truly representative of those in the original tissue. A subsequent study based on morphometry failed to show any rise in numbers of LL cells, as identified by electron microscopy, during the later part of the luteal phase of the sheep oestrous cycle (O'Shea *et al.*, 1986).

An increase in the numbers of LL cells and a decrease in SL cells in dissociated populations from cyclic sheep CL have also been demonstrated in response to luteotrophic stimulation by human chorionic gonadotrophin (hCG) (Niswender *et al.*, 1985). While these observations are also consistent with the idea of transformation of SL cells to LL cells, they remain subject to the general problems associated with quantitation based on dissociated cell populations.

Evidence based on luteinization of follicular cysts. The strongest evidence in sheep that theca-derived cells may be able to differentiate into LL cells is found in a study by Cran (1983): luteinization of follicular cysts, formed in response to PMSG, in which granulosa cells had largely degenerated, was accompanied by the formation of cells whose ultrastructure closely resembled that of LL cells.

Evidence based on monoclonal antibodies. The studies of Alila & Hansel (1984), referred to above, also provide evidence related to the possible transformation of SL cells to LL cells: the numbers of LL cells binding a granulosa-specific antibody declined progressively throughout the oestrous cycle, while the number binding a theca-specific antibody increased. The majority of the SL cells bound the theca-specific antibody, and no SL cells bound the granulosa-specific antibody after Day 6 of the cycle. The findings of this study, in which numbers of cells binding both of these antibodies were reduced progressively throughout pregnancy, obviously add considerable strength to the case for transformation of theca-derived cells, and presumably of SL cells, to LL cells in the normal cow CL. While the possibility that granulosa-derived cells might at a later stage in their life cycle express antigens originally limited to the theca cannot be wholly eliminated, Alila & Hansel (1984) argue strongly against this on the grounds that no cells were ever detected which expressed both granulosa-specific and theca-specific antigens at the same time. The concept that theca-derived LL cells replace the original granulosa-derived cells would be strengthened if direct evidence of the death and disposal of the original LL cells could be provided.

Implications of cellular transformation. Although granulosa and theca cells may share a common origin during ovarian histogenesis (Byskov, 1986), they are structurally and functionally distinct cell types at least from the time of folliculogenesis, and are separated by a complete basal lamina until close to the time of ovulation. Hence, subsequent convergent differentiation to a common, new type of specialized cell, the LL cell, would appear intrinsically surprising. Nonetheless, if the dual origin theory of LL cell formation can be confirmed to be correct, it remains an open question as to why LL cells of granulosa origin should need to be replaced by theca-derived cells. One possibility might be that granulosa-derived LL cells have a limited lifespan, perhaps sufficient for the oestrous cycle but inadequate for pregnancy. Disappearance of the original granulosa-derived cells might then serve as a stimulus for further differentiation of theca-derived cells, as appears to have happened in the luteinized follicles studied by Cran (1983). An additional possibility could be that, although superficially similar, theca-derived LL cells are functionally different from those of granulosa origin. For example, replacement might be necessary to provide a relaxin-synthetic LL cell type in place of an oxytocin-synthetic type, leading to the testable hypothesis that individual LL cells may be able to synthesize either oxytocin or relaxin, but not both.

Function

Knowledge as to the distinctive functions of LL and SL cells has been obtained primarily from in-vitro studies of enriched populations derived from enzymically dissociated luteal tissue. Emphasis here will be placed on data related to the individual functions of LL and SL cells, and to the possibility of functional interactions. While other functions of the CL as a whole have been identified, the major products which have been linked with one or other luteal cell type are progesterone, relaxin and oxytocin. Further discussion will therefore be limited to these hormones.

Progesterone

Control of synthesis. While luteinizing hormone (LH) appears to be the major luteotrophic hormone in domestic ruminants (Niswender *et al.*, 1985), evidence from several studies of sheep (Fitz *et al.*, 1982; Rodgers & O'Shea, 1982; Rodgers *et al.*, 1983a) and cattle (Ursely & Leymarie, 1979; Koos & Hansel, 1981) indicates a marked difference in response to LH by LL and SL cells *in vitro*. This evidence, which is consistent with the much higher numbers of LH receptors on SL cells (Fitz *et al.*, 1982), shows that the progesterone secretory response to LH is confined largely, if not wholly, to the SL cells. The response by SL cells is mediated by cAMP, intracellular concentrations of which are elevated in response to LH (Hoyer *et al.*, 1984). Dibutyryl cAMP (Fitz *et al.*, 1982; Rodgers *et al.*, 1983a) or activators of adenylate cyclase (Hoyer *et al.*, 1984) also stimulate progesterone production by these cells. Progesterone synthesis by SL cells is also stimulated by hCG, although quantitative aspects of the response differ from those to LH (Bourdage *et al.*, 1984). The only documented exception to the response of SL cells to LH is found in cows during the later stages of pregnancy (Chegini *et al.*, 1984; Weber *et al.*, 1984), and may represent a regressionary change.

The response to LH by LL cells is consistently less than that of SL cells, and in studies of sheep can be explained in terms of contamination of the LL cell populations by small numbers of LH-responsive SL cells (Fitz *et al.*, 1982; Rodgers *et al.*, 1983a). LL cells also fail to show any progesterone-secretory response to dibutyryl cAMP (Fitz *et al.*, 1982) or to activators of adenylate cyclase, although these activators do increase the synthesis of cAMP by LL cells (Hoyer *et al.*, 1984).

Studies *in vitro* also suggest the possibility that other hormones, including prostaglandins, oestradiol, catecholamines and oxytocin, may have physiological actions on the progesterone-secretory function of cells of ruminant CL. In relation to prostaglandins (PG), Fitz *et al.* (1984a) demonstrated a stimulatory effect of PGE-1 and PGE-2 on basal (i.e. without LH) progesterone production by ovine LL cells, in the absence of any changes in cAMP levels in the cells or media, while SL cells did not respond to these PGs. PGF-2α, known to be a luteolytic hormone in sheep, caused a decrease in progesterone production by the LL cells (Fitz *et al.*, 1984b). These findings are again consistent with the demonstrated distribution of receptors for PGE and PGF, which are confined largely to the LL cells (Fitz *et al.*, 1982). In contrast to the findings of Fitz *et al.* (1984a, b), Rodgers *et al.* (1985) found no effect of PGE-2 on basal progesterone production by either type of sheep luteal cell, and Koos & Hansel (1981) were unable to detect any effect of PGF-2α on progesterone production by LL or SL cells from cow CL, with or without LH stimulation.

The presence of a cytosolic receptor for oestradiol has been demonstrated in ovine luteal cells, primarily LL cells, by Glass *et al.* (1984). However, the physiological significance of this receptor is not yet clear. Rodgers *et al.* (1985) observed no effect of oestradiol on basal progesterone production by either type of ovine luteal cell *in vitro*, while Ursely & Leymarie (1979) reported a dose-dependent inhibitory effect on LH-stimulated progesterone secretion by bovine LL and SL cells, suggesting a possible physiological role in luteolysis.

β-Adrenergic stimulation of progesterone secretion by isoprenaline, and its blockage by the β-adrenergic antagonist propranolol, have been demonstrated in mixed populations of luteal cells from sheep (Jordan *et al.*, 1978) and cattle (Battista & Condon, 1986). Increased synthesis of cAMP was also observed in sheep (Jordan *et al.*, 1978), suggesting that SL cells rather than LL cells may have been responsible for the increased progesterone production. However, Rodgers *et al.* (1985), using separated populations of ovine LL and SL cells, observed that isoprenaline produced a significant stimulation of progesterone production only by LL cells. The reason for this apparent discrepancy is not clear, and the cellular location of β-adrenergic receptors in ruminant CL has not been investigated.

Evidence from mixed populations of bovine luteal cells (Tan *et al.*, 1982) suggests that oxytocin may produce a minor stimulatory action on progesterone production at low doses, and a marked suppression of hCG-stimulated progesterone production at higher doses. In sheep, oxytocin has been reported to inhibit LH-stimulated progesterone production by SL cells (Niswender *et al.*, 1985), although Rodgers *et al.* (1985) were unable to detect any effect on LL or SL cell progesterone synthesis. Various degrees of damage to receptors arising from variations in method of tissue dissociation could perhaps explain these findings.

In addition to hormonal influences, there is also evidence of the importance of intracellular levels of calcium in progesterone secretion by sheep luteal tissue (Higuchi *et al.*, 1976). The importance of calcium fluxes is also suggested by the observation that the Ca^{2+} channel-blocking agent verapamil substantially suppresses basal progesterone production by both LL and SL cells from sheep (Rodgers *et al.*, 1985).

Quantitative aspects of secretion. With the exception of the work of Chegini *et al.* (1984) on pregnant cows, all of the published studies comparing basal progesterone production by ruminant LL and SL cells have indicated that LL cells, on a per cell basis, produce substantially more progesterone than do SL cells. Quantitative comparisons become more complicated, however, when considering contributions to total luteal progesterone production. Two attempts (Rodgers *et al.*, 1983a; Niswender *et al.*, 1985) have been made to assess the relative contributions of LL and SL cells to total progesterone production by the sheep CL. To do this one needs quantitative data on basal secretion, on responses to LH, on absolute or relative numbers of the two cell types in luteal tissue, and on luteal (or blood) levels of LH *in vivo*. It is also necessary to extrapolate findings on the functional behaviour of LL and SL cells from an in-vitro to an in-vivo situation, clearly a hazardous undertaking. However, on these bases, and using different sets of data, Rodgers *et al.* (1983a) concluded that progesterone production by the SL cell type could exceed that from the LL cells, while Niswender *et al.* (1985) estimated an SL cell contribution of less than 22%. It seems unlikely that either of these estimates is very accurate, but they do suggest that the contribution of SL cells to luteal progesterone production may well be quantitatively significant and should never be ignored.

Morphological correlates of secretion. As indicated above, LL and SL cells possess features consistent with a steroidogenic function. However, discussion of the structural correlates of luteal progesterone secretion in ruminants has been dominated for over a decade by the suggestion that progesterone (presumably bound to a carrier protein) may be released in association with the exocytosis of the secretory granules of the LL cells. This suggestion, originally advanced by Corteel (1973) and Gemmell *et al.* (1974), has received wide support based primarily on a large amount of circumstantial evidence linking the timing and extent of granule release to that of progesterone secretion (Gemmell *et al.*, 1977; Sawyer *et al.*, 1979; Parry *et al.*, 1980; Heath *et al.*, 1983). Additional evidence derives from the demonstration that progesterone is concentrated in association with secretory granules after differential centrifugation of bovine luteal tissue homogenates (Quirk *et al.*, 1979). However, the granule theory of progesterone secretion has become less tenable since it has been established that SL cells also secrete progesterone, but without the exocytosis of granules. Furthermore, a recent attempt to demonstrate the presence of a progesterone-binding protein in the secretory granules of ovine LL cells was unsuccessful (Sernia *et al.*, 1982), and granule release can be induced in sheep by PGF-2α *in vivo* without any accompanying increase in ovarian venous blood levels of progesterone (Fairclough *et al.*, 1985).

LL–SL cell interaction. Using incubation of LL and SL cells separately and after recombination, and a superfusion system in which SL cells were incubated in series with, and upstream from, LL cells, Lemon & Mauléon (1982) demonstrated an interaction between pig LL and SL cells in the synthesis of progesterone. The data suggest that some product(s) of the SL cells was able to enhance progesterone production by the LL cells. Attempts to demonstrate a similar interaction in cow (Ursely & Leymarie, 1979) and sheep (Rodgers *et al.*, 1985) CL have so far been unsuccessful.

The possibility of an inhibitory interaction between ruminant LL and SL cells, in which oxytocin might be released from LL cells under the influence of PGF-2α and suppress the response of SL cells to LH, has also been postulated (Rodgers *et al.*, 1985; Niswender *et al.*, 1985). As discussed above, however, evidence of any inhibitory action of oxytocin on progesterone production *in vitro* is equivocal, and oxytocin receptors have not yet been demonstrated on SL cells.

Relaxin

Cattle. Fields *et al.* (1980) were able to demonstrate relaxin in bovine LL cells from the middle third of pregnancy by using the immunoperoxidase method with light microscopy. Although the polyclonal antibodies used were produced against pig relaxin, strong evidence from biochemical, immunological and bioassay studies supported the conclusion that the antigen localized was in fact relaxin. Reaction product was confined to the cytoplasm of the LL cells, but it was not possible to establish its subcellular localization. However, both by extrapolation from data from other species and in terms of known mechanisms of secretion of polypeptides by other cell types, storage and release of relaxin in granule form would be predicted.

Sheep. Evidence of the occurrence of relaxin in sheep CL is less convincing than that in the cow. Kruip *et al.* (1976) reported a positive identification of relaxin in CL from ewes in the last half of pregnancy, using immunofluorescence with an antibody produced against pig relaxin (NIH-R-P1), but also noted that sections treated without the antirelaxin serum were "not completely negative". No confirmatory evidence of the occurrence of relaxin in sheep CL was produced. Subsequent immunological localization studies in CL from pregnant ewes, using polyclonal (Renegar & Larkin, 1985) and monoclonal (J. Patterson, J. D. O'Shea, C. S. Lee & M. R. Brandon, unpublished data) antibodies to pig relaxin, have given wholly negative results.

Oxytocin

Sheep. In this species, many lines of evidence now indicate that oxytocin is synthesized, stored and secreted by LL cells. Localization of oxytocin and neurophysin I in LL cells, but not SL cells, was demonstrated with light microscopic immunocytochemistry by Watkins (1983), and has since been confirmed by electron microscopy (Theodosis *et al.*, 1986). Synthesis of oxytocin *in vitro* has also been shown to be restricted to LL cells (Rodgers *et al.*, 1983b). There is close correlation between the content and release of luteal oxytocin and the content and exocytosis of LL cell secretory granules. Granule loss of LL cells during luteal tissue dissociation (Rodgers & O'Shea, 1982) is accompanied by a substantial fall in the oxytocin content of the LL cells (Rodgers *et al.*, 1983b), while PGF-2α-induced release of oxytocin *in vivo* is associated with the exocytosis of large numbers of secretory granules (Fairclough *et al.*, 1985). Immune electron microscopy (Theodosis *et al.*, 1986) has now provided definitive proof of the granular localization of oxytocin and neurophysin I in ovine LL cells.

Cattle. Although data are less complete than for sheep, it is clear that the LL cells, but not the SL cells, of cyclic cows contain immunoreactive oxytocin and neurophysin (Guldenaar *et al.*, 1984; Kruip *et al.*, 1985). Secretory granule release in response to PGF-2α also occurs in cow CL (Heath *et al.*, 1983), but this has not yet been linked directly to oxytocin secretion.

Goats. Although morphological studies have shown that secretory granules are plentiful in LL cells through the cycle and into late pregnancy, and goat CL do contain oxytocin (Freeman & Currie, 1985), it has not yet been established whether the LL-cell granules contain oxytocin, relaxin, or some other product.

References

Alila, H.W. & Hansel, W. (1984) Origin of different cell types in the bovine corpus luteum as characterized by specific monoclonal antibodies. *Biol. Reprod.* **31**, 1015–1025.

Azmi, T.I. & Bongso, T.A. (1985) The ultrastructure of the corpus luteum of the goat. *Pertanika* **8**, 215–222.

Battista, P.J. & Condon, W.A. (1986) Serotonin-induced stimulation of progesterone production by cow luteal cells *in vitro*. *J. Reprod. Fert.* **76**, 231–238.

Bourdage, R.J., Fitz, T.A. & Niswender, G.D. (1984) Differential steroidogenic responses of ovine luteal cells to ovine luteinizing hormone and human chorionic gonadotropin. *Proc. Soc. exp. Biol. Med.* **175**, 483–486.

Byskov, A.G. (1986) Differentiation of mammalian embryonic gonad. *Physiol. Rev.* **66**, 71–117.

Chegini, N., Ramani, N. & Rao, Ch.V. (1984) Morphological and biochemical characterization of small and large bovine luteal cells during pregnancy. *Mol. cell. Endocr.* **37**, 89–102.

Corteel, M. (1973) Etude histologique de la transformation du follicle preovulatoire en corps jaune cyclique chez la brebis. *Annls Biol. anim. Biochim. Biophys.* **13**, 249–258.

Cran, D.G. (1983) Follicular development in the sheep after priming with PMSG. *J. Reprod. Fert.* **67**, 415–423.

Deane, H.W., Hay, M.F., Moor, R.M., Rowson, L.E.A. & Short, R.V. (1966) The corpus luteum of the sheep: relationships between morphology and function during the oestrous cycle. *Acta endocr., Copenh.* **51**, 245–263.

Donaldson, L. & Hansel, W. (1965) Histological study of bovine corpora lutea. *J. Dairy Sci.* **48**, 905–909.

Fairclough, R.J., Staples, L.D. & O'Shea, J.D. (1985) Ovarian oxytocin and progesterone secretion and degranulation of luteal cells in sheep given exogenous prostaglandins. *Proc. Aust. Soc. Reprod. Biol.* **17**, 55, Abstr.

Fields, M.J., Fields, P.A., Castro-Hernandez, A. & Larkin, L.H. (1980) Evidence for relaxin in corpora lutea of late pregnant cows. *Endocrinology* **107**, 869–876.

Fields, M.J., Dubois, W. & Fields, P.A. (1985) Dynamic features of luteal secretory granules: ultrastructural changes during the course of pregnancy in the cow. *Endocrinology* **117**, 1675–1682.

Fitz, T.A., Sawyer, H.R. & Niswender, G.D. (1981) Characterization of two steroidogenic cell types in the ovine corpus luteum. *Biol. Reprod.* **24**, Suppl. 1, 54A, Abstr.

Fitz, T.A., Mayan, M.H., Sawyer, H.R. & Niswender, G.D. (1982) Characterization of two steroidogenic cell types in the ovine corpus luteum. *Biol. Reprod.* **27**, 703–711.

Fitz, T.A., Hoyer, P.B. & Niswender, G.D. (1984a) Interactions of prostaglandins with subpopulations of ovine luteal cells. I. Stimulatory effects of prostaglandins E_1, E_2 and I_2. *Prostaglandins* **28**, 119–126.

Fitz, T.A., Mock, E.J., Mayan, M.H. & Niswender, G.D. (1984b) Interactions of prostaglandins with subpopulations of ovine luteal cells. II. Inhibitory effects of $PGF_{2\alpha}$ and protection by PGE_2. *Prostaglandins* **28**, 127–138.

Freeman, L.C. & Currie, W.B. (1985) Variations in the oxytocin content of caprine corpora lutea across the breeding season. *Theriogenology* **23**, 481–486.

Friend, D.S. & Gilula, N.B. (1972) A distinctive cell contact in the rat adrenal cortex. *J. Cell Biol.* **53**, 148–163.

Gemmell, R.T., Stacy, B.D. & Thorburn, G.D. (1974)

Ultrastructural study of secretory granules in the corpus luteum of the sheep during the estrous cycle. *Biol. Reprod.* **11**, 447–462.

Gemmell, R.T., Stacy, B.D. & Nancarrow, C.D. (1977) Secretion of granules by the luteal cells of the sheep and the goat during the estrous cycle and pregnancy. *Anat. Rec.* **189**, 161–168.

Glass, J.D., Fitz, T.A. & Niswender, G.D. (1984) Cytosolic receptor for estradiol in the corpus luteum of the ewe: variation throughout the estrous cycle and distribution between large and small steroidogenic cell types. *Biol. Reprod.* **31**, 967–974.

Guldenaar, S.E.F., Wathes, D.C. & Pickering, B.T. (1984) Immunocytochemical evidence for the presence of oxytocin and neurophysin in the large cells of the bovine corpus luteum. *Cell Tiss. Res.* **237**, 349–352.

Heath, E., Weinstein, P., Merritt, B., Shanks, R. & Hixon, J. (1983) Effects of prostaglandins on the bovine corpus luteum: granules, lipid inclusions and progesterone secretion. *Biol. Reprod.* **29**, 977–985.

Higuchi, T., Kaneko, A., Abel, J.H. & Niswender, G.D. (1976) Relationship between membrane potential and progesterone release in ovine corpora lutea. *Endocrinology* **99**, 1023–1032.

Hoyer, P.B., Fitz, T.A. & Niswender, G.D. (1984) Hormone-independent activation of adenylate cyclase in large steroidogenic ovine luteal cells does not result in increased progesterone secretion. *Endocrinology* **114**, 604–608.

Jordan, A.W., Caffrey, J.C. & Niswender, G.D. (1978) Catecholamine-induced stimulation of progesterone and adenosine 3′,5′-monophosphate production by dispersed ovine luteal cells. *Endocrinology* **103**, 385–392.

Kirsch, T.M., Vogel, R.L. & Flickinger, G.L. (1983) Macrophages: a source of luteal cybernins. *Endocrinology* **113**, 1910–1912.

Koos, R.D. & Hansel, W. (1981) The large and small cells of the bovine corpus luteum: ultrastructural and functional differences. In *Dynamics of Ovarian Function*, pp. 197–203. Eds N. B. Schwartz & M. Hunzicker-Dunn. Raven Press, New York.

Kruip, Th.A.M., Taverne, M.A.M. & van Beneden, Th.H. (1976) Demonstration of relaxin in the ovary of pregnant sheep by immunofluorescence. *Proc. 8th Int. Congr. Anim. Reprod. & A.I. Krakow*, vol. 3, 375–377.

Kruip, Th.A.M., Vullings, H.G.B., Schams, D., Jonis, J. & Klarenbeek, A. (1985) Immunocytochemical demonstration of oxytocin in bovine ovarian tissues. *Acta endocr., Copenh.* **109**, 537–542.

Le Beux, Y.J. (1969) An unusual ultrastructural association of smooth membranes and glycogen particles: the glycogen body. *Z. Zellforsch. mikrosk. Anat.* **101**, 433–437.

Lemon, M. & Mauléon, P. (1982) Interaction between two luteal cell types from the corpus luteum of the sow in progesterone synthesis *in vitro*. *J. Reprod. Fert.* **64**, 315–323.

McClellan, M.C., Diekman, M.A., Abel, J.H. & Niswender, G.D. (1975) Luteinizing hormone, progesterone and the morphological development of normal and superovulated corpora lutea in sheep. *Cell. Tiss. Res.* **164**, 291–307.

Mossman, H.W. & Duke, K.L. (1973) *Comparative Morphology of the Mammalian Ovary*, pp. 209–220. University of Wisconsin Press, Madison.

Niswender, G.D., Reimers, T.J., Diekman, M.A. & Nett, T.M. (1976) Blood flow: a mediator of ovarian function. *Biol. Reprod.* **14**, 64–81.

Niswender, G.D., Schwall, R.H., Fitz, T.A., Farin, C.E. & Sawyer, H.R. (1985) Regulation of luteal function in domestic ruminants: new concepts. *Recent Progr. Horm. Res.* **41**, 101–151.

O'Shea, J.D. & Wright, P.J. (1985) Regression of the corpus luteum of pregnancy following parturition in the ewe. *Acta anat.* **122**, 69–76.

O'Shea, J.D., Cran, D.G. & Hay, M.F. (1979) The small luteal cell of the sheep. *J. Anat.* **128**, 239–251.

O'Shea, J.D., Cran, D.G. & Hay, M.F. (1980) Fate of the theca interna following ovulation in the ewe. *Cell Tiss. Res.* **210**, 305–319.

O'Shea, J.D., Wright, P.J. & Davis, K.E. (1985) Numbers of granulosa cells in preovulatory follicles from ewes during the breeding season and after LHRH administration during seasonal anoestrus. *Proc. 12th Int. Anat. Congr., London*, A529, Abstr.

O'Shea, J.D., Rodgers, R.J. & Wright, P.J. (1986) Cellular composition of the sheep corpus luteum in the mid- and late luteal phases of the oestrous cycle. *J. Reprod. Fert.* **76**, 685–691.

Paavola, L.G. & Christensen, A.K. (1981) Characterization of granule types in luteal cells of sheep at the time of maximum progesterone secretion. *Biol. Reprod.* **25**, 203–215.

Parry, D.M., Willcox, D.L. & Thorburn, G.D. (1980) Ultrastructural and cytochemical study of the bovine corpus luteum. *J. Reprod. Fert.* **60**, 349–357.

Priedkalns, J. & Weber, A.F. (1968) Ultrastructural studies of the bovine graafian follicle and corpus luteum. *Z. Zellforsch. mikrosk. Anat.* **91**, 554–573.

Priedkalns, J., Weber, A.F. & Zemjanis, R. (1968) Qualitative and quantitative morphological studies of the cells of the membrana granulosa, theca interna and corpus luteum of the bovine ovary. *Z. Zellforsch. mikrosk. Anat.* **85**, 501–520.

Quirk, S.J., Willcox, D.L., Parry, D.M. & Thorburn, G.D. (1979) Subcellular location of progesterone in the bovine corpus luteum: a biochemical, morphological and cytochemical investigation. *Biol. Reprod.* **20**, 1133–1145.

Renegar, R.H. & Larkin, L.H. (1985) Relaxin concentrations in endometrial, placental and ovarian tissues and in sera from ewes during middle and late pregnancy. *Biol. Reprod.* **32**, 840–847.

Rodgers, R.J. & O'Shea, J.D. (1982) Purification, morphology, and progesterone production and content of three cell types isolated from the corpus luteum of the sheep. *Aust. J. biol. Sci.* **35**, 441–455.

Rodgers, R.J., O'Shea, J.D. & Findlay, J.K. (1983a) Progesterone production *in vitro* by small and large ovine luteal cells. *J. Reprod. Fert.* **69**, 113–124.

Rodgers, R.J., O'Shea, J.D., Findlay, J.K., Flint, A.P.F. & Sheldrick, E.L. (1983b) Large luteal cells the source of luteal oxytocin in the sheep. *Endocrinology* **113**, 2302–2304.

Rodgers, R.J., O'Shea, J.D. & Bruce, N.W. (1984) Morphometric analysis of the cellular composition of the ovine corpus luteum. *J. Anat.* **138**, 757–769.

Rodgers, R.J., O'Shea, J.D. & Findlay, J.K. (1985) Do

small and large luteal cells of the sheep interact in the production of progesterone? *J. Reprod. Fert.* **75,** 85–94.

Sawyer, H.R., Abel, J.H., McClellan, M.C., Schmitz, M. & Niswender, G.D. (1979) Secretory granules and progesterone secretion by ovine corpora lutea *in vitro. Endocrinology* **104,** 476–486.

Sernia, C., Thorburn, G.D. & Gemmell, R.T. (1982) Search for a progesterone-binding protein in secretory granules of the ovine corpus luteum. *Endocrinology* **110,** 2151–2158.

Singh, U.B. (1975) Structural changes in the granulosa lutein cells of pregnant cows between 60 and 240 days. *Acta anat.* **93,** 447–457.

Tan, G.J.S., Tweedale, R. & Biggs, J.S.G. (1982) Effects of oxytocin on the bovine corpus luteum of early pregnancy. *J. Reprod. Fert.* **66,** 75–78.

Theodosis, D.T., Wooding, F.B.P., Sheldrick, E.L. & Flint, A.P.F. (1986) Ultrastructural localisation of oxytocin and neurophysin in the ovine corpus luteum. *Cell Tiss. Res.* **243,** 129–135.

Ursely, J. & Leymarie, P. (1979) Varying response to luteinizing hormone of two luteal cell types isolated from bovine corpus luteum. *J. Endocr.* **83,** 303–310.

Watkins, W.B. (1983) Immunohistochemical localization of neurophysin and oxytocin in the sheep corpora lutea. *Neuropeptides* **4,** 51–54.

Weber, D.M., Roberts, R.F., Romrell, L. & Fields, M.J. (1984) Biochemically and morphologically different luteal cells in late pregnant cows. *Biol. Reprod.* **30,** Suppl. **1,** 91, Abstr.

J. Reprod. Fert., Suppl. **34** (1987), 87–99

Printed in Great Britain
© 1987 Journals of Reproduction & Fertility Ltd

Luteal peptides and intercellular communication

D. Schams

*Lehrstuhl für Physiologie der Fortpflanzung und Laktation, Technische Universität München,
8050 Freising—Weihenstephen, FRG*

Summary. The variety of peptides synthesized by the corpus luteum (relaxin, vaso-pressin, oxytocin and oxytocin-related neurophysin) and their possible intracellular effects are reviewed. After luteinization of the granulosa cells and in response to LH and FSH, the output of oxytocin is increased. In addition, insulin-like growth factor is a very potent stimulus of oxytocin secretion. Although luteal cells respond to gonado-trophins by increased production of progesterone, there is no further secretion of oxytocin. Oxytocin is localized in large luteal cells which seem not to be under the direct control of gonadotrophins. Synthesis of luteal oxytocin seems to occur during the early luteal phase according to measurements of oxytocin mRNA. Highest tissue concen-trations and secretion under in-vitro conditions were observed during the mid-luteal phase, and so synthesis, storage and secretion are unlikely to occur concomitantly.

Under in-vitro conditions, oxytocin is secreted concomitantly with neurophysin and progesterone, and there appears to be some form of communication between small and large luteal cells for the secretion of progesterone and oxytocin under in-vivo conditions.

Evidence has been obtained that oxytocin may have local effects in the ovary by inhibition of secretion (synthesis ?) of progesterone, especially during the early luteal phase. A mechanism can be suggested whereby, under physiological conditions, oxy-tocin may delay the increase of progesterone by inhibition of progesterone secretion and therefore delay down regulation of its own receptor. This would prolong the life-span of the CL and the oestrous cycle. Exogenous progesterone given on Days 1–4 shortens the cycle to about 12 days. The best evidence that oxytocin may be involved in controlling luteolysis comes from immunization experiments in ewes and goats, but there is no clear evidence of this type for cattle. Basal concentrations of oxytocin at the end of the luteal phase may interact with oxytocin receptors after the inhibitory effect of progesterone in the uterus is reduced, thus initiating synthesis of PGF-2α.

Introduction

The corpus luteum (CL) occupies a central position in the reproductive process of all mammals. Progesterone is the primary endocrine secretory product, although the CL also secretes prostaglan-dins and, in some species, oestradiol-17β, and a variety of protein and peptide hormones. The first peptide found in luteal tissue of pigs was relaxin (for review see Bryant-Greenwood, 1982). The primary structure of pig relaxin consists of two peptide chains (A and B) of 22 and 31 amino acids, respectively, covalently linked by two interchain disulphide bonds with an intradisulphide link in the A chain (M_r 6000). Oxytocin and vasopressin have been identified in luteal tissue from sheep (Wathes & Swann, 1982), women (Wathes *et al.*, 1982) and cows (Wathes *et al.*, 1983a). This paper is concerned primarily with luteal peptides in domestic ruminants. It focusses on their synthesis, storage, secretion and possible intercellular function.

Relaxin

Relaxin classically causes uterine quiescence during pregnancy and cervical dilatation before parturition. It may have also some activities during mammary gland development. The common denominator of these activities is its action on the remodelling of connective tissue. There are suggestions that relaxin is a local intrauterine hormone in some species, e.g. woman, and may have some effects during follicular growth and rupture. Our knowledge of relaxin in ruminants is rather limited due to the lack of specific and sensitive assays. Large luteal cells from cows in the middle third of pregnancy stained positively for relaxin with the immunoperoxidase method (Fields *et al.*, 1980). Relaxin has also been isolated from CL of late pregnant cows (Fields *et al.*, 1982).

Relaxin has not been detected immunocytochemically by light or electron microscopy in luteal tissue at the end of pregnancy in sheep. It appears that quantities, if any, are very low (Renegar & Larkin, 1985). Pig relaxin given into the cervical os or by intramuscular injection to beef heifers, beginning 4 days before expected parturition, induced significant dilatation of the cervix 8 and 16 h later and increased growth rate of the pelvic area (Perezgrovas & Anderson, 1982).

Vasopressin

Vasopressin has been extracted from CL of non-pregnant cows (Wathes *et al.*, 1983a, 1984). Concentrations (pg range/g wet weight) were highest during the mid-luteal phase, declined thereafter and were undetectable during pregnancy, following a trend similar to that of oxytocin. Ivell & Richter (1984) found 1000 times less vasopressin mRNA than oxytocin mRNA in luteal tissue.

Oxytocin and neurophysin I

Tissue concentrations

Although evidence for oxytocin in the CL of the goat was published as long ago as 1910 (Ott & Scott, 1910), it is only in the past 4 years that its presence has been confirmed in sheep (Wathes & Swann, 1982; Flint & Sheldrick, 1983a), women (Wathes *et al.*, 1983b) and cows (Wathes *et al.*, 1983a, 1984; Fields *et al.*, 1983; Schams *et al.*, 1984, 1985a, 1987). The material extracted displaced the tracer parallel to oxytocin in a radioimmunoassay, stimulated uterine contractions and eluted at the same position as oxytocin when tested by high-performance liquid chromatography (HPLC). Final confirmation was obtained by purification of oxytocin-like material from sheep CL and identification by sequence analysis (Watkins *et al.*, 1985). Further evidence was given by demonstration that the oxytocin gene is highly transcribed in the cow CL during the mid-luteal phase of the oestrous cycle. The active CL produced up to 250 times more oxytocin mRNA than did a single hypothalamus (Ivell & Richter, 1984). More detailed studies indicated that transcription is maximal at the time of ovulation and decreases rapidly thereafter (Ivell *et al.*, 1985; Schams *et al.*, 1987). Swann *et al.* (1984) measured incorporation of [^{35}S]cysteine into oxytocin and neurophysin I in dispersed cell cultures of CL from cows and sheep.

There is general agreement that oxytocin concentrations in tissue increase up to the mid-luteal phase and decrease thereafter, remaining low after luteal regression and during pregnancy. A positive correlation was found between oxytocin and progesterone until the mid-luteal phase, but not during the late luteal phase. There seem to be breed differences for luteal oxytocin. Maximal content ranged from 0·4 to 1·8 µg/g CL in cows and was about 2 µg/g CL in sheep. Levels of neurophysin I in cows were also maximal during the mid-luteal phase and declined afterwards to low values during pregnancy.

In-vivo studies

Secretion of luteal oxytocin and neurophysin I during the oestrous cycle

Sheep. Measurements of oxytocin during the oestrous cycle of the ewe show that the circulating levels increase and decrease synchronously with changes in progesterone concentration, declining to a minimum at oestrus (Sheldrick & Flint, 1981; Webb *et al.*, 1981; Mitchell *et al.*, 1982; Schams *et al.*, 1982; Flint & Sheldrick, 1983a). Oxytocin concentrations plateau about 2 days earlier than do progesterone values. After ovariectomy of cyclic ewes, circulating oxytocin concentrations fell to the limit of detection of the assay (Schams *et al.*, 1982). Intermittent pulses of the metabolite of prostaglandin (PG) F-2α and neurophysin I/II were measured concomitantly in blood during the period of luteal regression (Fairclough *et al.*, 1980).

Cow. Circulating oxytocin concentration during the oestrous cycle are lower than in the ewe but follow the same trend, with the highest values found in the early and mid-luteal phases and again with a short increase at the time of luteolysis (Schams, 1983; Schams *et al.*, 1985b). More detailed studies after frequent bleeding at different stages of the oestrous cycle and comparison of concentrations of oxytocin and neurophysin I in the posterior vena cava and jugular vein showed that concentrations of oxytocin were similar in both vessels during the non-luteal phase but were higher in the vena cava during the luteal phase, indicating the ovarian origin of oxytocin. Oxytocin was secreted in a pulsatile manner concomitantly with progesterone and neurophysin I, and concentrations of oxytocin also increased concomitantly with these two hormones. Oxytocin concentrations increased parallel with surges of PGF-2α at the time of luteolysis (Walters *et al.*, 1984; Walters & Schallenberger, 1984; Schallenberger *et al.*, 1984; Schams *et al.*, 1985b, c).

The resumption of oxytocin and progesterone secretion during the first luteal phase after parturition in intact as well as hysterectomized heifers shows that oxytocin concentrations increased faster than those of progesterone during the early luteal phase. In hysterectomized animals oxytocin decreased after the first 2 weeks of the luteal phase and remained at low concentrations whereas progesterone values remained high (Schams *et al.*, 1985b).

Goat. A similar pattern of ovarian oxytocin secretion probably occurs in the goat, in which circulating oxytocin concentrations decline from >40 to <10 pg/ml between Days 12 and 16 in non-pregnant and pregnant animals (Homeida & Cooke, 1983) and become undetectable after ovariectomy (Cooke *et al.*, 1984).

Stimulation of luteal oxytocin secretion

Prostaglandins. Injections of cloprostenol, PGF-2α or PGE-2 caused a significant increase of oxytocin secretion in cows (Schallenberger *et al.*, 1984; Schams *et al.*, 1985b) and ewes (Flint & Sheldrick, 1982, 1983b; Watkins *et al.*, 1984). The stimulation of oxytocin release by prostaglandin injection depends on the presence of the ovary and a functional CL. A persistent CL 53–83 days after hysterectomy (performed on Days 6–7 of the cycle) was not able to secrete oxytocin after an injection of cloprostenol (Sheldrick & Flint, 1983). However, a 35-day persistent CL of a hysterectomized heifer released some oxytocin after a prostaglandin challenge, even when unstimulated basal secretion had ceased for weeks. The posterior pituitary seems to release only minute amounts of oxytocin after the injection of prostaglandins.

Gonadotrophins. After infusion of a crude FSH preparation (10 mg) or i.v. injection of 100 i.u. hCG, a simultaneous release of progesterone and oxytocin was observed (Schams *et al.*, 1985b). A high correlation between the number of CL and the increase in oxytocin and immunoreactive neurophysin I was found after stimulation of superovulation with PMSG or pituitary FSH in heifers. Even when PMSG exerted a more pronounced luteotrophic effect on progesterone secretion than did FSH, there was no further stimulation of oxytocin or neurophysin I secretion (Schams *et al.*, 1985c).

In-vitro studies

Follicular oxytocin

There is evidence that the follicle is able to secrete small amounts of oxytocin. Immunoreactive oxytocin has been measured in cow follicular fluid. Concentrations (on average 48–108 pg/ml) increased significantly with the size of the follicle and were significantly higher in histologically verified cysts (190 pg/ml). After culture of follicles the amount of oxytocin released into the medium increased, indicating de-novo synthesis. The granulosa cells were the main source of follicular oxytocin. Secretion increased during luteinization, indicating that luteinization is an important step for production of oxytocin in ovaries (Schams *et al.*, 1985a). Geenen *et al.* (1985) obtained similar results and further demonstrated the production of neurophysin I by bovine granulosa cells. The production of immunoreactive oxytocin was significantly greater in cultures of granulosa cells harvested from large follicles than in those derived from small follicles, but immunoreactive oxytocin was demonstrated immunocytochemically in the granulosa cells of small and large follicles (Kruip *et al.*, 1985). Low levels of oxytocin mRNA were detected in bovine follicles. Contrary to the previous results, Jungclas & Luck (1986) could not measure oxytocin in follicular fluid. They agree that luteinization is the main stimulus for secretion of oxytocin.

To clarify the role of oxytocin for ovarian function, in-vitro studies were undertaken using bovine granulosa and luteal cells. The aim was to find out what stimulates oxytocin, and whether the stimuli effective for progesterone secretion were also operative. Some of the results have been published by Schams *et al.* (1987). Ovaries were collected from the slaughter house.

Stimulation was performed with 10 and 100 ng/ml medium of highly purified sheep LH and FSH (kindly supplied by Dr O. D. Sherwood, U.S.A.).

The effect of growth factors was tested additionally in granulosa cells. IGF-I (somatomedin-C) was kindly supplied by Dr C. H. Li (U.S.A.). Fibroblast growth factor (IGF-II) and nerve cell growth factor were purchased from Sigma, St Louis, MO, U.S.A.

Culture of granulosa cells

Concentrations of progesterone increased in the medium during culture of bovine granulosa cells. LH and FSH exhibited a clear stimulating effect with a 2–8-fold increase over controls. The effect of FSH was more pronounced, especially after 120 h of incubation. Concentrations of oxytocin increased in controls only at the end of incubation (120 h). FSH was clearly more effective especially after 120 h (Fig. 1). FSH stimulated over a 10-fold increase in the output of oxytocin as compared to controls (Table 1). Increase of oestradiol-17β synthesis by addition of aromatase substrate alone (40 ng androstendione) or in combination with FSH was uneffective.

As shown in Fig. 2 insulin-like growth factor-I increased concentrations of progesterone and oxytocin in a dose-dependent manner to values higher than with FSH. IGF + FSH gave no additional effect. The other growth factors used exerted no effect on secretion of progesterone or oxytocin.

Luteal cells

In controls, concentrations of progesterone increased in the medium during incubation of cells obtained from CL of cows at different stages of the oestrous cycle. FSH, and particularly LH, stimulated a further output of progesterone (Table 1). However, although the concentration of oxytocin in the medium increased during incubation no stimulating effect of LH or FSH was observed. In an experiment in which CL from heifers were collected at specific times of the oestrous cycle (Days 3–4, 6–7, 10–12, 15–17 and 19–21), no significant difference was observed. Luteal tissue concentrations of oxytocin mRNA, immunoreactive oxytocin and progesterone were also

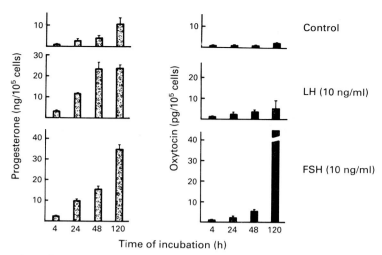

Fig. 1. Secretion of progesterone and oxytocin (mean ± s.d., 4 dishes/dose) during culture of bovine granulosa cells without and with stimulation of gonadotrophins (M 199 + 0·5% BSA). (Modified from Schams *et al.*, 1987.)

Table 1. Effect of stimulation with gonadotrophins (10 ng/ml), expressed as a factor over controls, on hormone secretion by cultured granulosa (120 h) and luteal (4 h) cells of cows

Cell type	Progesterone		Oxytocin	
	LH	FSH	LH	FSH
Granulosa	3·0 ± 1·5 (13)	6·3 ± 3·3 (13)	4·0 ± 3·3 (11)	10·0 ± 10·8 (11)
Luteal	2·2 ± 1·5 (24)	1·4 ± 0·3 (24)	1·0 ± 0·1 (25)	−1·0 ± −0·2 (25)

Values are mean ± s.d. for the no. of experiments indicated in parentheses.

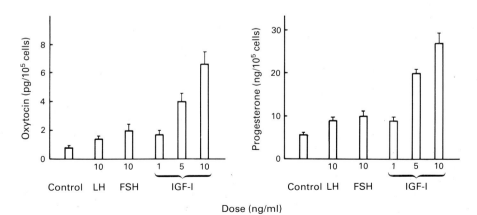

Fig. 2. Culture of bovine granulosa cells in M 199 + 0·5% BSA for 120 h. Effects of LH, FSH and insulin-like growth factor (IGF-I) on secretion of oxytocin and progesterone. Values are mean ± s.d., 4 dishes/dose.

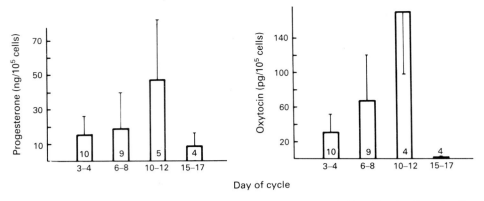

Fig. 3. Secretion of progesterone and oxytocin after incubation (4 h) of bovine luteal cells at different stages of the oestrous cycle. Values are mean ± s.d. for the no. of observations indicated.

measured. Highest oxytocin mRNA levels were found on Day 1 or Days 3–4 with a marked decrease afterwards. In contrast, immunoreactive oxytocin content was highest during Days 10–12 and decreased thereafter. Progesterone content followed the same trend, but was still high on Days 15–17 (Schams *et al.*, 1987). The secretion rate of oxytocin from unstimulated luteal cells showed a similar pattern (Fig. 3).

Conclusions

The in-vitro data clearly demonstrate that luteinization of granulosa cells seems to be the main stimulus for the secretion and, most probably, the synthesis of ovarian oxytocin. This is an agreement with data from Jungclas & Luck (1986), who also found an intercellular communication between theca and granulosa cells. Oxytocin but not progesterone output could be consistently increased by addition of pieces of theca interna tissue, or theca-conditioned medium, to the cultures. In cultured granulosa cells an increase of oxytocin mRNA could not be clearly measured (R. Ivell, unpublished observations). Both gonadotrophins have a stimulating effect, but FSH appears to play a dominating role. Insulin-like growth factor (IGF-I), acting as a potent stimulator of the production of progesterone and oxytocin of granulosa luteal cells is obviously also important. The mechanism involved remains uncertain. Receptors for insulin-like growth factor have been demonstrated in pig granulosa luteal cells (Barranao & Hammond, 1984; Veldhuis & Furlanetto, 1985). The tissue contents of defined luteal material confirmed earlier observations (Ivell *et al.*, 1985; Schams *et al.*, 1987). Oxytocin mRNA content suggests that most of the oxytocin must be synthesized during the early luteal phase and secreted at a high rate. The high oxytocin content during the mid-luteal phase is more likely to reflect storage than synthesis. In the CL of the non-pregnant ewe (Watkins, 1983; Sawyer *et al.*, 1986) and cow (Guldenaar *et al.*, 1984; Kruip *et al.*, 1985), specific staining for both oxytocin and neurophysin I has been demonstrated only in the large luteal cells, which are known to contain secretory granules (Gemmell *et al.*, 1977; Parry *et al.*, 1980). The exocytosis of these granules accelerates during the mid-luteal phase (Gemmell *et al.*, 1977; Gemmell & Stacy, 1979; Quirk *et al.*, 1979). Rice & Thorburn (1985) showed that the oxytocin in ovine CL was associated with a particulate fraction, sedimenting at a density of 1054–1061 g/ml, which contained electron-dense granules with a diameter of 200–250 nm. Theodosis *et al.* (1986) have shown specific staining of similar granules with antisera to oxytocin and neurophysin, using an immunogold labelling technique. Large cells from sheep CL are most active in oxytocin synthesis (Rodgers *et al.*, 1983). The oxytocin tissue content correlates with the number of large luteal cells in sheep (Schwall *et al.*, 1986). There was a 400% increase in the total number of cells between Days 4 and 8 but no change between Days 8 and 12, and a 70% decrease at Day 16. Alila & Hansel (1984) demonstrated that in the cow the large luteal cells initially develop from the granulosa layer but are soon replaced by the growth of theca-derived cells. Granulosa-derived cells disappear during early pregnancy, while cells of thecal origin persist throughout pregnancy. Large cells from sheep CL contained more receptors for prostaglandins, while the small cells contained more receptors for LH (Fitz *et al.*, 1982). The lack of LH and possibly FSH receptors in large cells may explain why secretion of oxytocin could no longer be stimulated with gonadotrophins. The secretion of oxytocin from large cells seems not to be under the direct control of gonadotrophins. In in-vitro studies using slices of bovine CL, adding prostaglandins or blocking their synthesis by indomethacin did not increase or decrease secretion of oxytocin.

Intra-ovarian effects of oxytocin on steroidogenesis

The first evidence for a direct action of oxytocin on steroidogenesis was obtained by Tan *et al.* (1982a, b; Tan & Biggs, 1984), using short-term incubations of dispersed luteal cells from cows and women. In both cases there was a tendency for lower doses of oxytocin (4–40 mU/ml) to stimulate progesterone production whereas higher doses (>400 mU/ml) inhibited both basal and hCG-stimulated progesterone release. However, Richardson & Masson (1985) were unable to demonstrate such an effect with human luteal cells and Mukhopadhyay *et al.* (1984) also found no effect of a range of oxytocin doses from 10^{-8} to 4×10^{-6} M on basal of hCG-stimulated progesterone production by luteal cells from pseudopregnant rats. Oxytocin added to large and small luteal cell fractions obtained from fully developed sheep CL did not affect progesterone production by either fraction (Rodgers *et al.*, 1985).

We were not able to demonstrate an inhibitory effect on LH-stimulated progesterone production by granulosa cells undergoing luteinization of oxytocin in doses ranging from 0·1 to 100 ng/ml

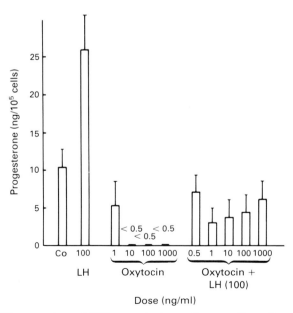

Fig. 4. Effect of LH, oxytocin and LH + oxytocin on the secretion of progesterone by cells obtained from the CL of cows at Days 4–5 of the cycle and incubated for 4 h. Values are mean ± s.d. for 4 dishes.

(Schams *et al.*, 1987). In preliminary experiments FSH-stimulated progesterone production by granulosa cells was depressed with 0·5 ng/ml oxytocin after culture for 120 h. In short-term incubations (4 h) of bovine luteal cells only doses of oxytocin from 1–1000 ng/ml showed a clear inhibitory effect, especially in CL obtained at Day 4–5 of the cycle. LH-stimulated progesterone production was depressed with as little as 0·5 ng oxytocin (see Fig. 4). The data indicate an inhibitory effect of oxytocin on progesterone secretion especially during the early luteal phase. It can be presumed that luteal cells have receptors for oxytocin, but this has not been confirmed by our preliminary experiments.

Luteolysis and ovarian oxytocin

Exogenous oxytocin

The luteolytic action of exogenous oxytocin was first described by Armstrong & Hansel (1959), who found that daily injections of 50–100 units oxytocin in heifers during the first 7 days of the oestrous cycle led to a significant decrease in oestrous cycle length. Hansel & Wagner (1960) reported that oxytocin was only luteolytic between Days 3 and 6 of the cycle. In the goat, oxytocin treatment on Days 3–6 of the cycle also led to a premature return to oestrus (Cooke & Knifton, 1981; Cooke & Homeida, 1982). Exogenous oxytocin, however, does not appear to cause luteal regression in ewes, although it may be followed by a slight depression in progesterone secretion (Milne, 1963; Hatjiminaoglou *et al.*, 1979).

Oxytocin treatment decreased the concentrations of progesterone in jugular venous blood on Day 8 and increased uterine venous PGF-2α concentrations in heifers (Milvae & Hansel, 1980; Oyedipe *et al.*, 1984).

An increase of PGF-2α in response to oxytocin was maximal on Day 3 of the cycle in heifers (Newcomb *et al.*, 1977). Exogenous oxytocin also stimulated the release of PGF-2α from the uterine endometrium in sheep (Roberts & McCracken, 1976) and goats (Cooke & Homeida, 1982). It appears that the luteolytic action of oxytocin is normally mediated by this response, since it can be prevented by hysterectomy in the cow (Armstrong & Hansel, 1959; Anderson *et al.*, 1965; Ginther *et al.*, 1967) and by simultaneous treatment with a PG synthetase inhibitor in the goat (Cooke & Knifton, 1981; Cooke & Homeida, 1983).

Immunization

The best evidence that oxytocin may be involved in controlling luteal regression comes from the active immunization of cyclic ewes against oxytocin (Sheldrick *et al.*, 1980; Schams *et al.*, 1983) and goats (Cooke & Homeida, 1985) in which luteal regression was delayed. However, passive immunization of heifers did not prolong the oestrous cycle (D. Schams, unpublished observations).

Interaction of oxytocin and uterine endometrium

Concentrations of endometrial oxytocin receptor change during the oestrous cycle, reaching a maximum at oestrus and declining to almost undetectable levels during the mid-luteal phase before increasing again on Days 14–15 of the cycle in ewes (Roberts *et al.*, 1976; Sheldrick & Flint, 1985a) or Days 18–19 of the cycle in cows (Schams *et al.*, 1987). Levels of the uterine oxytocin receptor are under the control of steroid hormones. Oestradiol induces and progesterone reduces the formation of receptors for oxytocin (Soloff, 1975; Nissenson *et al.*, 1978). As demonstrated in sheep (for review see McCracken *et al.*, 1984), oestradiol induces the formation of receptors for oxytocin in about 6 h. During the luteal phase, progesterone inhibits the action of oestradiol by blocking the nuclear accumulation of oestrogen receptor and hence the ability of oestradiol to maintain synthesis of oxytocin receptors. However, after about 10 days the action of progesterone in the

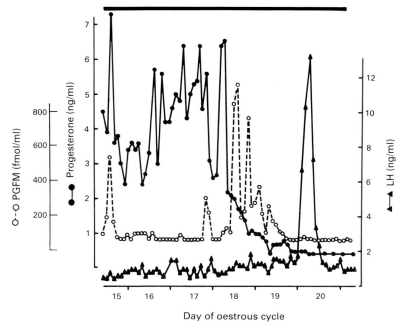

Fig. 5. Concentrations of progesterone, metabolite of PGF-2α (PGFM) and LH in jugular vein plasma during continuous infusion of oxytocin (12 μg/100 kg/day) in a heifer (peripheral oxytocin concentration 25–30 pg/ml plasma). Closed bar indicates time of infusion. Blood was sampled at 2-h intervals.

uterus begins to diminish (possibly due to progesterone-induced loss of its own receptor), and so oestradiol is able to induce synthesis of oxytocin receptors again and also permits oxytocin-induced secretion of PGF-2α. At this time, however, PGF-2α secretion in response to oxytocin is 100-fold greater than before the influence of progesterone. It is suggested that the massive discharge of PGF-2α from the uterus towards the end of the luteal phase is stimulated by oxytocin release, thereby initiating luteolysis. However, it has been shown by Moore *et al.* (1986) that concentrations of PGF-2α in utero-ovarian vein samples in ewes begin to increase before the concentrations of oxytocin and oxytocin-associated neurophysin. This suggests that uterine PGF-2α initiates the release of ovarian oxytocin and oxytocin-associated neurophysin pulses during luteolysis in ewes. This is consistent with the inhibition of pulsatile oxytocin-associated neurophysin release in ewes (Watkins *et al.*, 1984) and oxytocin release in goats (Cooke & Homeida, 1984) after systemic treatment with indomethacin. Pulsatile release of oxytocin was also absent in hysterectomized cows bearing a persistent CL (Schams *et al.*, 1985b). It appears therefore that basal concentrations of oxytocin interact with uterine oxytocin receptors, thus initiating PGF-2α release which induces further release of oxytocin and hence amplifies the release of PGF-2α from the uterus. Earlier suggestions indicate that the pulsatile secretion of PGF-2α may be due to down-regulation of the oxytocin receptor (McCracken *et al.*, 1984; Flint & Sheldrick, 1985). Constant infusion of oxytocin beginning on Day 13 caused prolongation of the oestrous cycle. However, we could not confirm these observations in cattle. Continuous infusion of oxytocin (12 μg/100 kg/day) starting on Day 15 did not prevent induction of oxytocin receptors or prevent luteal regression in cyclic heifers (J. Kotwica & D. Schams, unpublished observations; Fig. 5). Results recently obtained in ewes indicate that the pulsatile secretion of PGF-2α reflects a post-receptor failure of response, possibly due to depletion of prostaglandin precursors (Sheldrick & Flint, 1985b).

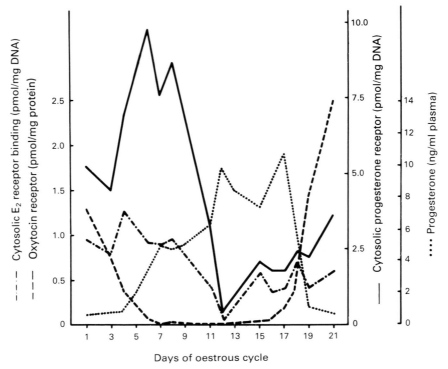

Fig. 6. Concentrations of bovine endometrial receptors for oxytocin, cytosolic progesterone and oestrogen and peripheral blood concentrations for progesterone. Heifers (4/group) were slaughtered at known stages of the oestrous cycle on Days 1–2, 3–4, 6–7, 10–12, 16–17, 18–19.

It may be that the relationships between oxytocin and PGF-2α during luteal regression are different in cattle and sheep. Our studies in heifers (T. Mittermeier, H. H. D. Meyer & D. Schams, unpublished data) give results that are comparable with those in sheep in that initially progesterone may depress the concentrations of receptors for oxytocin, oestrogen and eventually its own receptor. Afterwards, the action of progesterone in the uterus begins to diminish, and so oestradiol may be able to stimulate synthesis of oxytocin receptors and secretion of PGF-2α. The profile of receptors for oxytocin, cytosolic progesterone and oestrogen in the endometrium of heifers slaughtered at defined times (N = 4 on Days 3–4, 6–7, 10–12, 16–17, 18–19 and 21) of the oestrous cycle are given in Fig. 6. The responsiveness of the endometrium in heifers to oxytocin varies with the stage of the oestrous cycle. Experiments with incubated minced endometrial tissue obtained from heifers indicate that basal secretion is lowest during the mid-luteal phase and increases at the time of luteolysis, with the highest sensitivity at oestrus. Exogenous oxytocin stimulated PGF-2α only after luteolysis, especially at oestrus after long-term incubation for 24 h (J. Kotwica, D. Schams & H. H. D. Meyer, unpublished observations).

References

Alila, H.W. & Hansel, W. (1984) Origin of different cell types in the bovine corpus luteum as characterized by specific monoclonal antibodies. *Biol. Reprod.* **31,** 1015–1025.

Anderson, L.L., Bowerman, A.M. & Melampy, R.M.

(1965) Oxytocin on ovarian function in cycling and hysterectomized heifers. *J. Anim. Sci.* **24,** 964–968.

Armstrong, D.T. & Hansel, W. (1959) Alteration of the bovine estrous cycle with oxytocin. *J. Dairy Sci.* **42,** 533–542.

Barranao, J.L.S. & Hammond, J.M. (1984) Comparative effects of insulin and insulin-like growth factors on DNA synthesis and differentiation of porcine granulosa cell. *Biochem. Biophys. Res. Commun.* **124**, 484–490.

Bryant-Greenwood, G.D. (1982) Relaxin as a new hormone. *Endocr. Rev.* **3**, 62–90.

Cooke, R.G. & Homeida, A.M. (1982) Plasma concentrations of 13,14-dihydro-15-keto-prostaglandin $F_{2\alpha}$ and progesterone during oxytocin-induced oestrus in the goat. *Theriogenology* **18**, 453–460.

Cooke, R.G. & Homeida, A.M. (1983) Prevention of the luteolytic action of oxytocin in the goat by inhibition of prostaglandin synthesis. *Theriogenology* **20**, 363–365.

Cooke, R.G. & Homeida, A.M. (1984) Delayed luteolysis and suppression of the pulsatile release of oxytocin after indomethacin treatment in the goat. *Res. vet. Sci.* **36**, 48–51.

Cooke, R.G. & Homeida, A.M. (1985) Suppression of prostaglandin F-2α release and delay of luteolysis after active immunization against oxytocin in the goat. *J. Reprod. Fert.* **75**, 63–68.

Cooke, R.G. & Knifton, A. (1981) Oxytocin-induced oestrus in the goat. *Theriogenology* **16**, 95–97.

Cooke, R.G., Homeida, A.M. & Watkins, W.B. (1984) Simultaneous release of neurophysin and ovarian oxytocin during luteolysis in the goat. *J. Physiol., Lond.* **354**, 97 *P, Abstr.*

Fairclough, R.J., Moore, L.G., McGowan, L.T., Peterson, A.J., Smith, J.F., Tervit, H.R. & Watkins, W.B. (1980) Temporal relationship between plasma concentrations of 13,14-dihydro-15-keto-prostaglandin F and neurophysin I/II around luteolysis in sheep. *Prostaglandins* **20**, 199–208.

Fields, M.J., Fields, P.A., Castro-Hernandez, A. & Larkin, L.H. (1980) Evidence for relaxin in corpora lutea of late pregnant cows. *Endocrinology* **107**, 869–876.

Fields, M.J., Roberts, R. & Fields, P.A. (1982) Octadecylsilica and carboxymethyl cellulose isolation of bovine and porcine relaxin. *Ann. N.Y. Acad. Sci.* **380**, 36–46.

Fields, P.A., Eldridge, R.K., Fuchs, A.R., Roberts, R.F. & Fields, M.J. (1983) Human placental and bovine corpora luteal oxytocin. *Endocrinology* **112**, 1544–1546.

Fitz, T.A., Mayan, M.H., Sawyer, H.R. & Niswender, G.D. (1982) Characterization of two steroidogenic cell types in the ovine corpus luteum. *Biol. Reprod.* **27**, 703–711.

Flint, A.P.F. & Sheldrick, E.L. (1982) Ovarian secretion of oxytocin is stimulated by prostaglandin. *Nature, Lond.* **297**, 587–588.

Flint, A.P.F. & Sheldrick, E.L. (1983a) Secretion of oxytocin by the corpus luteum in sheep. *Prog. Brain Res.* **60**, 521–530.

Flint, A.P.F. & Sheldrick, E.L. (1983b) Evidence for a systemic role for ovarian oxytocin in luteal regression in sheep. *J. Reprod. Fert.* **67**, 215–225.

Flint, A.P.F. & Sheldrick, E.L. (1985) Continuous infusion of oxytocin prevents induction of uterine oxytocin receptor and blocks luteal regression in cyclic ewes. *J. Reprod. Fert.* **75**, 623–631.

Geenen, V., Legros, J.J., Hazée-Hagelstein, M.T., Louis-Kohn, F., Lecomte-Yerna, M.J., Demoulin, A. & Franchimont, P. (1985) Release of immunoreactive oxytocin and neurophysin I by cultured luteinizing bovine granulosa cells. *Acta endocr., Copenh.* **110**, 263–270.

Gemmell, R.T. & Stacy, B.D. (1979) Effect of cycloheximide on the ovine corpus luteum: the role of granules in the secretion of progesterone. *J. Reprod. Fert.* **57**, 87–89.

Gemmell, R.T., Stacy, B.D. & Nancarrow, C.D. (1977) Secretion of granules by the luteal cells of the sheep and the goat during the oestrous cycle and pregnancy. *Anat. Rec.* **189**, 161–168.

Ginther, O.J., Woody, C.O., Mahajan, S., Janakiraman, K. & Casida, L.E. (1967) Effect of oxytocin administration of the oestrous cycle on unilaterally hysterectomized heifers. *J. Reprod. Fert.* **14**, 225–229.

Guldenaar, S.E.F., Wathes, D.C. & Pickering, B.T. (1984) Immunocytochemical evidence for the presence of oxytocin and neurophysin in the large cells of the bovine corpus luteum. *Cell Tissue Res.* **237**, 349–352.

Hansel, W. & Wagner, W.C. (1960) Luteal inhibition in the bovine as a result of oxytocin injections, uterine dilatation, and intrauterine infusions of seminal and preputial fluids. *J. Dairy Sci.* **43**, 796–805.

Hatjiminaoglou, I., Alifakiotis, T. & Zervas, N. (1979) The effect of exogenous oxytocin on estrous cycle-length and corpus luteum cysts in ewes. *Annls Biol. anim. Biochim. Biophys.* **19**, 355–365.

Homeida, A.M. & Cooke, R.G. (1983) Suppression of pulsatile release of oxytocin during early pregnancy in the goat. *Prostaglandins* **26**, 103–109.

Ivell, R. & Richter, D. (1984) The gene for the hypothalamic peptide hormone oxytocin is highly expressed in the bovine corpus luteum: biosynthesis, structure and sequence analysis. *EMBO J.* **3**, 2351–2354.

Ivell, R., Brackett, K., Fields, M.J. & Richter, D. (1985) Ovulation triggers oxytocin gene expression in the bovine ovary. *FEBS Lett.* **190**, 263–267.

Jungclas, B. & Luck, M.R. (1986) Evidence for granulosa-theca interaction in the secretion of oxytocin by bovine ovarian tissue. *J. Endocr.* **109**, R_1–R_4.

Kruip, T.A.M., Vullings, H.G.B., Schams, D., Jonis, J. & Klarenbeck, A. (1985) Immunocytochemical demonstration of oxytocin in bovine ovarian tissues. *Acta endocr., Copenh.* **109**, 537–542.

McCracken, J.A., Schramm, W. & Okulicz, W.C. (1984) Hormone receptor control of pulsatile secretion of $PGF_{2\alpha}$ from the ovine uterus during luteolysis and its abrogation in early pregnancy. *Anim. Reprod. Sci.* **7**, 31–55.

Milne, J.A. (1963) Effects of oxytocin on the oestrous cycle of the ewe. *Aust. vet. J.* **39**, 51–52.

Milvae, R.A. & Hansel, W. (1980) Concurrent uterine venous and ovarian arterial prostaglandin F concentrations in heifers treated with oxytocin. *J. Reprod. Fert.* **60**, 7–15.

Mitchell, M.D., Kraemer, D.L., Brennecke, S.P. & Webb, R. (1982) Pulsatile release of oxytocin during the estrous cycle, pregnancy and parturition in sheep. *Biol. Reprod.* **27**, 1169–1173.

Moore, L.G., Choy, V.J., Elliot, R.L. & Watkins, W.B. (1986) Evidence for the pulsatile release of PGF-2α inducing the release of ovarian oxytocin during luteolysis in the ewe. *J. Reprod. Fert.* **76**, 159–166.

Mukhopadhyay, A.K., Kumar, A., Tams, R., Bohnet, H.G. & Leidenberger, F.A. (1984) Oxytocin and vasopressin have no effect on progesterone production and cyclic AMP accumulation by rat luteal cells *in vitro*. *J. Reprod. Fert.* **72**, 137–141.

Newcomb, R., Booth, W.D. & Rowson, L.E.A. (1977) The effect of oxytocin treatment on the levels of prostaglandin F in the blood of heifers. *J. Reprod. Fert.* **49**, 17–24.

Nissenson, R., Flouret, G. & Hechter, O. (1978) Opposing effects of estradiol and progesterone on oxytocin receptors in rabbit uterus. *Proc. natn. Acad. Sci. U.S.A.* **75**, 2044–2048.

Ott, I. & Scott, J.C. (1910) The galactogogue action of the thymus and corpus luteum. *Proc. Soc. exp. Biol. Med.* **8**, 49.

Oyedipe, E.O., Gustafsson, B. & Kindahl, H. (1984) Blood levels of progesterone and 15-keto-13,14-dihydro prostaglandin $F_{2\alpha}$ during the estrous cycle of oxytocin treated cows. *Theriogenology* **22**, 329–339.

Parry, D.M., Willcox, D.L. & Thorburn, G.D. (1980) Ultrastructural and cytochemical study of the bovine corpus luteum. *J. Reprod. Fert.* **60**, 349–357.

Perezgrovas, R. & Anderson, L.L. (1982) Effect of porcine relaxin on cervical dilatation, pelvic area and parturition in beef heifers. *Biol. Reprod.* **26**, 765–776.

Quirk, S.J., Willcox, D.L., Parry, D.M. & Thorburn, G.D. (1979) Subcellular location of progesterone in the bovine corpus luteum: a biochemical, morphological and cytochemical investigation. *Biol. Reprod.* **20**, 1133–1145.

Renegar, R.H. & Larkin, L.H. (1985) Relaxin concentrations in endometrial placental and ovarian tissues in sera from ewes during middle and late pregnancy. *Biol. Reprod.* **32**, 840–847.

Rice, G.E. & Thorburn, G.D. (1985) Subcellular localization of oxytocin in the ovine corpus luteum. *Can. J. Physiol. Pharmacol.* **63**, 309–314.

Richardson, M.C. & Masson, G.M. (1985) Lack of direct inhibitory action of oxytocin on progesterone production by dispersed cells from human corpus luteum. *J. Endocr.* **104**, 149–151.

Roberts, J.S. & McCracken, J.A. (1976) Does prostaglandin $F_{2\alpha}$ released from the uterus by oxytocin mediate the oxytocic action of oxytocin? *Biol. Reprod.* **15**, 457–463.

Roberts, J.S., McCracken, J.A., Gavagan, J.E. & Soloff, M.S. (1976) Oxytocin stimulated release of prostaglandin $F_{2\alpha}$ from ovine endometrium in vitro: correlation with estrous cycle and oxytocin-receptor binding. *Endocrinology* **99**, 1107–1114.

Rodgers, R.J., O'Shea, J.D., Findlay, J.K., Flint, A.P.F. & Sheldrick, E.L. (1983) Large luteal cells: the source of luteal oxytocin in the sheep. *Endocrinology* **113**, 2302–2304.

Rodgers, R.J., O'Shea, J.D. & Findlay, J.K. (1985) Do small and large cells of the sheep interact in the production of progesterone? *J. Reprod. Fert.* **75**, 85–94.

Sawyer, H.R., Moeller, C.L. & Kozlowski, G.P. (1986) Immunocytochemical localization of neurophysin and oxytocin in ovine corpora lutea. *Biol. Reprod.* **34**, 543–548.

Schallenberger, E., Schams, D., Bullermann, B. &

Walters, D.L. (1984) Pulsatile secretion of gonadotrophins, ovarian steroids and ovarian oxytocin during prostaglandin-induced regression of the corpus luteum in the cow. *J. Reprod. Fert.* **71**, 493–501.

Schams, D. (1983) Oxytocin determination by radioimmunoassay III. Improvement to subpicogram sensitivity and application to blood levels in cyclic cattle. *Acta endocr., Copenh.* **103**, 180–132.

Schams, D., Lahlou-Kassi, A. & Glatzel, P. (1982) Oxytocin concentrations in peripheral blood during the oestrous cycle and after ovariectomy in two breeds of sheep with high and low fecundity. *J. Endocr.* **92**, 9–13.

Schams, D., Prokopp, S. & Barth, D. (1983) The effect of active and passive immunization against oxytocin on ovarian cyclicity in ewes. *Acta endocr., Copenh.* **103**, 337–344.

Schams, D., Meyer, H.H.D., Schallenberger, E., Breitinger, H., Enzenhöfer, G. & Legros, J.J. (1984) Ovarian oxytocin and its possible role during the estrous cycle in cattle. *Proc. 10th Int. Congr. Anim. Reprod. & A.I.*, Urbana-Champaign, Vol. 3, Abstr. 514.

Schams, D., Kruip, T.A.M. & Koll, R. (1985a) Oxytocin determination in steroid producing tissues and in vitro production in ovarian follicles. *Acta endocr., Copenh.* **109**, 530–536.

Schams, D., Schallenberger, E., Meyer, H.H.D., Bullermann, B., Breitinger, H.-J., Enzenhöfer, G. Koll, R., Kruip, T.A.M., Walters, D.L. & Karg, H. (1985b) Ovarian oxytocin during the estrous cycle in cattle. In *Oxytocin, Clinical and Laboratory Studies*, pp. 317–334. Eds J. A. Amico & A. G. Robinson. Elsevier Biomedical, Amsterdam.

Schams, D., Schallenberger, E. & Legros, J.J. (1985c) Evidence for the secretion of immunoreactive neurophysin I in addition to oxytocin from the ovary in cattle. *J. Reprod. Fert.* **73**, 165–171.

Schams, D., Koll, R., Ivell, R., Mittermeier, Th. & Kruip, T.A.M. (1987) The role of oxytocin in follicular growth and luteal function. In *Follicular Growth and Ovulation Rate in Farm Animals* (in press). Eds J. Roche & D. O'Callaghan. Martinus Nijhoff, The Hague.

Schwall, R.H., Sawyer, H.R. & Niswender, G.D. (1986) Differential regulation by LH and prostaglandins of steroidogenesis in small and large luteal cells of the ewe. *J. Reprod. Fert.* **76**, 821–829.

Sheldrick, E.L. & Flint, A.P.F. (1981) Circulating concentrations of oxytocin during the estrous cycle and early pregnancy in sheep. *Prostaglandins* **22**, 631–636.

Sheldrick, E.L. & Flint, A.P.F. (1983) Regression of the corpora lutea in sheep in response to cloprostenol is not affected by loss of luteal oxytocin after hysterectomy. *J. Reprod. Fert.* **68**, 155–160.

Sheldrick, E.L. & Flint, A.P.F. (1985a) Endocrine control of uterine oxytocin receptors in the ewe. *J. Endocr.* **106**, 249–258.

Sheldrick, E.L. & Flint, A.P.F. (1985b) Luteal oxytocin and the control of prostaglandin-$F_{2\alpha}$ secretion during luteolysis in ewes. *J. Endocr.* **107**, Suppl. 89, Abstr.

Sheldrick, E.L., Mitchell, M.D. & Flint, A.P.F. (1980) Delayed luteal regression in ewes immunized against oxytocin. *J. Reprod. Fert.* **59**, 37–42.

Soloff, M.S. (1975) Uterine receptor for oxytocin: effects

of estrogen. *Biochem. Biophys. Res. Commun.* **65**, 205–212.

Swann, R.W., O'Shaughnessy, P.J., Birkett, S.D., Wathes, D.C., Porter, D.G. & Pickering, B.T. (1984) Biosynthesis of oxytocin in the corpus luteum. *FEBS Lett.* **174**, 262–266.

Tan, G.J.S. & Biggs, J.S.G. (1984) Progesterone production by dispersed luteal cells of non-pregnant cows: effects of oxytocin and oestradiol. *Anim. Reprod. Sci.* **7**, 441–445.

Tan, G.J.S., Tweedale, R. & Biggs, J.S.G. (1982a) Effects of oxytocin on the bovine corpus luteum of early pregnancy. *J. Reprod. Fert.* **66**, 75–78.

Tan, G.J.S., Tweedale, R. & Biggs, J.S.G. (1982b) Oxytocin may play a role in the control of the human corpus luteum. *J. Endocr.* **95**, 65–70.

Theodosis, D.T., Wooding, F.B.P., Sheldrick, E.L. & Flint, A.P.F. (1986) Ultrastructural localization of oxytocin and neurophysin in the ovine corpus luteum. *Cell Tissue Res.* **243**, 129–135.

Veldhuis, J.D. & Furlanetto, F.W. (1985) Trophic actions of human somatomedin C insulin-like growth factor I on ovarian cells: in vitro studies with swine granulosa cells. *Endocrinology* **116**, 1235–1242.

Walters, D.L. & Schallenberger, E. (1984) Pulsatile secretion of gonadotrophins, ovarian steroids and ovarian oxytocin during the periovulatory phase of the oestrous cycle in the cow. *J. Reprod. Fert.* **71**, 503–512.

Walters, D.L., Schams, D. & Schallenberger, E. (1984) Pulsatile secretion of gonadotrophins, ovarian steroids and ovarian oxytocin during the luteal phase of the oestrous cycle in the cow. *J. Reprod. Fert.* **71**, 479–491.

Wathes, D.C. & Swann, R.W. (1982) Is oxytocin an ovarian hormone? *Nature, Lond.* **297**, 225–227.

Wathes, D.C., Swann, R.W., Pickering, B.T., Porter, D.G., Hull, M.G.R. & Drife, J.O. (1982) Neurohypophysial hormones in the human ovary. *Lancet* **ii**, 410–412.

Wathes, D.C., Swann, R.W., Birkett, S.D., Porter, D.G. & Pickering, B.T. (1983a) Characterization of oxytocin, vasopressin and neurophysin from the bovine corpus luteum. *Endocrinology* **113**, 693–698.

Wathes, D.C., Swann, R.W., Hull, M.G.R., Drife, J.O., Porter, D.G. & Pickering, B.T. (1983b) Gonadal sources of the posterior pituitary hormones. *Prog. Brain Res.* **60**, 513–520.

Wathes, D.C., Swann, R.W. & Pickering, B.T. (1984) Variation in oxytocin, vasopressin and neurophysin concentrations in the bovine ovary during the oestrous cycle and pregnancy. *J. Reprod. Fert.* **71**, 551–557.

Watkins, W.B. (1983) Immunohistochemical localization of neurophysin and oxytocin in the sheep corpora lutea. *Neuropeptides* **4**, 51–54.

Watkins, W.B., Moore, L.G., Flint, A.P.F. & Sheldrick, E.L. (1983) Secretion of neurophysin I/II and oxytocin by the ovary in sheep. *Peptides* **5**, 61–64.

Watkins, W.B., Moore, L.G., Fairclough, R.J., Peterson, A.J. & Tervit, H.R. (1984) Possible role for ovarian prostaglandin $F_{2\alpha}$ in stimulating luteal oxytocin release in ewes at luteolysis. *J. Steroid Biochem.* **20**, 1507, Abstr.

Watkins, W.B., Choy, V.J., Chaiken, I.M. & Spiess, J. (1985) Isolation and sequence analysis of oxytocin from the sheep corpus luteum. *Neuropeptides* **7**, 87–95.

Webb, R., Mitchell, M.D., Falconer, J. & Robinson, J.S. (1981) Temporal relationships between peripheral plasma concentrations of oxytocin, progesterone and 13,14,dihydro-15-keto-prostaglandin $F_{2\alpha}$ during the oestrous cycle and early pregnancy in the ewe. *Prostaglandins* **22**, 443–454.

J. Reprod. Fert., Suppl. **34** (1987), 101–114

Printed in Great Britain
© 1987 Journals of Reproduction & Fertility Ltd

Spermatogenesis and Sertoli cell numbers and function in rams and bulls

M. T. Hochereau-de Reviers, C. Monet-Kuntz and M. Courot

I.N.R.A., Reproductive Physiology, 37380 Nouzilly, France

Summary. The two main types of cellular associations (type I, 2 generations of spermatocytes + 1 of spermatids; type II, 1 of spermatocytes and 2 of spermatids) occupy, respectively, more than half and about a third of the seminiferous epithelium cycle in rams and bulls. However, the duration of the cycle of the seminiferous epithelium and that of spermatogenesis differ between the species. A_1 spermatogonia and Sertoli cell total numbers are highly correlated in adult rams and bulls. Mitosis in Sertoli cells occurs mostly *in utero* but may still occur for a short period after birth. Between birth and puberty there is about a 5-fold increase in the number of Sertoli cells. After that there are no seasonal- or age-related increases in the number of adult Sertoli cells. Some factors (season of birth; nutrition; genetics; hormones) affect mitosis of Sertoli cells in prepubertal animals. Sertoli cells differentiate after cessation of mitosis. Their differentiation is affected by cryptorchidism, nutrition, genetics and hormones. Their adult function is only poorly known. ABP and rete testis fluid secretions and nuclear Sertoli volume fluctuate under the influence of the same factors, but they are not always linked together. This reinforces the need for more knowledge of Sertoli cell secretions and function.

Introduction

Development and function of the germinal epithelium are linked to the development of the somatic elements of the testis. The Sertoli cells are permanent elements of the seminiferous epithelium which originate from the mesonephros (Zamboni & Upadhyay, 1982). In the male fetus primordial germ cells divide but do not differentiate in the male gonad, although when they occasionally enter the fetal adrenal, they initiate spermatogonial multiplications and undergo meiosis (Upadhyay & Zamboni, 1982). However, after birth and differentiation of Sertoli cells, germ cells enter spermatogenesis.

The Sertoli cells could play different roles in inhibiting germinal differentiation in the fetal testis, inducing germ cell multiplication and differentiation during and after puberty and probably influencing sperm quality (Hochereau-de Reviers & Courot, 1978). The different factors involved in such complex phenomena are not yet known. More than 80 proteins are secreted by rat Sertoli cells in culture (Wright *et al.*, 1981). Very few of these proteins have been identified in rams and bulls. Androgen-binding protein (ABP) is one of them (Jegou *et al.*, 1979). Clusterin, which is a cell aggregating factor, has been identified in testes of sheep (Blaschuk *et al.*, 1983) but no physiological analysis has been done. Tissue-type plasminogen activor is secreted by bovine Sertoli cells in culture (Jenkins & Ellison, 1986). The existence of bovine anti-Müllerian hormone (Josso, 1973) and its structure (Cate *et al.*, 1986) and its homology with inhibin (Mason *et al.*, 1985) have been reported. The object of this review is (1) to summarize the similarities and dissimilarities of the seminiferous epithelium in rams and bulls and (2) to analyse the control of Sertoli cell multiplication and function in both species.

Spermatogenesis in rams and bulls

Two main types of cellular associations have been distinguished: type I with two generations of primary spermatocytes and a single generation of spermatids and type II with only one generation of primary spermatocytes and two of spermatids. Types I and II represent 50–60% and 30–40% respectively in the seminiferous epithelial cycle of bulls and rams, and in both species, spermatozoa are released before the new generation of preleptotene primary spermatocytes is initiated. The two types of cellular associations can be subdivided further according to the arrangement and shape of the germ cells (Ortavant, 1958; Cupps & Laben, 1960; Amann, 1962; Hochereau, 1967; Guraya & Bilaspuri, 1976; Bilaspuri & Guraya, 1986) or the stages of development of the acrosome (Clermont & Leblond, 1955; Kramer, 1960; Berndtson & Desjardins, 1974). The relative frequencies of cellular association vary according to the classification method. However, equivalence can be drawn and comparisons can be made (Courot et al., 1970).

There are no significant variations in frequencies between regions in the same testis, between testes or between individuals provided a sufficient number of cross-sections of tubules are analysed (Amann, 1962; Hochereau, 1963). If not enough tubules are counted, local variations in grouping of tubules at the same stage are observed, indicating a local control of onset of spermatogenesis as in the mouse (Redi, 1986). This local assembly of tubules at the same stages could result in apparent variation of relative frequencies of the cellular association (Kramer, 1960). This phenomenon has to be taken into account if analyses are performed on small biopsy specimens.

Spermatogonial divisions and stem cell renewal

The number of generations between A_1 and preleptotene spermatocytes has been analysed by different complementary methods: (1) incorporation of radiolabelled precursors of DNA (Hochereau et al., 1964; Hochereau, 1967; Hochereau-de Reviers, 1970; Hochereau-de Reviers et al., 1976b); (2) the morphological appearance of nuclei in the spermatogonia including their nuclear volume (Ortavant, 1958; Kramer, 1960; Amann, 1962; Bilaspuri & Guraya, 1986); and (3) the evolution of their number per cross-section of tubules during the seminiferous epithelial cycle (Ortavant, 1958; Amann, 1962).

Six spermatogonial divisions have been observed after labelling with [^{3}H]thymidine in the ram and the bull. They occur at the same stages in the seminiferous epithelial cycle. Three type A, one intermediate and two type B spermatogonial generations have been observed in rams and bulls (Hochereau, 1967; Berndtson & Desjardins, 1974; Hochereau-de Reviers et al., 1976b; Bilaspuri & Guraya, 1986). However, the duration of the seminiferous epithelial cycle differs in the two species (Table 1).

The origin of the cycling stem cells and the significance of A_0 spermatogonia (round and pale type A) are still disputed (Hochereau-de Reviers, 1981; Lok et al., 1982). In adult rams and bulls, the ratio between A_0 and A_1 (ovoid pale with a central dark nucleolus) spermatogonia differs markedly: for example, this ratio, $A_0/(A_0 + A_1)$, is about 15% in adult bulls (Hochereau-de Reviers, 1970) and about 50% in adult rams in the breeding season (Hochereau-de Reviers et al., 1976b). In the sheep testis it varies with photoperiod (Hochereau-de Reviers et al., 1985) and endocrinological status, the A_1 spermatogonia disappearing after hypophysectomy and being at least partly related to FSH secretion (Courot et al., 1979). A_0 spermatogonia could represent the first step of the cell cycle of A_1 spermatogonia (G_0 or beginning of G_1: Hochereau-de Reviers, 1981). The A_0 spermatogonia which are isolated or single (A_s) could represent the basic stem cell, the A paired (A_p) and aligned (A_{al}) spermatogonia being the multiplicating cells which ensure the formation of new A_1 spermatogonia (Lok et al., 1982). In the ram the total numbers of $A_s + A_p + A_{al}$ (Lok et al., 1982) nearly double between stages 2 and 5 of the classification of Roosen-Runge & Giesel (1950) in which three peaks of [^{3}H]thymidine labelling corresponding to A_2, A_3 and intermediate spermatogonia are observed (Hochereau-de Reviers et al., 1976b). At two

Table 1. Comparison of duration of seminiferous epithelial cycle in sheep and cattle

Species		Reference	Method	Duration (days)
Sheep	(*Ovis aries*)	Ortavant (1958)	^{32}P	10·4
		Hochereau *et al.* (1964)	[^{3}H]thymidine	10·4
Cattle	(*Bos taurus*)	Orgebin (1961)	^{32}P	13·4
	(*Bos taurus*)	Hochereau *et al.* (1964)	[^{3}H]thymidine	13·5
	(*Bos indicus* × *Bos taurus*)	Salim & Entwistle (1982)	[^{3}H]thymidine	13·4
	(*Bos indicus*)	Cardoso & Godinho (1983)	[^{3}H]thymidine	14·0
	(*Bubalus bubalis*)	Sharma & Gupta (1980)	[^{3}H]thymidine	8·6
		Bilaspuri & Guraya (1980)	[^{3}H]thymidine	8·5

divisions after labelling, most of the new A_1 labelled spermatogonia in rams and bulls arise from precursor mother cells labelled as A_1 and A_2 spermatogonia (Hochereau-de Reviers, 1970; Hochereau-de Reviers *et al.*, 1976b). In bulls, A_1 spermatogonia at stages 7 and 8 are present as single (25%) or grouped (75%) cells (Hochereau-de Reviers, 1970). This suggests segregation of the precursor cells of A_1 spermatogonia earlier than the A_2 divisions (Hochereau-de Reviers, 1971).

Nevertheless, A_1 spermatogonia and Sertoli cell total numbers per testis are highly and positively correlated ($r > +0.65$) in rams and bulls. This is not observed for the A_0 spermatogonia population (de Reviers & Courot, 1976) and we conclude that A_1 spermatogonia are the first step of the spermatogenic cycle and are clearly dependent on Sertoli cell function.

Relations between Sertoli and germ cell populations

The existence of a correlation between Sertoli and A_1 spermatogonia indicates a control of spermatogenesis by the Sertoli population very early in the spermatogenic cycle, the end point of which is control of daily sperm production (de Reviers *et al.*, 1980). The establishment of a Sertoli cell population is therefore a primordial factor controlling sperm production.

Sertoli cell multiplications

Sertoli cell multiplications occur mostly during fetal life. In the Ile-de-France lamb shortly after sexual differentiation (40 days of fetal life) the total number of future Sertoli cells (supporting cells) is about 1×10^6 per testis (Courot, 1971). It increases 300- to 400-fold until birth (Table 2). Between birth and the post-pubertal phase of testicular growth, total number of Sertoli cells still increases 5- to 10-fold in rams according to breed and by a factor of 5 in Normand bulls (Table 2). The age at which mitosis of Sertoli cells is arrested varies from 40 to 80 days in different breeds of sheep. No further increase is observed after the prepubertal period in sheep and cattle (Table 2). No variation in total numbers of Sertoli cell is observed between breeding and non-breeding season in adult rams (Table 3).

After puberty the Sertoli cell population, estimated by the same technique, does not vary in numbers and so there is a quantitatively stable population of Sertoli cells in adult rams and bulls. However, the following factors can influence the Sertoli cell multiplications.

Genetic factors

Between breeds, variation in Sertoli cell populations has been reported (Hochereau-de Reviers *et al.*, 1984a) for rams and bulls (Table 4). These variations affect at least partly those of daily

Table 2. Comparative development of total numbers of Sertoli cells (corrected number for nuclear size) according to age in rams and bulls of different breeds

Sheep	Ile-de-France (adapted from Courot, 1971; Kilgour et al., 1985)	Romanov (Lafortune et al., 1984; and unpublished data)	Dorset Horn × Finn (J. R. McNeilly, unpublished data)	Romanov × Prealpes × Ile-de-France (Monet-Kuntz et al., 1984, 1987)
Birth	3·1	3·1	4·30 ± 0·4	—
60–70 days of age	17·0 ± 6·5	—	—	22·5 ± 3·0
100–200 days of age	36·2 ± 3·0	20·6 ± 1·4	20·5 ± 1·1	25·6 ± 2·0
18 months	28·0 ± 3·3	19·8 ± 1·1	20·3 ± 1·1	—
5 years	33·7 ± 2·5	—	—	—

Cattle	Française Frisonne Pie noire (M.T. Hochereau-de Reviers, unpublished data)	Normand (adapted from Attal & Courot, 1963)
Birth	—	9·3 ± 0·4
120 days	—	46·8 ± 7·1
240 days	—	44·2 ± 4·4
18 months	35·9 ± 1·2	—
3 years	—	54·6 ± 3·6
6 years	36·6 ± 2·7	—

Values are mean ± s.e.m. × 10^{-8}.

Table 3. Comparison of total numbers (corrected for nuclear size) of Sertoli cells per testis in different breeds of sheep according to season

Breed of sheep	Reference	Rams/ group	Non-breeding season	Breeding season
Île-de-France	Hochereau-de Reviers et al. (1984a)	8	27·9 ± 7·7	33·2 ± 9·2
	B. D. Schanbacher (unpublished)	5	31·4 ± 6·7	31·2 ± 2·7
Soay	Hochereau-de Reviers et al. (1985)	5	15·4 ± 0·6	12·9 ± 0·6

Values are mean ± s.e.m. × 10^{-8}.

production of spermatozoa. In Montbeliard bulls there is a tendency for the animals classified as 'good' or 'intermediate' for their sperm characteristics to have a higher number of Sertoli cells per testis (30 ± 7 and 29 ± 4 × 10^8 respectively) than those classified as 'bad' (22 ± 1 × 10^8; Abdel Malak, 1983).

Environmental factors

Nutrition. Severe restriction of food, inducing a reduction of mean daily weight increase (136 versus 280 g/day), from the first week of age results in a decrease of the Sertoli cell population at 100 days of age (21 *vs* 39 × 10^8 Sertoli cells/testis) in cross-bred Romanov × Limousin lambs (Brongniart et al., 1985). However, the Sertoli cell multiplications can be maintained for a longer period as rapid testis growth is delayed and starts just before 100 days of age in underfed lambs. Therefore, no conclusion can be reached on the presence or absence of a decrease in the adult Sertoli cell population.

Table 4. Breed differences in Sertoli cell populations (corrected numbers) and daily sperm production in sheep and cattle

	Reference	Breed	Sertoli cells (total no./testis $\times 10^{-8}$)	Daily sperm production ($\times 10^{-9}$)
Sheep	Hochereau-de Reviers *et al.* (1984a)	Soay	13·9 ± 1·4	2·1 ± 0·2
		Romanov	19·8 ± 1·3	2·5 ± 0·2
		Prealpes-du-Sud	24·9 ± 1·4	4·1 ± 0·2
		Ile-de-France	36·2 ± 3·1	4·1 ± 0·3
	J. R. McNeilly (unpublished)	Dorset Horn × Finn	20·3 ± 1·1	2·6 ± 0·2
	M. Seck (unpublished)	Merinos d'Arles	24·7 ± 1·4	2·8 ± 0·2
Cattle	Abdel Malak (1983)	Montbeliard	27·2 ± 1·8	2·7 ± 0·2
	Lafortune (1983)	Française Frisonne Pie Noire	23·0 ± 2·4	2·7 ± 0·2
		Zebu	21·0 ± 1·5	2·7 ± 0·1
	M. T. Hochereau-de Reviers (unpublished)	Holstein × FFPN	36·6 ± 2·7	—
	Attal & Courot (1963)	Normand	54·6 ± 3·6	3·4 ± 0·2

Values are mean ± s.e.m.

Season of birth. A higher number of Sertoli cells after puberty is observed in the testes of rams born in the summer: +25% in Ile-de-France adult rams (de Reviers *et al.*, 1980) and +50% in Finn × Dorset Horn cross breds, 6 months old (Hochereau-de Reviers *et al.*, 1984b). The photoperiod modifies the gonadotrophin secretion (Courot *et al.*, 1975; Lafortune *et al.*, 1984) during the prepubertal period in lambs. Gonadotrophin binding per testis (Barenton & Pelletier, 1983) and increased secretions of testosterone and ABP (Jegou *et al.*, 1979) are evidence of seasonal variations with a maximum during the summer months and a minimum during the winter ones.

Experimental situations

Unilateral castration. Unilateral castration of prepubertal lambs or calves resulted in a hyperplasia of the Sertoli cell population (Hochereau-de Reviers, 1976; de Reviers *et al.*, 1980; Waites *et al.*, 1985). Such a numerical increase is not obtained after unilateral castration of pubertal or adult animals (Hochereau-de Reviers *et al.*, 1984b).

Cryptorchidism. In rams and bulls testicular descent into the scrotum occurs early in fetal life (Hullinger & Wensing, 1985), around the end of the second third of gestation. Lambs have been rendered experimentally cryptorchid at birth and orchidopexy occurs at onset of prepubertal rapid testicular growth. After 2 months of cryptorchidism, a significant increase in the Sertoli cell population is observed compared to normal lambs (Monet-Kuntz *et al.*, 1987).

Hypophysectomy. Hypophysectomy of 50-day-old Ile-de-France lambs results in a decrease in Sertoli cell numbers (Courot, 1971; Table 5): 15 days after hypophysectomy total numbers are reduced by 40% compared to values of controls and by 60% compared to that of postpubertal animals of the same breed. Treatment with LH or FSH alone prevents Sertoli cell numbers from decreasing. However, treatment with both FSH and LH results in a synergistic action on Sertoli cell multiplications such that the normal adult number in that breed is restored (Table 5).

Table 5. Hormonal control of Sertoli cell population and of seminiferous tubule mean diameter in prepubertal Ile-de-France lambs (corrected number) (adapted from Courot (1971) and Kilgour *et al.* (1984))

	Sertoli cell		Seminiferous tubule mean diam. (μm)
	Total no. ($\times 10^{-8}$)	Nuclear area (μm²)	
Control, 50 days	$14\cdot8 \pm 2\cdot3$	$22\cdot5 \pm 0\cdot7$	$55\cdot1 \pm 1\cdot7$
Hypox. + 15 days	$10\cdot3 \pm 3\cdot1$	$17\cdot5 \pm 0\cdot4$	$36\cdot2 \pm 1\cdot2$
Hypox. + FSH	$14\cdot9 \pm 2\cdot2$	$18\cdot5 \pm 0\cdot3$	$39\cdot4 \pm 2\cdot5$
Hypox. + LH	$17\cdot3 \pm 2\cdot2$	$24\cdot5 \pm 0\cdot4$	$58\cdot0 \pm 3\cdot9$
Hypox. + LH + FSH	$30\cdot6 \pm 3\cdot8$	$27\cdot2 \pm 0\cdot5$	$63\cdot5 \pm 3\cdot1$
Control, 100 days	$38\cdot2 \pm 5\cdot5$	$33\cdot1 \pm 1\cdot3$	$104\cdot0 \pm 13\cdot2$
Passively immunized against LH	$20\cdot7 \pm 2\cdot8$	$30\cdot3 \pm 1\cdot3$	$104\cdot9 \pm 10\cdot7$
Control, 100 days	$32\cdot8 \pm 0\cdot3$	$30\cdot4 \pm 2\cdot2$	$125\cdot2 \pm 5\cdot7$
Passively immunized against FSH	$20\cdot1 \pm 2\cdot7$	$28\cdot9 \pm 1\cdot3$	$85\cdot2 \pm 9\cdot9$

Values are mean ± s.e.m.

Immunization against gonadotrophins. Continuous passive immunizations since birth until 100 days of age with antibodies against either FSH or LH decrease the total number of Sertoli cells per testis at 100 days of age in lambs; the value observed at birth (3×10^8) has not been obtained (Table 5; Kilgour *et al.*, 1984). Multiplications of Sertoli cells are not completely arrested.

Culture in vitro. Ovine Sertoli cells cultured *in vitro* incorporate [³H]thymidine into DNA whatever the age, between 2 and 12 weeks of age. This incorporation is not stimulated but is decreased by FSH treatment, probably by an increase of their maturation (A. S. Speight, J. M. Clifford & G. M. H. Waites, unpublished data). A seminiferous growth factor has been isolated from calf and mouse testes; it stimulates Sertoli cell proliferation *in vitro* (Bellvé & Feig, 1984).

Functional differentiation of Sertoli cells

At the end of the prepubertal period, Sertoli cells progressively differentiate. One of the first morphological signs is an increase in cytoplasmic and nuclear cross-sectional area (Monet-Kuntz *et al.*, 1984). In the lamb, mean cellular and nuclear volumes increase 3- and 1·5-fold respectively, between 25 and 100 days of age. Between 6 weeks and 18 months of age Sertoli cell nuclear volume increases about 5-fold. In the Normand bull, the mean Sertoli nuclear volume increases by a factor of 3·5 between 4 months and 3 years of age.

In the lamb, the total number of FSH binding sites per testis increases about 150-fold between 10 and 120 days (Barenton *et al.*, 1983b), and those of androgen cytoplasmic and nuclear binding sites increase 16- and 12-fold between 25 and 100 days of age respectively (Monet-Kuntz *et al.*, 1984). In fact, between cessation of mitosis of Sertoli cells and puberty the total numbers of FSH and androgen-binding sites increase about 10-fold. ABP testicular content increases by a factor of 3·5 from 50 to 120 days of age (Carreau *et al.*, 1979). Similarly, an increase in FSH binding sites of 13-fold has been reported for calves between 100 days and 2·5 years of age (Dias & Reeves, 1982).

During pubertal testicular growth, inhibin secretion increases with Sertoli differentiation in the ram (Blanc *et al.*, 1981).

The factors which can alter Sertoli cell differentiation are as follows.

Genetic factors

In lambs from two established lines selected for high (H) and low (L) testicular growth between 6 and 20 weeks of age (McNeilly *et al.*, 1986; unpublished data), the total numbers of Sertoli cells

Table 6. Changes of testicular values in two fixed lines of Dorset Horn × Finn rams selected for their high (H) and low (L) rate of testicular growth (J. R. McNeilly, unpublished data)

		Testis weight (g)	Total no. of Sertoli cells/testis ($\times 10^{-8}$)*	Sertoli cell nuclear area (μm^2)	Seminiferous tubule diam. (μm)	Daily production of round spermatids ($\times 10^{-9}$)
Birth	H	0.7 ± 0.07^a	4.30 ± 0.4^a	17.4 ± 0.4^a	36.0 ± 0.4^a	—
	L	0.79 ± 0.16^a	4.4 ± 0.7^a	17.7 ± 0.2^d	34.0 ± 1.3^a	—
6 weeks	H	1.97 ± 0.23^b	6.64 ± 0.3^b	22.3 ± 0.8^b	61.7 ± 2.8^b	—
	L	1.47 ± 0.18^b	5.82 ± 0.8^b	20.4 ± 1.0^b	51.0 ± 2.8^c	—
20 weeks	H	106.6 ± 4.00^d	19.3 ± 1.0^c	45.1 ± 1.0^d	203.3 ± 4.7^d	1.57 ± 0.08^b
	L	74.6 ± 11.10^e	21.4 ± 0.5^c	39.2 ± 0.6^c	172.4 ± 11.7^e	0.97 ± 0.24^a
Adult	H	218.7 ± 8.10^f	20.3 ± 1.1^c	64.15 ± 1.6^e	242.0 ± 7.0^f	2.62 ± 0.16^c
	L	222.8 ± 15.80^f	23.7 ± 2.8^c	63.93 ± 1.6^e	237.0 ± 4.0^f	2.75 ± 0.28^c

Values are mean ± s.e.m.
*Corrected number.
Within columns, values with different letters indicate significant differences ($P < 0.05$).

did not differ at any age between lines. However, their function could be different. The mean diameter of the seminiferous tubules, which reflects Sertoli cellular volume variation, and the nuclear area of Sertoli cells are greater at 6 and 20 weeks of age respectively in the H line than in the L line sheep. Seminiferous tubule diameter and daily production of round spermatids are also increased in the H line at 20 weeks of age (Table 6), but in adult rams no differences are observed. The two lines of sheep are therefore distinguished by a transitory difference in Sertoli and germ cell characteristics. Plasma FSH concentrations are significantly lowered in the H line sheep at 16 and 20 weeks of age (McNeilly *et al.*, 1986), and this could reflect a more precocious increase in FSH binding sites in the H than in the L line animals.

In Ile-de-France, Prealpes-du-Sud and Romanov lambs, at 8 months of age and during the non-breeding season, variations in FSH binding sites per testis or per Sertoli cell reflect variations in precocity and/or seasonality (Barenton *et al.*, 1983a).

Sertoli cell function during pubertal testis growth could therefore result in variations of testicular characteristics and possibly in earlier puberty.

Nutritional factors

In sheep that are underfed during the prepubertal period, there is reduced Sertoli cell nuclear size (mean nuclear area: 17.1 versus 28.4 μm^2) resulting in a decrease of 63% in volume (I. Brongniart, unpublished data).

Experimental situations

Cryptorchidism. In sheep experimental cryptorchidism at birth inhibits the normal differentiation of the Sertoli cells during the pubertal process. Nuclear volume of Sertoli cells is reduced by 30% by cryptorchidism. Total ABP (Table 7) content per testis, total numbers of FSH and cytoplasmic androgen binding sites per testis are greatly reduced (Monet-Kuntz *et al.*, 1987). After orchidopexy at 2 months of age the daily production of round spermatids, 5 months later, is only partly restored, due to the presence of seminiferous tubules empty of germ cells. However, the ABP content of testis and the total numbers of FSH and cytoplasmic androgen binding sites are restored (Monet-Kuntz *et al.*, 1987; Table 7). This indicates that restoration of Sertoli cell function in terms of ABP secretion

Table 7. Effect of experimental cryptorchidism on lamb testicular parameters (from Monet-Kuntz *et al.*, 1987), expressed as % of the normal values at 7 months of age

	Cryptorchid	Cryptorchid + orchidopexy at 2 months of age
Testis weight (g)	14	84
% of empty tubules*	100	16·5
Sertoli cell total number/testis ($\times 10^{-8}$)	88	149
FSH binding (pmol/testis)	3·6	90
Cytoplasmic androgen binding (pmol/testis)	14·3	96
ABP (pmol/testis)	19·2	100
Daily production of round spermatids ($\times 10^{-9}$)	0	45

*Empty tubules in normal lambs equals 1·9%.

or of FSH and androgen binding sites does not depend on that of spermatogenesis in the whole testis.

Hypophysectomy. Hypophysectomy of 50-day-old lambs provokes a decrease in cellular (as indicated by the variation in seminiferous tubule diameter) and nuclear size of Sertoli cells (Courot, 1971; Table 5). FSH treatment does not support these measures. LH supplementation maintains the initial cytoplasmic and nuclear size and LH + FSH treatment promotes the Sertoli cell development. ABP production is restored after hypophysectomy, mainly by testosterone treatment (Carreau *et al.*, 1980).

Immunization against gonadotrophins. Passive immunizations of sheep from birth until 100 days of age with antibodies against either LH or FSH do not modify Sertoli cell nuclear development significantly (Table 5).

Culture in vitro. Sertoli cells of prepubertal sheep and cattle have been cultured to assess their ability to be stimulated by FSH and/or androgen. In calves, 5–11 months old, FSH or testosterone treatments change the overall rate of protein synthesis and secretion without the detection of specific qualitative changes and age effects (Hayes & Brooks, 1985). In cultures of Sertoli cells from 6–8-month-old calves, FSH but not LH induces a synthesis of cAMP and [³H]leucine-labelled proteins (Smith & Griswold, 1981). FSH stimulates the synthesis of tissue-type plasminogen activator of Sertoli cells of prepubertal calves but this response is abolished by dexamethasone, which induces a specific protease inhibitor (Jenkins & Ellison, 1986). In the lamb, from 2 to 12 weeks of age, Sertoli cells cultured *in vitro* demonstrate a significant age-dependent increase in the proportion of [³H]leucine incorporation into Sertoli cell secreted proteins (Waites *et al.*, 1985).

Adult Sertoli cell functions

Cyclic variations of Sertoli cell nuclei according to the seminiferous epithelium stages have been observed in rams, with maximum development during the stages when elongation of spermatids takes place (Hochereau-de Reviers *et al.*, 1985). However, the separation of functional seminiferous tubules in bull and ram testis is not possible, due to the importance of connective fibres in inter-tubular tissue. A stage-dependent analysis of seminiferous epithelium function to compare with that of the rat (Parvinen, 1982) has not yet been possible.

Sertoli cell functions of adults may vary according to the factors below.

Table 8. Comparisons of testicular values and hormonal binding in 18-month-old adult Romanov and Ile-de-France rams during the breeding season (C. Monet-Kuntz, unpublished data)

	Romanov (N = 6)	Ile-de-France (N = 5)
Testis weight (g)	172·8 ± 26·9	255·2 ± 45·8*
Sertoli cell total no./testis × 10^{-8}	13·6 ± 2·7	28·0 ± 2·3*
FSH receptors/Sertoli cell (pmol)	13·90	12·25
Sertoli nuclear cross-sectional area (μm^2)	70·4 ± 1·3	65·9 ± 1·1*
FSH binding (pmol/testis)	189·0 ± 38	343·0 ± 66·0*
Cytoplasmic androgen binding (pmol/testis)	52·7 ± 19	74·5 ± 22·0
Daily production of round spermatid (× 10^{-9})	1·89 ± 0·35	3·34 ± 0·48*
DSP/Sertoli cell (× 10^{-7})	13·9	11·9
Cross-sectional area of spermatid cell (μm^2)	53·3 ± 0·9	58·7 ± 1·7*
Rete testis fluid secretion (ml/h)	1·03 ± 0·14	1·35 ± 0·35

Values are mean ± s.e.m.
*Significantly different from value for Romanov rams, $P < 0.05$.

Genetic factors

In sheep, breed differences have been demonstrated in rete-testis fluid flow and compared to that of Sertoli cell numbers per testis (Dacheux *et al.*, 1981). Romanov as compared to Ile-de-France rams have 50% less of the number of Sertoli cells, of FSH binding sites and of daily spermatid production per testis, 30% less of testis weight, of androgen binding sites and of rete testis fluid flow rate (Table 8). However, the mean Sertoli cell nuclear cross-sectional area, the daily production of spermatids and the FSH binding sites are slightly greater in Romanov than in Ile-de-France rams, while cross-sectional area of spermatid cell is smaller.

Seasonal factors

In adult rams, seasonal variations in nuclear cross-sectional area of Sertoli cells have been observed with a maximum during the breeding season (Hochereau-de Reviers *et al.*, 1976a, 1985). Total numbers of FSH binding sites per testis (Barenton & Pelletier, 1983), ABP concentration in the rete testis fluid (Jegou *et al.*, 1979) and rete testis fluid flow rate (Dacheux *et al.*, 1981) exhibit seasonal variations. The maximum of rete testis fluid flow precedes by 1·5 months that of sperm production, and this delay corresponds approximately to the duration of one spermatogenic cycle and suggests that the increase of Sertoli cell function, indicated by rete testis flow rate, induces and/or is correlated with, that of spermatogenic functions.

Experimental situations

In adult Ile-de-France rams, hypophysectomy induces a decrease in Sertoli cell nuclear volume which is restored nearly completely by PMSG or hCG treatments but not by testosterone (Courot *et al.*, 1979).

In sheep, active immunization against oestradiol-17β provokes an increase in testicular volume (Schanbacher, 1984) which is accompanied by an increase in LH, FSH and testosterone plasma concentrations (Schanbacher *et al.*, 1986). During the breeding season, this testicular increase is related to an increase in sperm production and of ABP concentration in the rete testis fluid which is

Table 9. Effect of active immunization against oestradiol-17β on testicular values in Ile-de-France adult rams (B. D. Schanbacher, unpublished data)

	Non-breeding season		Breeding season	
	Control	Immunized	Control	Immunized
Testis weight (g)	173 ± 26^a	219 ± 28^b	275 ± 20^c	369 ± 29^d
Total no. of Sertoli cells/testis ($\times 10^{-8}$)	31.4 ± 2.9^a	36.8 ± 3.2^a	30.3 ± 4.5^a	35.5 ± 2.4^a
FSH binding (pmol/testis)	20.7 ± 3.7	25.9 ± 3.4	161 ± 12^a	212 ± 18^b
Cytoplasmic androgen binding (pmol/testis)	117 ± 25^a	37.1 ± 30^b	125 ± 12^a	82 ± 18^c
ABP concentration in rete-testis fluid (10^{-9} M)	—	—	10.7 ± 1.4^a	20.0 ± 6.4^a
Daily production of spermatids ($\times 10^{-9}$)	1.95 ± 0.14^a	2.44 ± 0.26^a	3.33 ± 0.55^b	5.02 ± 0.38^c

Values are mean ± s.e.m.
Within rows, values with different letters are significantly different, $P < 0.05$.

nearly doubled. In parallel, the total number of FSH binding sites per testis is increased while that of androgen binding sites and androgen concentrations in the rete testis, are decreased (unpublished data; Table 9).

Passive immunization against FSH depresses Sertoli nuclear cross-area (by 14%) and sperm production (Courot *et al.*, 1984; unpublished data).

Discussion

The major events of spermatogenesis in ram and bull testes are now relatively well known. However, the determination of daily sperm production requires a precise knowledge of duration of the seminiferous epithelial processes in all domestic ruminants.

The relative proportion of associations of Types I and II is quite different from that observed in rodents in which Types I and II represent about 25% and 70% respectively of the seminiferous epithelial cycle, and in human male in which the two types are nearly equal. This reinforces the need for a comparative study between species in which spermiation, entrance of meiotic cells into the adluminal compartment and the division of type A_1 spermatogonia are concomitant or not in the seminiferous tubules. Compared to rodents or monkeys (*Cercopithecus aethiops*), the number of spermatogonial generations is similar but the ratio of type A and B spermatogonia differs (see review by Courot *et al.*, 1970).

As in rats, the mode of stem cell renewal is still disputed for sheep. However, the main problem is to understand the physiological significance of A_0 stem cells and of the different ratio of $A_0/(A_0 + A_1)$ in rams and bulls. These differences could be related to the presence of seasonal variation in sheep, but this needs to be tested. Total numbers of A_0 and A_1, spermatogonia vary inversely in the ovine testis and the A_0 population is negatively correlated to FSH plasma concentrations. Furthermore, A_1 spermatogonia and Sertoli cell population per testis are highly correlated in ram and bull testes. Firstly, this indicates that Sertoli cells control the very early step of spermatogenesis (inhibiting and/or permissive action) and not only the adluminal compartment. Secondly, the establishment of the Sertoli cell population before puberty could have specific consequences on sperm production in the adult. This reinforces the need for a better knowledge of the control of Sertoli cell multiplications in the fetus and neonate.

In sheep and cattle, most Sertoli cell multiplications occur during fetal life as observed in the rat (Orth, 1982) but multiplication does continue until the onset of prepubertal testicular growth. In rams and bulls between birth and adulthood the Sertoli cell population increases by a factor of 5–10 before the onset of spermatogenesis, while in rodents the two phenomena are concomitant. In these adult animals no numerical variations with age, after puberty or with season are observed. This is different from the conclusion of Johnson & Thompson (1983) for the stallion. However, these authors observe modifications in total volume of Sertoli nuclei according to age or season without variation in individual Sertoli nuclear volume and draw conclusions on variations in number. This is not confirmed by Jones & Berndtson (1986) who report a decline with age of the Sertoli cell population. Results from sheep and cattle testes support variations in nuclear volume of Sertoli cells with age, hormonal status, season and genetics without variations in numbers and correspond to the observation of Lino (1971).

In rats, Sertoli cell multiplications are under FSH control (Orth, 1984), but FSH alone appears to be unable to promote Sertoli cell multiplications in hypophysectomized lambs or *in vitro*. A synergistic effect of LH and FSH *in vivo* is necessary to ensure normal multiplications (Courot, 1971) and, as far as morphology of the cell is concerned, LH could be necessary to maintain normal Sertoli cells, possibly allowing further response to FSH. No such study has been done for the calf.

Before puberty, environmental factors, via variation in hormonal secretion, hormone binding to target cells and specific secretions induced by hormones, affect Sertoli cell multiplications and differentiation. Variation in precocity could be partly explained by variations in Sertoli function, possibly by those of FSH binding and further stimulation, as induced in Sertoli cells cultured *in vitro*. Hormonal control of Sertoli secretion in prepubertal or adult bull and rams is poorly understood as the only proteinaceous productions to be analysed are ABP and plasminogen activator. ABP is mainly testosterone dependent in the lamb (Carreau *et al.*, 1980) and fluctuates with season (Jegou *et al.*, 1979). Plasminogen activator secretion appears to be FSH dependent (Jenkins & Ellison, 1986). Inhibin variations had been suspected in normal and cryptorchid rams (Blanc *et al.*, 1978, 1981), but the relationship between anti-Müllerian hormone (Josso *et al.*, 1980) and inhibin production, their hormonal control and their respective role on germ cell multiplications or differentiation have to be investigated. Moreover, the respective roles of the EGF domain of tissue-type plasminogen activator (Patthy, 1985) and of the transforming growth factor β domain of inhibin (Mason *et al.*, 1985) and of anti-Müllerian hormone (Cate *et al.*, 1986) are probably of prime importance for the regulation of spermatogonial multiplications as for other cell types (Roberts *et al.*, 1985).

Furthermore, the relation between quality of spermatozoa and Sertoli cell secretion has been poorly investigated in cattle and sheep. The role of factors such as transferrin (Foresta *et al.*, 1986) or clusterin (Blaschuk *et al.*, 1983) has to be investigated.

In conclusion, the numerical variation of Sertoli cells and their relation with sperm production are now relatively well understood, but their secretions and hormonal control are still poorly understood in sheep and cattle, despite the economic importance of these species.

We thank Dr M. de Reviers for discussions throughout this work; Dr M. Thibier and Dr G. Abdel Malak for discussion of the results with Montbeliard bulls; Dr J. Findlay for suggestions to improve the text; Mrs C. Perreau, C. Pisselet and I. Fontaine for help in preparing the material and the analysis of results; and Mrs G. Ploux and D. Tanzi for typing the manuscript.

References

Abdel-Malak, G. (1983) *Relations quantitatives entre paramètres spermatiques et hormonaux du taurillon post-pubère (Bos taurus).* Thèse, Dr. es Sciences, Paris VI.

Amann, R.P. (1962) Reproduction capacity of dairy bulls. IV. Spermatogenesis and testicular germ cell degeneration. *Am. J. Anat.* **110,** 69–78.

Attal, J. & Courot, M. (1963) Développement testiculaire et établissement de la spermatogenèse chez le taureau. *Annls Biol. anim. Biochim. Biophys.* **3,** 219–241.

Barenton, B. & Pelletier, J. (1983) Seasonal changes in testicular gonadotropin receptors and steroid content in the ram. *Endocrinology* **112,** 1441–1446.

Barenton, B., Hochereau-de Reviers, M.T. & Perreau, C.

(1983a) Breed differences in testicular histology and numbers of LH and FSH receptors in the lamb. *IRCS Med. Sci.* **11**, 471.

Barenton, B., Hochereau-de Reviers, M.T., Perreau, C. & Saumande, J. (1983b) Changes in testicular gonadotropin receptors and steroid content through postnatal development until puberty in the lamb. *Endocrinology* **112**, 1447–1453.

Bellvé, A.R. & Feig, L.A (1984) Cell proliferation in the mammalian testis. Biology of the seminiferous growth factor (SGF). *Recent Prog. Horm. Res.* **40**, 531–567.

Berndtson, W.E. & Desjardins, C. (1974) The cycle of the seminiferous epithelium and spermatogenesis in the bovine testis. *Am. J. Anat.* **140**, 167–180.

Bilaspuri, G.S. & Guraya, S.S. (1980) Quantitative studies on spermatogenesis in buffalo (*Bubalus bubalis*). *Reprod. Nutr. Develop.* **20**, 975–982.

Bilaspuri, G.S. & Guraya, S.S. (1986) The seminiferous epithelial cycle and spermatogenesis in rams (*Ovis aries*). *Theriogenology* **25**, 485–505.

Blanc, M.R., Cahoreau, C., Courot, M., Dacheux, J.L., Hochereau-de Reviers, M.T. & Pisselet, C. (1978) Plasma follicle stimulating hormone (FSH) and luteinizing hormone (LH) suppression in the cryptorchid ram by a non-steroid factor (Inhibin) from ram rete testis fluid. *Intern. J. Androl., Suppl.* **2**, 139–146.

Blanc, M.R., Hochereau-de Reviers, M.T., Cahoreau, C., Courot, M. & Dacheux, J.L. (1981) Inhibin: effects on gonadotropin secretion and testis function in the ram and the rat. In *Intragonadal Regulation of Reproduction*, pp. 299–326. Eds P. Franchimont & C. P. Channing. Academic Press, London.

Blaschuk, O., Burdzy, K. & Fritz, I.B. (1983) Purification and characterization of a cell aggregating factor (clusterin) the major glycoprotein in ram rete testis fluid. *J. biol. Chem.* **258**, 7714–7720.

Brongniart, I., Theriez, M., Perreau, C. & Hochereau-de Reviers, M.T. (1985) Testis parameters, LH and testosterone secretion in underfed lambs. *J. Androl.* **6**, 64P, Abstr.

Cardoso, F.M. & Godinho, H.P. (1983) Cycle of the seminiferous epithelium and its duration in the Zebu, *Bos indicus*. *Anim. Reprod. Sci.* **5**, 231–245.

Carreau, S., Drosdowsky, M.A. & Courot, M. (1979) Age related effects on androgen binding protein (ABP) in sheep testis and epididymis. *Int. J. Androl.* **2**, 49–61.

Carreau, S., Drosdowsky, M.A., Pisselet, C. & Courot, M. (1980) Hormonal regulation of androgen binding protein in lamb testes. *J. Endocr.* **85**, 443–448.

Cate, R.L., Mattaliano, R.J., Hession, C., Tizard, R., Farber, N.M., Cheung, A., Ninfa, E.G., Frey, A.Z., Gash, D.J., Chow, E.P., Fisher, R.A., Bertonis, J.M., Torres, G., Wallner, B.P., Ramachandran, K.L., Ragin, R.C., Manganaro, T.F., MacLaughlin, D.T. & Donahoe, P.K. (1986) Isolation of the bovine and human genes for Müllerian inhibiting substance and expression of the human gene in animal cells. *Cell* **45**, 685–698.

Clermont, Y. & Leblond, C.P. (1955) Spermiogenesis of man, monkey, ram, and other mammals as shown by the 'periodic acid Schiff' technique. *Am. J. Anat.* **96**, 229–255.

Courot, M. (1971) *Etablissement de la spermatogenése chez l'agneau* (Ovis aries). *Etude expérimentale de son contrôle gonadotrope; importance de la lignée sertolienne.* Thèse, Dr. es Sciences, Paris CNRS No 6317.

Courot, M., Hochereau-de Reviers, M.T. & Ortavant, R. (1970) Spermatogenesis. In *The Testis*, vol. I, pp. 339–432. Eds A. D. Johnson, W. R. Gomes & N. L. VanDemark. Academic Press, New York.

Courot, M., de Reviers, M.M. & Pelletier, J. (1975) Variations in pituitary and blood LH during puberty in the male lamb. Relation to time of birth. *Annls Biol. anim. Biochim. Biophys.* **15**, 509–516.

Courot, M., Hochereau-de Reviers, M.T., Monet-Kuntz, C., Locatelli, A., Pisselet, C., Blanc, M.R. & Dacheux, J.L. (1979) Endocrinology of spermatogenesis in the hypophysectomized ram. *J. Reprod. Fert., Suppl.* **26**, 165–173.

Courot, M., Hochereau-de Reviers, M.T., Pisselet, C., Kilgour, R.J., Dubois, M.P. & Sairam, M.R. (1984) Effect of passive immunization against ovine FSH on spermatogenesis in the ram. In *The Male in Farm Animal Reproduction*, pp. 75–79. Ed. M. Courot. Martinus Nijhoff, The Hague.

Cupps, P.T. & Laben, R.C. (1960) Spermatogenesis in relation to spermatozoa concentration in bovine semen. *J. Dairy Sci.* **43**, 782–786.

Dacheux, J.L., Pisselet, C., Blanc, M.R., Hochereau-de Reviers, M.T. & Courot, M. (1981) Seasonal variations in rete testis fluid secretion and sperm production in different breeds of ram. *J. Reprod. Fert.* **61**, 363–371.

de Reviers, M., Hochereau-de Reviers, M.T., Blanc, M.R., Brillard, J.P., Courot, M. & Pelletier, J.P. (1980) Control of Sertoli and germ cell populations in the cock and sheep testis. *Reprod. Nutr. Develop.* **20**, 241–249.

de Reviers, M.T. & Courot, M. (1976) Stem spermatogonia and Sertoli cells in the bull and ram. *Proc. 8th Int. Congr. Anim. Reprod. & A. I., Cracow*, pp. 85–88.

Dias, J.A. & Reeves, J.J. (1982) Testicular FSH receptor numbers and affinity in bulls of various ages. *J. Reprod. Fert.* **66**, 39–45.

Foresta, C., Manoni, F., Businaro, V., Donadel, C., Indino, M. & Scandellari, C. (1986) Possible significance of transferrin levels in seminal plasma of fertile and infertile men. *J. Androl.* **7**, 77–82.

Guraya, S.S. & Bilaspuri, G.S. (1976) Stages of seminiferous epithelial cycle in the buffalo (*Bos bubalis*). *Annls Biol. anim. Biochim. Biophys.* **16**, 137–144.

Hayes, H.K. & Brooks, D.E. (1985) Hormone effects on protein synthesis and secretion by Sertoli cells in the developing bovine testis. *Proc. Aust. Soc. Reprod. Biol.* **17**, 59, Abstr.

Hochereau, M.T. (1963) Constance des fréquences relatives des stades du cycle de l'épithélium séminifère chez le taureau et chez le rat. *Annls Biol. anim. Biochim. Biophys.* **3**, 93–102.

Hochereau, M.T. (1967) Synthèse de l'ADN au cours des multiplications et du renouvellement des spermatogonies chez le taureau. *Archs Anat. microsc. Morph. exp.* **56**, 85–96.

Hochereau, M.T., Courot, M. & Ortavant, R. (1964) Marquage des cellules germinales du bélier et du taureau par injection de thymidine tritiée dans l'artère spermatique. *Annls Biol. anim. Biochim. Biophys.* **4**, 157–161.

Hochereau-de Reviers, M.T. (1970) *Etude des divisions spermatogoniales et du renouvellement de la spermato-*

gonie souche chez le taureau. Thèse, Dr. es. Sciences, Paris, CNRS No. 3976.

Hochereau-de Reviers, M.T. (1971) Etude cinétique des spermatogonies chez les mammifères. In *La Cinétique de Prolifération Cellulaire,* pp. 189–216. INSERM, Paris.

Hochereau-de Reviers, M.T. (1976) Variation in the stock of testicular stem cells and in the yield of spermatogonial divisions in ram and bull testes. *Andrologia* **8,** 137–146.

Hochereau-de Reviers, M.T. (1981) Control of spermatogonial multiplications. In *Reproductive Process and Contraception,* pp. 307–332. Ed. K. W. McKerns. Plenum Press, New York.

Hochereau-de Reviers, M.T. & Courot, M. (1978) Sertoli cells and development of seminiferous epithelium. *Annls Biol. anim. Biochim. Biophys.* **18,** 573–583.

Hochereau-de Reviers, M.T., Loir, M. & Pelletier, J. (1976a) Seasonal variations in the response of the testis and LH levels to hemicastration of adult rams. *J. Reprod. Fert.* **46,** 203–209.

Hochereau-de Reviers, M.T., Ortavant, R. & Courot, M. (1976b) Type A spermatogonia in the ram. *Prog. Reprod. Biol.* **1,** 13–19.

Hochereau-de Reviers, M.T., Bindon, B.M., Courot, M., Lafortune, E., Land, R.B., Lincoln, G.A. & Ricordeau, G. (1984a) Number of Sertoli cells in the ram testis. In *The Male in Farm Animal Reproduction,* pp. 69–74, Ed. M. Courot. Martinus Nijhoff, The Hague.

Hochereau-de Reviers, M.T., Land, R.B., Perreau, C. & Thompson, R. (1984b) Effect of season of birth and of hemicastration on the histology of the testis of 6-month-old lambs. *J. Reprod. Fert.* **70,** 157–163.

Hochereau-de Reviers, M.T., Perreau, C. & Lincoln, G.A. (1985) Photoperiodic variations in somatic and germ cell populations in the Soay ram testis. *J. Reprod. Fert.* **74,** 329–334.

Hullinger, R.L. & Wensing, C.J.G. (1985) Descent of the testis in the fetal calf. A summary of the anatomy and process. *Acta Anat.* **121,** 63–68.

Jegou, B., Dacheux, J.L., Garnier, D.H., Terqui, M., Colas, G. & Courot, M. (1979) Biochemical and physiological studies of androgen binding protein in the reproductive tract of the ram. *J. Reprod. Fert.* **57,** 311–318.

Jenkins, N. & Ellison, J. (1986) Suppression of tissue plasminogen activator by corticosteroids in the bovine Sertoli cell. In *Proc. 4th European Workshop on Molecular and Cellular Endocrinology of the Testis,* Abstr. A7, p. 45.

Johnson, L. & Thompson, D.L. (1983) Age-related and seasonal variation in the Sertoli cell population, daily sperm production and serum concentrations of FSH, LH and testosterone in stallions. *Biol. Reprod.* **29,** 777–789.

Jones, L.S. & Berndtson, W.E. (1986) A quantitative study of Sertoli cell and germ cell populations as related to sexual development and aging in stallion. *Biol. Reprod.* **35,** 138–148.

Josso, N. (1973) *In vitro* synthesis of müllerian inhibiting hormone by seminiferous tubules isolated from the calf fetal testis. *Endocrinology* **93,** 829–834.

Josso, N., Picard, J.Y. & Tran, D. (1980) A new testicular glycoprotein: anti-müllerian hormone. In *Testicular Development, Structure and Function,* pp. 21–31. Eds

A Steinberger & E. Steinberger. Raven Press, New York.

Kilgour, R.J., Pisselet, C., Dubois, M.P., Courot, M. & Sairam, M.R. (1984) The role of FSH in the establishment of spermatogenesis in the lamb. *Proc. 10th Int. Cong. Anim. Reprod. & A. I.* Urbana-Champaign, **2,** 42, Abstr.

Kramer, M.F. (1960) *Spermatogenesis bij de stier.* Thesis, Rijkuniversiteit, Utrecht.

Lafortune, E. (1983) *Etablissement de la fonction sexuelle chez le taureau en milieu tropical.* Thèse Dr. 3° cycle, Université Paris VI.

Lafortune, E., Blanc, M.R., Pelletier, J., Perreau, C., Terqui, M. & Hochereau-de Reviers, M.T. (1984) Variations in the plasma levels of gonadotrophin and testosterone and in Leydig and Sertoli cell populations between birth and adulthood in Romanov lambs born in spring or autumn. *Reprod. Nutr. Develop.* **24,** 937–946.

Lino, B.F. (1971) Cell count correction factors for the quantitative histological analysis of the germinal epithelium of the ram. *Anat. Rec.* **170,** 413–420.

Lok, D., Weenk, D. & de Rooij, D.G. (1982) Morphology, proliferation and differentiation of undifferentiated spermatogonia in the Chinese hamster and the ram. *Anat. Rec.* **203,** 83–99.

McNeilly, J.R., Fordyce, M., Land, R.B., Lee, G.J. & Webb, R. (1986) Endocrine differences in rams after genetic selection for testis size. *J. Reprod. Fert.* **76,** 131–140.

Mason, A.J., Hayflick, J.S., Ling, N., Esch, F., Ueno, N., Ying, S.Y., Guillemin, R., Niall, H. & Seeburg, P.H. (1985) Complementary DNA sequences of ovarian follicular fluid inhibin show precursor structure and homology with transforming growth factor-β. *Nature, Lond.* **318,** 659–663.

Monet-Kuntz, C., Hochereau-de Reviers, M.T. & Terqui, M. (1984) Variations in testicular androgen receptors and histology of the lamb testis from birth to puberty. *J. Reprod. Fert.* **70,** 203–210.

Monet-Kuntz, C., Barenton, B., Locatelli, A., Fontaine, I., Perreau, C. & Hochereau-de Reviers, M.T. (1987) Effects of experimental cryptorchidism and subsequent orchidopexy on seminiferous tubule functions in the lamb. *J. Androl.* (in press).

Orgebin, M.C. (1961) Etude du transit épididymaire des spermatozoïdes de taureau marqués à l'aide du [32]P. *Annls Biol. anim. Biochim. Biophys.* **1,** 117–120.

Ortavant, R. (1958) *Le cycle spermatogénétique chez le bélier.* Thèse, Dr. es Sciences, Paris, CNRS A3118.

Orth, J.M. (1982) Proliferation of Sertoli cells in fetal and postnatal rats: a quantitative autoradiographic study. *Anat. Rec.* **203,** 485–492.

Orth, J.M. (1984) The role of FSH in controlling Sertoli cell proliferation in testes of fetal rats. *Endocrinology* **115,** 1248–1252.

Patthy, L. (1985) Evolution of the proteases of blood coagulation and fibrinolysis by assembly from modules. *Cell* **41,** 657–663.

Parvinen, M. (1982) Regulation of the seminiferous epithelium. *Endocrine Review* **3,** 404–417.

Redi, C.A. (1986) Relative positions of the spermatogenic stages in the mouse testis: morphological evidences for their non-randomized adjacences. *Andrologia* **18,** 25–32.

Roberts, A.B., Anzano, M.A., Wakefield, L.M., Roche, N.S., Stern, D.F. & Sporn, M.B. (1985) Type β transforming growth factor: A bifunctional regulator of cellular growth. *Proc. natn. Acad. Sci., U.S.A.* **82**, 119–123.

Roosen-Runge, E.C. & Giesel, L.O. (1950) Quantitative studies on spermatogenesis in the albino rat. *Am. J. Anat.* **87**, 1–30.

Salim, B. & Entwistle, K.W. (1982) Duration of the seminiferous epithelial cycle in hybrid *Bos indicus* × *Bos taurus* bulls. *J. Reprod. Fert.* **66**, 729–734.

Schanbacher, B.D. (1984) Regulation of luteinizing hormone secretion in male sheep by endogenous estrogen. *Endocrinology* **115**, 944–950.

Schanbacher, B.D., Pelletier, J. & Hochereau-de Reviers, M.T. (1986) Follicle stimulating hormone, luteinizing hormone and testicular Leydig cell responses to estradiol immunization in Ile-de-France rams. *J. Androl.* (in press).

Sharma, A.K. & Gupta, R.C. (1980) Duration of seminiferous epithelial cycle in Buffalo bulls (*Bubalus bubalis*). *Anim. Reprod. Sci.* **3**, 217–224.

Smith, B.C. & Griswold, M.D. (1981) Primary culture of supporting cells from bovine testis. *In Vitro* **17**, 612–618.

Upadhyay, S. & Zamboni, L. (1982) Ectopic germ cells: natural model for the study of germ cell sexual differentiation. *Proc. natn. Acad. Sci., U.S.A.* **79**, 6584–6588.

Waites, G.M.H., Speight, A.C. & Jenkins, N. (1985) The functional maturation of the Sertoli cell and Leydig cell in the mammalian testis. *J. Reprod. Fert.* **75**, 317–326.

Wright, W.W., Musto, N.A., Mather, J.P. & Bardin, C.W. (1981) Sertoli cells secrete both testis specific and serum proteins. *Proc. natn. Acad. Sci., U.S.A.* **78**, 7565–7567.

Zamboni, L. & Upadhyay, S. (1982) The contribution of the mesonephros to the development of the sheep fetal testis. *Am. J. Anat.* **165**, 339–356.

J. Reprod. Fert., Suppl. **34** (1987), 115–131

Function of the epididymis in bulls and rams

R. P. Amann

Animal Reproduction Laboratory, Colorado State University, Fort Collins, Colorado 80523, U.S.A.

Introduction

Spermatozoa undergo a series of remarkable transformations during the 5–6 weeks between when they originate as spherical spermatids in the germinal epithelium and, assuming natural mating, encounter an oocyte(s) within the female reproductive tract. Reproductive biologists tend to consider spermatozoa from the perspective of their own narrow interests such as epididymal function, characteristics of ejaculated spermatozoa, in-vitro fertilization, or sperm transport within the female reproductive tract. This approach ignores the fact that events occurring at any point during the 5–6-week life-span of the spermatid/spermatozoon must influence subsequent events. Since the epididymis has a crucial role in controlling sperm function, an understanding of epididymal function is important for all gamete biologists.

At a first level of consideration, functions of the epididymis are: (a) maturation of spermatozoa so they have a maximal potential for fertilizing oocytes with a minimal loss of embryos, (b) maintenance of mature spermatozoa in optimum condition until ejaculated or voided from the excurrent duct system, and (c) transport of spermatozoa through the epididymal duct. Available evidence (Lino & Braden, 1972; Amann *et al.*, 1974) supports the conclusion that the epididymis is not a major site for dissolution or removal of spermatozoa from the excurrent duct system in a normal male; most spermatozoa produced by the testes are voided during ejaculation or micturition.

Testicular spermatozoa of bulls and rams (Amann & Griel, 1974; Fournier-Delpech *et al.*, 1979), are infertile whereas spermatozoa from the distal cauda epididymidis have a fertilizing potential equivalent to that of ejaculated spermatozoa. Spermatozoa should not be considered as mature until they have acquired normal motility, fertility, and ability to induce normal embryonic development. However, spermatozoa from the cauda epididymidis, which apparently are functionally mature, are further altered by admixture with seminal plasma (Killian & Amann, 1973).

The sites within the epididymis where different aspects of sperm maturation occur probably are similar for rams (Fig. 1) and bulls. Spermatozoa from the distal caput epididymidis bind to about 50% of sheep oocytes during in-vitro incubation, but are immotile and do not fertilize oocytes after artificial insemination into the uterine horn. However, for samples from the distal corpus epididymidis both the motility and fertilizing ability of spermatozoa are markedly greater (Fig. 1); the majority of fertilized oocytes result in lambs (69%) whereas oocytes fertilized by spermatozoa from the central corpus epididymidis invariably underwent early embryonic death. There is an improvement in sperm function between the distal corpus and the proximal cauda epididymidis and cells from either the proximal or distal cauda are equivalent to ejaculated spermatozoa (Fournier-Delpech *et al.*, 1979). The unusually high embryonic death rate associated with oocytes fertilized by spermatozoa from the distal corpus epididymidis is a consequence of delayed early cleavage divisions (Fournier-Delpech *et al.*, 1981a) which presumably is an intrinsic cause of embryo death or leads to improper synchrony between embryo development and the environment afforded by the oviduct and uterus. From Fig. 1, it is obvious that sperm maturation must involve a series of complex changes in the spermatozoa, some of which are discussed by Hammerstedt & Parks (1987), which result from a sequence of events occurring at different points within the caput and corpus epididymidis.

Epididymal function is androgen-dependent, although extensive data are not available for domestic ruminants. The electrophoretic profile of proteins in fluid taken by micropuncture at several sites of the epididymis is different in intact and orchidectomized rams (Fournier-Delpech

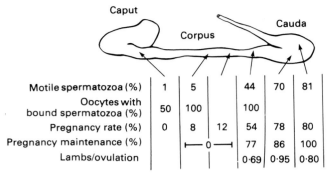

Fig. 1. Sites in the ram epididymis of progressive changes in sperm function associated with sperm maturation. Data for binding of spermatozoa to oocytes are based on in-vitro studies with sheep oocytes and pregnancy data are for spermatozoa inseminated into the uterine horn. From data reported by Fournier-Delpech *et al.* (1979, 1983).

et al., 1981b). Also, differentiation of the initial segment of bulls and rams (Abdel-Raouf, 1960; Nilnophakoon, 1978) does not occur until testicular development has progressed to a stage when the rete testis fluid probably has normal concentrations of androgens and androgen-binding protein (Carreau *et al.*, 1984b). Benoit (1926) noted that castration altered morphology, and presumably function, of the epididymal epithelium in mice and guinea-pigs. He postulated that epididymal secretions were needed to permit the acquisition of motility and maintain the vitality of spermatozoa. Although sperm maturation requires <4 days (Amann, 1981), data for laboratory animals (Bedford, 1975; Orgebin-Crist *et al.*, 1975) support the conclusion that ageing spermatozoa in a specific location in the epididymis for several days is insufficient to induce maturation. The sequential exposure of spermatozoa to specific environments produced in progressively more distal sites within the epididymal duct is necessary. Furthermore, the process of sperm maturation is dependent upon secretions of the epididymal epithelium, some of which are androgen dependent (Orgebin-Crist & Jahad, 1978; Brooks, 1981, 1983; Jones *et al.*, 1982).

Maintenance of fertile spermatozoa in the cauda epididymidis also is androgen-dependent. For orchidectomized rams given 0, 60 or 300 mg implants of testosterone, the function of spermatozoa taken 9 days after orchidectomy from the distal cauda epididymidis was dependent on androgen concentration in blood (Fournier-Delpech *et al.*, 1984).

This paper focusses on epididymal function in rams and bulls. It is likely that the underlying processes are similar in other domestic ruminants and laboratory animals. However, comparisons will emphasize the fact that certain functions may be located in different regions of the epididymis, depending on the species.

Morphology of the epithelium

The epididymis consists of three distinct compartments which are: (a) the ductal lumen, containing spermatozoa and epididymal plasma; (b) the epithelium lining the lumen; and (c) the connective tissue, vascular elements, smooth muscle and nerves which comprise the extraductal tissue. Emphasis will be focussed on the epithelium and interpretation of analyses of the luminal fluid to deduce epithelial functions; the spermatozoa and extraductal tissue will not be considered.

Cell types

Throughout the epididymis, the ductal epithelium consists of tall, narrow columnar cells, termed principal cells, and small spherical cells, termed basal cells. As discussed below, the cytoplasmic characteristics of principal cells differ in different regions of the epididymis (Nicander, 1958; Tröger, 1969; Sinowatz, 1981; Wagley *et al.*, 1984; Goyal, 1985). Basal cells have a scant cytoplasm containing only a few organelles, and are similar throughout the duct. Their function is unknown. The epithelium in the caput epididymidis (regions M_1 and M_2 described below) also contains apical cells. The basal half of these columnar cells is very narrow, but the apical half is of normal width and contains the oval nucleus, many mitochondria, and some ribosomes. Apical cells are rarely found in other parts of the ductus epididymidis (Sinowatz, 1981; Goyal, 1985). In addition, macrophages are found in the basal area of the epithelium and lymphocytes (although these may be monocytes) are found both basally and apically; both of these immunocompetent cell types are found along the entire length of the epididymal duct (Sinowatz, 1981). Detailed morphometric analyses of the bovine or ovine epididymis have not been reported, so the relative abundance of principal cells and basal cells is unknown.

In contrast to bulls and rams, the epididymal epithelium in rodents and rabbits also contains clear cells which are found in increasing numbers throughout the duct distal to the initial segment (Robaire & Hermo, 1987). Since rodent sperm typically lose their cytoplasmic droplet while in the cauda epididymidis, and membranous elements similar to those within a sperm cytoplasmic droplet are found free in the lumen of the epididymal duct and in vesicles within the clear cells, it has been proposed (Robaire & Hermo, 1987) that clear cells internalize elements of sperm cytoplasmic droplets. The majority of bull and ram spermatozoa retain their cytoplasmic droplet in the cauda epididymidis (Ortavant, 1953; Amann & Almquist, 1962), and neither bulls nor rams have clear cells within the epididymal epithelium.

Mitosis of principal cells in the epididymal epithelium of domestic ruminants has not been studied using [^{3}H]thymidine-labelling of cells in the S-phase, followed by radioautography. However, mitosis is rare (Sinowatz, 1981). Based on the labelling indices of cells in the rat epididymis, Clermont & Flannery (1970) concluded that the epithelium was static. This is generally interpreted as evidence that the epithelial cells rarely divide. However, based on several observations [Clermont & Flannery (1970) for rats, Pabst & Schick (1979) for rabbits, and Wagley *et al.* (1984) for cultured ram principal cells], it is likely that principal cells in the epididymal epithelium have a life span of 2–3 months. There is no evidence to support the concept that basal cells are precursors of principal cells. Indeed, from studies of the ontogeny of the epididymal epithelium in bulls and rodents, Abdel-Raouf (1960), Benoit (1926) and Sun & Flickinger (1979) concluded that columnar cells differentiate into principal cells and basal cells before puberty; Nilnophakoon (1978) reported similar data for rams, but felt that origin of the basal cells was obscure.

Regions of the epididymis

Regional differences in morphology of the epithelium lining the epididymal duct of bulls, and the concept that these reflected differences in function, were described by von Lanz (1924) and Benoit (1926). Benoit (1926) introduced the term 'initial segment'. He noted that there were few spermatozoa in the lumen and that the epithelium had histological characteristics which he, and others, incorrectly interpreted as evidence of secretory activity. The detailed description by Nicander (1958) of the regional histology and cytochemistry of the bull and ram epididymis is generally used to demarcate regions (Fig. 2a).

From a combined morphological and physiological perspective, it is advantageous to consider the epididymis as an initial segment involved in adsorption, a middle segment in which sperm maturation occurs, and a terminal segment in which fertile spermatozoa are stored (Glover & Nicander, 1971). Therefore designation of the middle and terminal segments as a maturation region and a fertility region respectively is appropriate. The designations in Fig. 2(b) are based on this concept, sperm function (Fig. 1), and the electron microscopic observations summarized below. The term initial segment has been retained and the maturation region has three morphologically distinct regions (M_1, M_2 and M_3). Region M_3 is subdivided into regions M_{3A} and M_{3B} which,

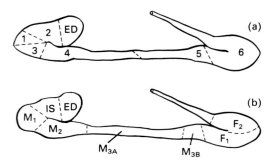

Fig. 2. Approximate demarcations between regions of the bull or ram epididymis (a) as defined by Nicander (1958) or (b) as proposed and defined herein.

although morphologically similar, differ in that spermatozoa from region M_{3A} rarely are motile and are infertile whereas many spermatozoa from region M_{3B} are fertile and result in offspring. The distinction between regions F_1 and F_2 of the fertility region is only in diameter of the duct, since the epithelium is similar throughout the cauda (Sinowatz, 1981; Goyal, 1985) and functional ability of the spermatozoa is similar (Fig. 1). The functional and morphological regions depicted in Fig. 2(b) are appropriate for the rabbit except that region M_{3B} probably should start more proximally.

Ultrastructure of principal cells

This description of regional differences in principal cells of the ram and bull is based on ultrastructural studies reported by Nicander (1979), Sinowatz (1981), Wagley et al. (1984) and Goyal (1985), although their descriptions are consistent with earlier light microscopic observations of von Lanz (1924), Benoit (1926), Nicander (1958) and Tröger (1969). General reviews by Hamilton (1972, 1975) and Robaire & Hermo (1987) on the mammalian epididymis should also be consulted. The epididymal epithelium is apparently similar in morphological characteristics and regional distribution in rams and bulls. Figure 3 depicts a synopsis of these data.

Principal cells in the initial segment are characterized by extensive long narrow, or dilated, cisternae of endoplasmic reticulum and a paucity of vesicles between the Golgi apparatus and their apical face. The apical face is covered by long, narrow microvilli among which are coated pits underlain by coated vesicles and a few large membrane-bound vesicles. In the central zone of the Golgi apparatus, small vesicles and cisternae of smooth endoplasmic reticulum are found and on the cis face of the Golgi there are occasional arrays of rough endoplasmic reticulum. The supranuclear area contains a few dense bodies. The oval nucleus is in the lower one-third of the cell and is surrounded by rough endoplasmic reticulum. Mitochondria are concentrated in the basal region of the cell.

Principal cells in region M_1 also have long cisternae of endoplasmic reticulum, with few ribosomes, in the apical portion of the cell, but there are more membrane-bound vesicles, vacuoles, and multi-vesicular bodies present than in principal cells in the initial segment. The apical face has many coated pits and apical vesicles and the microvilli are shorter. Below the prominent Golgi apparatus, vesicles, dense bodies and a few multi-vesicular bodies fill much of the cytoplasm. The nucleus tends to be invaginated and is associated with rough endoplasmic reticulum. The infranuclear region contains vesicles, and most of the mitochondria. In contrast to cells in the initial segment, principal cells in region M_1 have more apical vesicles, a larger Golgi apparatus, more supranuclear dense bodies and lipid vacuoles, and far more basal vesicles.

In region M_2, the apical face of the principal cells has numerous coated pits and underlying vesicles, and they appear to be less prevalent than in region M_1. The apical two-thirds of the cells are filled with large and intermediate sized multi-vesicular bodies, vesicles that vary in size and

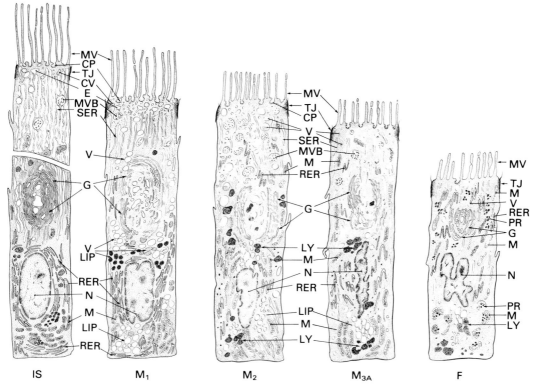

Fig. 3. Drawings illustrating the characteristics and differences in principal cells from specific regions (see Fig. 2) of the ram or bull epididymis. The cells (from left to right) typify principal cells in the initial segment (IS) and regions M_1, M_2, M_{3A} and F. Principal cells are 100–110 μm tall in the initial segment, 65–70 μm in regions M_1 and M_2, and progressively shorter distally to about 50 μm in region F. Structures are identified as: CP, coated pit; CV, coated vesicle; E, endosome; G, Golgi apparatus; LIP, lipid droplet; LY, lysosome; MV, microvilli; MVB, multi-vesicular body; N. nucleus; M, mitochondria; P, polyribosomes; RER, rough endoplasmic reticulum; SER, smooth endoplasmic reticulum; TJ, tight junction; V, vacuole. Based on unpublished observations and those reported by Goyal (1985), Nicander (1979), Sinowatz (1981) and Wagley *et al.* (1984).

electron density, and lysosomes in addition to a large Golgi apparatus. Mitochondria, rough endoplasmic reticulum, and free ribosomes are found throughout this region. The nucleus is narrow and invaginated and associated with mitochondria and rough endoplasmic reticulum. The basal portion of the cells contains numerous lipid droplets, mitochondria and smooth endoplasmic reticulum. In contrast to Sinowatz (1981) and Wagley *et al.* (1984), Goyal (1985) considered the principal cells in regions M_1 and M_2 to be similar, but distinct from those in the initial segment and region M_3.

Principal cells in regions M_{3A} and M_{3B} are similar in morphology, although there is a slight diminution in height progressing distally. The apical face contains a few coated pits and there are coated and uncoated small vesicles in the apical region of the cytoplasm. In all portions of the cells, multi-vesicular bodies, vesicles and lysosomes are less prevalent than in region M_2. The Golgi apparatus is relatively small and the nucleus is elongated and invaginated. Small dense bodies are found above the nucleus. Clusters of lipid droplets and dense bodies are found below, and occasionally above, the nucleus. Although mitochondria are found throughout the cells, orderly arrays of mitochondria are conspicuous near the basal membrane.

Principal cells are similar in morphological characteristics throughout region F, although they are progressively shorter passing distally. There are few coated pits or apical vesicles, although numerous polyribosomes and short profiles of rough endoplasmic reticulum are found above the relatively small Golgi apparatus. Many mitochondria are found above the irregularly-shaped nucleus and lysosomes are found in the basal portion of the cell. In contrast to mitochondria located above the nucleus of principal cells in region F, or at all locations in principal cells in more proximal regions, mitochondria located below the nucleus in principal cells in region F are cup-shaped and surround a small cluster of ribosomes.

The microvilli become progressively shorter, and broader, passing distally along the epididymal duct, and histochemical studies (Sinowatz, 1981; Goyal & Vig, 1984) revealed that they probably differ in function. In the initial segment and region M_1, the microvilli stain positive for alkaline phosphatase, but only occasionally for Mg/Ca-dependent ATPase. In contrast, the microvilli of principal cells in regions M_2, M_3 and F are devoid of alkaline phosphatase, but stain intensely for Mg/Ca-dependent ATPase. Acid phosphatase is distributed throughout the apical cytoplasm in principal cells in the initial segment and regions M_1 and M_2, and localized with high intensity in the basal portion of cells in region M_3.

Blood–epididymis barrier

A functional blood–epididymis barrier in bulls and rams is indicated by apically located tight junctions between adjacent principal cells (Sinowatz, 1981; Wagley *et al.*, 1984; Fig. 3), differences in the composition of luminal fluid at different sites along the bovine epididymis (Crabo, 1965), and differences between the luminal contents and blood. More extensive and direct evidence for the blood–epididymis barrier in other species is discussed by Hinton (1985) and Robaire & Hermo (1987). The actual site of the blood–epididymis barrier is the tight junctions found at the apical face of the epithelium.

Functions of the epithelium

Major functions of the epididymal epithelium are to maintain an environment appropriate for maturation of spermatozoa in the caput and corpus and for maintenance of fertile spermatozoa in the cauda epididymidis. With respect to the principal cells, however, three functions might be considered: transport of small molecules (e.g. amino acids, ions, sugars, water), transport of large molecules (e.g. proteins), and metabolism or synthesis. Transport of all sizes of molecules involves one, or usually both, of two coupled transport phenomena–adsorption and secretion. Adsorption of components from the lumen of the duct is coupled with secretion of material from the lateral or basal faces of the principal cells (below the blood–epididymis barrier) to allow eventual passage into the blood. Similarly, secretion requires adsorption of components from the extraductal or luminal environment, optional processing of the material by the cells (metabolism or synthesis), and secretion into the duct lumen. In this review, metabolism refers to modification of molecules (e.g. testosterone to dihydrotestosterone) in the principal cells, and synthesis to de-novo proteins or steroids; the terms do not refer to energy metabolism or to synthesis necessary for cell maintenance.

Transport of small molecules

The earliest quantitative evidence of micromolecule transport by the bull epididymis was provided by Crabo (1965) who used micropuncture techniques, and direct biochemical analyses, to study regional differences in the content of the luminal fluid. His data were largely confirmed by Bech & Koefoed-Johnsen (1973). Extensive data for bulls and rams (Amann *et al.*, 1974; Waites, 1977) show that about 40 ml/day enter the epididymis as rete testis fluid but <1 ml/day leaves as cauda epididymal plasma. Assuming that spermatozoa were neither made nor lost in the epididymis, Crabo (1965) used the concentration of spermatozoa in successive regions of the epididymis to

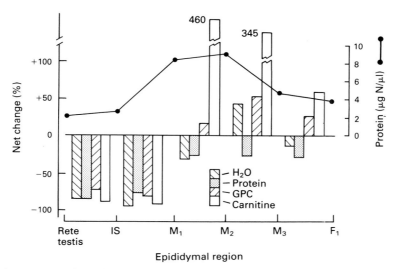

Fig. 4. The concentration of protein in luminal fluid from specific sites of the ram epididymis (●—●) and net changes in the amount of water, protein, glycerophosphocholine (GPC), and carnitine between rete testis fluid and luminal fluid from successive sites (see Fig. 2) in the bull epididymis. Net changes in fluid components, in terms of gain or loss between successive sites, were calculated using (a) the change in sperm concentration to estimate water gain or loss and (b) calculating the total amount of a component per unit volume of fluid at site 1 and the total amount of the component at site 2 for a unit volume adjusted for water gain or loss between site 1 and site 2; the difference was expressed as a percentage of the value at site 1. Calculated from data reported by Crabo (1965) and Besancon *et al.* (1985).

calculate water transport from or to the lumen. Sperm concentration is higher in the initial segment (360×10^6/ml) than in rete testis fluid (60×10^6/ml) and markedly higher in regions M_1 and M_2 (4500 and 6400×10^6/ml), although concentrations are lower in regions M_3 and F_1 (4500 and 5200×10^6/ml). These data, and the concentrations of components, allowed calculations of changes in electrolyte and protein concentrations along the epididymal duct.

Figures 4 and 5 show the net changes, in terms of entrance or loss into or from the luminal fluid of a number of molecules. These values for net change in concentration are, however, not equivalent to transport flux within a region of the epididymis. Measurements of flux must also consider residence time (i.e. length of duct and flow rate). Adsorption of water and glycerophosphocholine occurs between the initial segment and region M_1 (Fig. 4), but more distally, especially between regions M_2 and M_3, there is a net increase in water and glycerophosphocholine. Based on data for rabbits (Hammerstedt & Rowan, 1979), glycerophosphocholine is a synthetic product of the epididymal epithelium derived from blood lipoproteins. Although most of the glycerophosphocholine is apparently added between regions M_2 and M_3, some is also added to the ductal fluid in region M_{3B}.

Carnitine is an example of a compound transported directly from blood into the ductal lumen without modification in the principal cells. Transport of carnitine is maximal between regions M_1 and M_{3A} (Fig. 4), even though the concentration is maximal in region F_2 because of water removal (> 10 mM; Hinton *et al.*, 1979; Besancon *et al.*, 1985).

The net change in electrolyte composition of the luminal fluid (Fig. 5) reveals a continuous removal of Na, K, Cl, and P from the luminal fluid in amounts greater than would be equivalent to water loss. However, between regions M_2 and M_3, where water is added back into the ductal lumen (Fig. 4), there is a net increase in all four ions, especially Cl and P (Fig. 5). Nevertheless, as reflected in the ratio of Na to K (Fig. 5), Crabo (1965) found a continuous decline in Na concentration along the ductus epididymidis (from 121 to 33 mequiv./l) whereas the concentrations of K were higher in

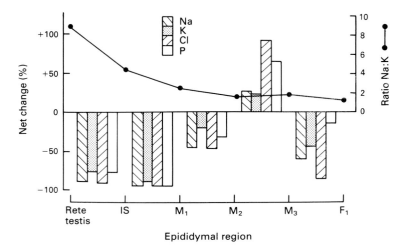

Fig. 5. The Na:K ratio in luminal fluid from specific sites of the ram epididymis (●—●) and net changes in the amount of Na, K, Cl, and P between rete testis fluid and luminal fluid from successive sites (see Fig. 2) in the bull epididymis. Net changes in fluid components were calculated as for Fig. 4 from data reported by Crabo (1965).

regions M_1, M_2 and M_3 than in the initial segment or region F_1 (about 45 vs 28 mequiv./l). Concentrations of Cl were reasonably constant between the initial segment and region M_3, but were much lower in region F_1 (60–80 *vs* 13 mequiv./l) and P concentrations were similar in luminal fluid from each region (38–46 mequiv./l).

The concentration of inositol is probably maximal in region M_1 (>20 mM), based on analyses of bovine rete testis fluid, cauda epididymal plasma and epididymal tissue (Voglmayr & Amann, 1973) and of fluid from several sites of the ram epididymis (Hinton *et al.*, 1980). This probably reflects the concentration of inositol, entering in rete testis fluid, as water is adsorbed (Voglmayr & Amann, 1973).

Transport and synthesis of large molecules

The uptake of carbon particles (India ink) by the epithelium of the proximal ram epididymis (Gunn, 1936) was the first proof of transport of large molecules. The concentration of protein in luminal fluid from the bovine epididymis is higher in region M_1 and M_2 than in more proximal or distal sites (Fig. 4), but there is apparently net loss of protein in all regions of the epididymis. Electrophoretic and immunological analyses of bovine rete testis fluid and cauda epididymal plasma (Amann *et al.*, 1973; Killian & Amann, 1973), and immunofluorescent analyses of the bovine epididymal epithelium (Barker & Amann, 1971), revealed profound changes in the proteins bathing spermatozoa and entering and leaving the epididymis. Many of these changes reflect macromolecular transport into and from the luminal fluid. Cauda epididymal plasma of rams and bulls (Alumot *et al.*, 1971; Killian & Amann, 1973) contains a variety of immunoglobulins as well as some blood serum antigens which were postulated (Killian & Amann, 1973) to enter the luminal fluid via the epididymal epithelium rather than in rete testis fluid. If these immunoglobulins and blood serum antigens are not contaminants of the collection process, their presence is evidence that the blood–epididymis barrier might be 'leaky'. However, most macromolecules transported into the lumen of the epididymal duct are synthesized by principal cells.

Adsorption of specific, identified proteins by the epididymal epithelium has not been demonstrated for rams or bulls, but 60% of the androgen-binding protein entering the epididymis daily in

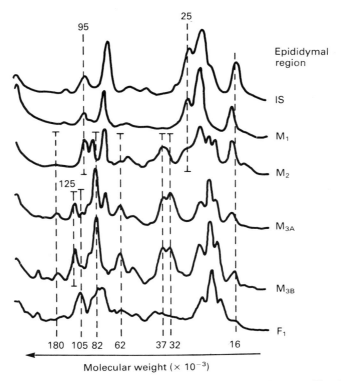

Fig. 6. Electrophoretic profiles of luminal proteins from specific sites (see Fig. 2) of the ram epididymis. The vertical lines designate proteins apparently not present in all fluids. Although not shown, the spectrum of proteins in rete testis fluid is very different from that for fluid from the initial segment (IS). Modified from Dacheux & Voglmayr (1983).

rete testis fluid is undetectable in cauda epididymal plasma. Selective adsorption of androgen-binding protein, α_2-macroglobulin and transferrin, entering the epididymis in rete testis fluid, has been demonstrated in rats. All three proteins are taken up, especially in the caput epididymidis, by receptor-mediated endocytosis via coated pits, and passed through coated vesicles and endosomes to multi-vesicular bodies (see Robaire & Hermo, 1987). The fate of androgen-binding protein has not been established. However, α_2-macroglobulin was transferred to lysosomes while transferrin was recycled back to the epididymal lumen.

The most common approach for study of macromolecule synthesis and transport has been electrophoresis of proteins obtained by micropuncture of the epididymal duct. Huang & Johnson (1975) found at least 6 proteins in the epididymal fluid of rams which were absent in serum. All were detectable in fluid from the initial segment, and one was present in progressively lower concentrations distally. Unfortunately, rete testis fluid or spermatozoa were not analysed and therefore could not be excluded as sources of the detected proteins.

Dacheux & Voglmayr (1983) compared proteins in epididymal fluid with those in rete testis fluid or on the surface of spermatozoa. Fluid from the initial segment lacked the major proteins of M_r 27 000, 32 000, 40 000 and 56 000 present in rete testis fluid. At least two proteins (M_r 25 000 and 95 000) in fluid from the initial segment (but absent in rete testis fluid), were present at decreasing concentrations through region M_2, but were absent in regions M_3 and F_1 (Fig. 6). At least 5 proteins (M_r 32 000, 37 000, 62 000, 82 000 and 180 000) first detectable in region M_2 were present in higher concentrations in region M_{3A} or M_{3B}, but at lower concentrations in region F_1. In

124   *R. P. Amann*

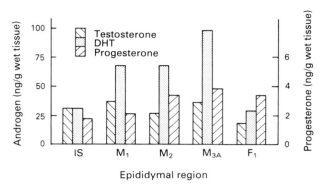

Fig. 7. Concentrations of testosterone, dihydrotestosterone (DHT), and progesterone in whole tissue from specific sites (see Fig. 2) of the bull epididymis. From data reported by Ganjam & Amann (1976).

contrast, a protein of M_r 105 000 first detectable in region M_{3A} was present at increasing concentrations passing distally, and a protein of M_r 125 000 was detectable only in regions M_{3A} and M_{3B}.

Dacheux & Voglmayr (1983) cautioned, as had previous workers (Barker & Amann, 1970; Amann *et al.*, 1973; Killian & Amann, 1973), that apparent changes in the electrophoretic pattern of proteins (or in immunological characteristics) could result from (1) adsorption of specific proteins from the tubule lumen; (2) modification within the lumen of existing proteins by cleavage, congregation, cross-linkage or glycosylation; (3) accumulation to detectable concentration of proteins lost from spermatozoa; or (4) by active secretion of new proteins by the epididymal epithelium. All four of these processes apparently occur.

Data for acid α-glucosidase illustrate the fallacy of making conclusions about luminal concentrations of a compound based on tissue concentration, although the comparison is confounded by species and technique. Besancon *et al.* (1985) measured acid α-glucosidase activity in epididymal plasma from rams and Jauhiainen & Vanha-Perttula (1985) measured the enzyme in bull tissue, using slightly different methods. In ram epididymal plasma, activity of α-glucosidase was low in the initial segment and regions M_1, M_2 and M_{3A}, but was 3-fold greater in region M_{3B}; values in regions F_1 and F_2 were even greater. In whole epididymal bull tissue, however, activity from region M_1 was almost 2-fold greater than for any other region.

Metabolism

Rete testis fluid and cauda epididymal plasma contain a diversity of steroids, including testosterone (> 100 nM), in concentrations much higher than in blood (Ganjam & Amann, 1976; Waites, 1977; Voglmayr *et al.*, 1977). However, based on analyses of epididymal tissue (Fig. 7), dihydrotestosterone must be produced from testosterone in region M_1, and possibly region M_{3A}. Principal cells, but not basal cells, metabolize testosterone to dihydrotestosterone (Klinefelter & Amann, 1980), and *in vitro* (Wagley *et al.*, 1984) testosterone is metabolized most actively by principal cells from region M_1 of rams (Fig. 8). Principal cells from the initial segment are virtually devoid of 5α-reductase activity. This is in stark contrast to data for rat principal cells (Brown & Amann, 1984) or tissue homogenates (Robaire *et al.*, 1981), for which the highest activity of 5α-reductase is in the initial segment. A comparison of testosterone metabolism by cultured principal cells and minced epididymal tissue (Fig. 9) confirmed that the initial segment of the ram epididymis contains little 5α-reductase activity.

Although cultured ram principal cells, especially from region M_1, rapidly metabolize testosterone

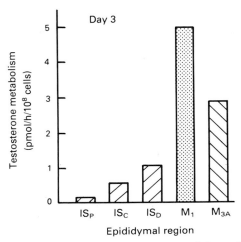

Fig. 8. Metabolism of testosterone to dihydrotesterone and 5α-androstane-3α,17β-diol by intact, cultured ram principal cells from specific sites of the epididymis (see Fig. 2) on Day 3 of culture ($n = 3$–12; D. V. Brown & R. P. Amann, unpublished data). For this study, the initial segment was subdivided into proximal (IS_P), central (IS_C), and distal (IS_D) portions before isolation of principal cells. Principal cells were isolated and cultured as outlined by Wagley *et al.* (1984) and testosterone metabolism was measured essentially as described by Brown & Amann (1984).

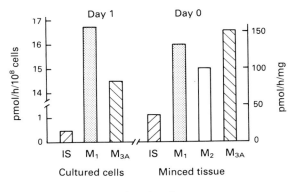

Fig. 9. Metabolism of testosterone to dihydrotestosterone and 5α-androstane-3α,17β-diol by intact and cultured ram principal cells on Day 1 of culture (left) and by fresh minced tissue (right) from specific segments (see Fig. 2) of the same ram epididymides ($n = 3$; D. V. Brown & R. P. Amann, unpublished data).

to dihydrotestosterone, there is slight ($< 10\%$) metabolism of dihydrotestosterone to 5α-androstane-3α,17β-diol; the 3β-epimer and other 5α-reduced metabolites are produced only in trace amounts (R. P. Amann, S. R. Marengo & D. V. Brown, unpublished data). The 5α-reductase in ram principal cells also metabolizes progesterone to 5α-pregnane-3,17-dione and androstenedione to 5α-androstane-3,17-dione; both products are further metabolized. Consequently, progesterone and androstenedione present in rete testis fluid (Ganjam & Amann, 1976) could serve as competing substrates for epididymal 5α-reductase.

R. P. Amann

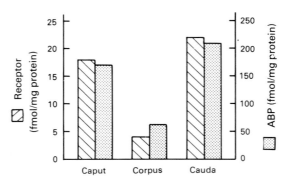

Fig. 10. Concentrations of cytosolic androgen receptor and androgen-binding protein (ABP) in whole tissue from the caput, corpus and cauda of the ram epididymis. From data reported by Carreau *et al.* (1984b).

It is unlikely that ram principal cells synthesize testosterone since, for several regions of the epididymis, they did not convert pregnenolone to progesterone, or 5-androstene-3β,17β-diol or androstenedione to testosterone (R. P. Amann & S. R. Marengo, unpublished data). The androgenic stimulus to the epididymal epithelium must therefore be derived from the blood or the almost 2 μg testosterone entering the epididymis daily in rete testis fluid (Ganjam & Amann, 1976).

Control of epididymal function

If the epididymis contains an androgen-dependent epithelium, the principal cells, or basal cells, should contain androgen receptors which evoke a response. Ram epididymal cytosol binds androgens (Carreau *et al.*, 1984a,b; F. R. Tekpetey & R. P. Amann, unpublished data) and binding proteins for several hormones have been found in rat and rabbit epididymal tissue (Robaire & Hermo, 1987). Although it would be desirable to quantify receptors in principal cells, basal cells and extraductal tissue, only data for whole tissue are available. In mouse tissue, however, [³H]dihydrotestosterone is localized to nuclei of principal cells, especially in the proximal caput and proximal cauda epididymidis (Schleicher *et al.*, 1984). This is consistent with data for the putative androgen receptor in low-salt cytosol from ram epididymal tissue (Fig. 10) and our preliminary data. However, only about 50% of the total androgen receptor per gram of tissue is extracted using low-salt buffer (F. R. Tekpetey & R. P. Amann, unpublished) and so the values reported in Fig. 10 are probably low. Ram epididymal cytosol also contains a protein which specifically binds oestradiol.

The regional distribution of the putative androgen receptor parallels that for androgen-binding proteins (Fig. 10). Androgen-binding protein was quantified by steady-state electrophoresis (Carreau *et al.*, 1984b) and was clearly separated from androgen receptor. Data in Fig. 10, and the 70 nм-dihydrotestosterone present in cauda epididymal plasma (Ganjam & Amann, 1976; Voglmayr *et al.*, 1977), lend credence to the speculation that dihydrotestosterone has a role in function of the cauda epididymidis as well as the distal caput.

Although there are discrepancies in available data (Ganjam & Amann, 1976; Voglmayr *et al.*, 1977; Carreau *et al.*, 1979, 1984b; Jégou *et al.*, 1979) for bulls and rams, the concentration of testosterone plus dihydrotestosterone in rete testis fluid exceeds that of androgen-binding protein (80–155 vs 20–30 nм) so that most of the androgen is presumably transported on the albumin-like protein secreted by Sertoli cells (Amann *et al.*, 1973). In cauda epididymal plasma, however, the concentration of androgen binding protein (> 80 nм) is similar to that of testosterone plus dihydrotestosterone. Androgen-binding protein therefore probably conserves dihydrotestosterone, and testosterone, in the epididymal fluid to maintain concentrations exceeding those in blood, as is true in rats (Turner *et al.*, 1984). Concentrations of androgen-binding protein and steroids in the luminal fluid at different sites along the epididymis are unavailable for domestic ruminants.

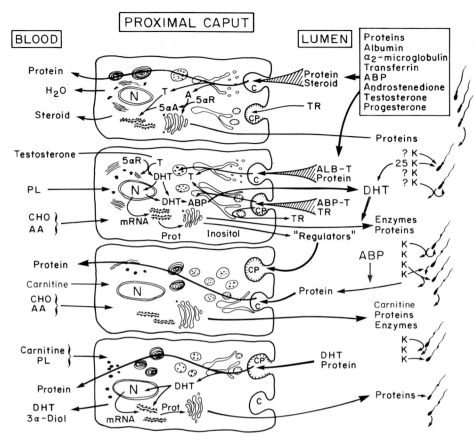

Fig. 11. A hypothetical model depicting the regional and sequential distribution of epithelial functions in the epididymis. See text for explanation.

Integrated hypothesis on regional function

Observations on the morphology of the epididymal epithelium and biochemical analyses of tissue and fluid, together with data for other species (Robaire & Hermo, 1987) are integrated into a conceptualized model (Fig. 11) depicting the possible regionalization of function. It would be naive to assume that all details of the model are correct or that other mechanisms are not involved in regulating epididymal function.

The initial segment is specialized for bulk removal of water, proteins, steroids, electrolytes and other components of rete testis fluid passing through the efferent ducts (Figs 4 & 5). There is a conspicuous difference in the electrophoretic characteristics of proteins in rete testis fluid and luminal fluid from the initial segment (Dacheux & Voglmayr, 1983), and so secretion must also occur in the initial segment (or efferent ducts). These proteins probably initiate sperm maturation. Since the initial segment is proximal to the sites of dihydrotestosterone synthesis (region M_1 and distally), it is unlikely that dihydrotestosterone regulates function of the initial segment in domestic ruminants; a role for testosterone cannot be excluded. The mode of endocytosis in these cells is not established, but principal cells of the initial segment are characterized by unique long arrays of smooth or sparsely-granulated endoplasmic reticulum, which is certainly involved in adsorption.

In region M_1, adsorption of water, proteins, small molecules and electrolytes continues, although the volumes removed by endocytosis are much less than in the initial segment; there is a 30% increase in sperm concentration between region M_1 and M_2 as contrasted with a >90% increase between the initial segment and region M_1. Reduction of the Na:K ratio is almost complete (Fig. 5) and osmotic pressure increases to 370 mosmol/kg (Bech & Koefoed-Johnsen, 1973). There is not a profound change in the spectrum of proteins in luminal fluid between the initial segment and region M_1 (Fig. 6), although the protein concentration is about 3-fold greater (Fig. 4). The principal cells in region M_1 are the most proximal, and primary, site of significant production of dihydrotestosterone (Figs 8 & 9). Thus, adsorption of diverse compounds and especially synthesis of dihydrotestosterone are major functions of region M_1.

Secretion of diverse products into luminal fluid occurs in regions M_2 and M_{3A} (Figs 4, 5 & 6) and may continue in regions M_{3B} or F. Comparing differences between fluids from regions M_1 and M_2, it is evident that the latter contains an extraordinary concentration of carnitine (Fig. 4) and at least five new proteins (Fig. 6). It is speculated that androgen receptor in the caput epididymidis (Fig. 10) actually is localized in region M_2, and present at low concentrations in region M_1. If this is correct, androgen binding protein may help to transport the dihydrotestosterone produced by principal cells in region M_1 to more distal sites such as regions M_2 or F, where it may provide the androgen stimulus necessary for epithelial function. Secretion of specific proteins and glycerophosphocholine, as well as transport of carnitine, are therefore functions of region M_2.

Between regions M_2 and M_3, there is a net loss of protein, but addition of water, electrolytes, glycerophosphocholine and carnitine (Figs 4 & 5). The concentration of protein decreases by almost 50% and osmotic pressure starts to decline. The apical one-third of principal cells in regions M_2 and M_3 contains numerous vesicles of varied electron density, and multi-vesicular bodies, and it is likely that these cells selectively adsorb components from the lumen in addition to their secretory roles. The abundance of lipid droplets in the basal portion of principal cells in regions M_2 and M_3 (Fig. 3) might be related to their uptake of blood lipoproteins which are processed (Hammerstedt & Rowan, 1979) and give rise to secretion of glycerophosphocholine (Fig. 4). Synthesis of dihydrotestosterone occurs in regions M_2 and M_{3A}, but it is also possible that preformed dihydrotestosterone is provided to these cells. Unique proteins appear in luminal fluid from region M_{3A} (Fig. 6), but given the apparent paucity of androgen receptors in the corpus epididymidis secretion of these proteins may not be androgen-dependent. Alternatively, this apparent discrepancy may reflect sub-regional differences in epithelial function and subtle differences in sampling sites in different laboratories.

Between regions M_{3A} and F_1, there is a net loss of water and protein, as well as electrolytes, but some entrance of glycerophosphocholine and carnitine into the ductal fluid (Figs 4 & 5). There may be specific adsorption of selected proteins, since the spectra of proteins in regions M_{3B} and F_1 differ (Fig. 6), or the secretory flux of proteins common with region M_{3B} is sufficient to dilute out some proteins (i.e. proteins of M_r 32 000 and 37 000). The culmination of these events is the maturation of the spermatozoa.

Unique attributes of the environment of the cauda epididymidis which allow prolonged maintenance of fertile spermatozoa have not been identified. The roles of the apparently high concentrations of androgen receptors (Fig. 10), dihydrotestosterone and progesterone (Ganjam & Amann, 1976), and certain proteins (Fig. 6) remain to be established.

Conclusions

During the past 20 years, reproductive biologists have used electron microscopy and microanalytical techniques to improve the descriptions of the epididymal epithelium and its functions, relative to those published 60 years ago. Unfortunately, there has been little improvement in our understanding of the biology of sperm maturation and no attempt to learn how fertile spermatozoa are maintained in the cauda epididymidis. Although it could be argued that epididymal malfunction is not a major problem in domestic ruminants, it is equally defensible to promote the ram as useful for study of epididymal function with the goal of developing a contraceptive for the human male. Regardless of the goal, future research should focus on control mechanisms and biological endpoints rather than phenomenology and data gathering. Only in this way can we hope to overcome our ignorance concerning how the epididymal epithelium, via its secretions, enables spermatozoa to become mature and remain fertile.

Unpublished research reported herein, and preparation of the manuscript, were supported by Grant HD-14,501 from NICHD and Cooperative Agreement CR-812725-01 with the US EPA.

References

Abdel-Raouf, M. (1960) The postnatal development of the reproductive organs in bulls with special reference to puberty. *Acta endocr., Copenh.* **34**, *Suppl.* **49**, 1–109.

Alumot, E., Lensky, Y. & Schindler, H. (1971) Separation of proteins in the epididymal fluid of the ram. *J. Reprod. Fert.* **25**, 349–353.

Amann, R.P. (1981) A critical review of methods for evaluation of spermatogenesis from seminal characteristics. *J. Androl.* **2**, 37–58.

Amann, R.P. & Almquist, J.O. (1962) Reproductive capacity of dairy bulls. VII. Morphology of epididymal sperm. *J. Dairy Sci.* **45**, 1516–1526.

Amann, R.P. & Griel, L.C., Jr (1974) Fertility of bovine spermatozoa from rete testis, cauda epididymidis, and ejaculated semen. *J. Dairy Sci.* **57**, 212–219.

Amann, R.P., Killian, G.J. & Benton, A.W. (1973) Differences in the electrophoretic characteristics of bovine rete testis fluid and plasma from the cauda epididymidis. *J. Reprod. Fert.* **35**, 321–330.

Amann, R.P., Kavanaugh, J.F., Griel, L.C., Jr & Voglmayr, J.K. (1974) Sperm production of Holstein bulls determined from testicular spermatid reserves, after cannulation of rete testis or vas deferens, and by daily ejaculation. *J. Dairy Sci.* **57**, 93–99.

Barker, L.D.S. & Amann, R.P. (1970) Epididymal physiology. I. Specificity of antisera against bull spermatozoa and reproductive fluids. *J. Reprod. Fert.* **22**, 441–452.

Barker, L.D.S. & Amann, R.P. (1971) Epididymal physiology. II. Immunofluorescent analyses of epithelial secretion and absorption, and of bovine sperm maturation. *J. Reprod. Fert.* **26**, 319–332.

Bech, J. & Koefoed-Johnsen, H.H. (1973) Spermiemorfologi og plasmas sammensaetning i bitestikler fra normale tyre samt to tyre med abnormt saedbillede. *Aarsberetning Inst. for Sterilitetsforskning*, pp. 9–30 Kgl. Vet.-og Landbohøjskole, Copenhagen.

Bedford, J.M. (1975) Maturation, transport, and fate of spermatozoa in the epididymis. In *Handbook of Physiology*, Section 7: *Endocrinology*, Vol. 5; *Male Reproductive System*, pp. 303–317. Eds R. O. Greep & E. B. Astwood. American Physiological Society, Washington, D.C.

Benoit, J. (1926) Recherches anatomiques, cytologiques et histophysiologiques sur les voies excrétrices du testicule chez les mammifères. *Archs Anat. Histol. Embryol.* **5**, 175–412.

Besancon, J., Dacheux, J.-L., Paquin, R. & Tremblay, R.R. (1985) Major contribution of epididymis to α-glucosidase content of ram seminal plasma. *Biol. Reprod.* **33**, 296–301.

Brooks, D.E. (1981) Metabolic activity in the epididymis and its regulation by androgens. *Physiol. Rev.* **61**, 515–555.

Brooks, D.E. (1983) Epididymal functions and their hormonal regulation. *Aust. J. biol. Sci.* **36**, 205–221.

Brown, D.V. & Amann, R.P. (1984) Inhibition of testosterone metabolism in cultured rat epididymal principal cells by dihydrotestosterone and progesterone. *Biol. Reprod.* **30**, 67–73.

Carreau, S., Drosdowsky, M.A. & Courot, M. (1979) Age related effects on androgen binding protein (ABP) in sheep testis and epididymis. *Int. J. Androl.* **2**, 49–61.

Carreau, S., Drosdowsky, M.A. & Courot, M. (1984a) Androgen-binding proteins in sheep epididymis: characterization of a cytoplasmic androgen receptor in the ram epididymis. *J. Endocr.* **103**, 273–279.

Carreau, S., Drosdowsky, M.A. & Courot, M. (1984b) Androgen-binding proteins in sheep epididymis: age-related effects on androgen-binding protein, cytosolic androgen receptor and testosterone concentrations. Correlations with histological studies. *J. Endocr.* **103**, 281–286.

Clermont, Y. & Flannery, J. (1970) Mitotic activity in the epithelium of the epididymis in young and old rats. *Biol. Reprod.* **3**, 283–292.

Crabo, B. (1965) Studies on the composition of epididymal content in bulls and boars. *Acta vet. scand.* **6**, Suppl. **5**, 1–94.

Dacheux, J.-L. & Voglmayr, J.K. (1983) Sequence of sperm cell surface differentiation and its relationship to exogenous fluid proteins in the ram epididymis. *Biol. Reprod.* **29**, 1033–1046.

Fournier-Delpech, S., Colas, G., Courot, M., Ortavant, R. & Brice, G. (1979) Epididymal sperm maturation in the ram: motility, fertilizing ability and embryonic survival after uterine artificial insemination in the ewe. *Annls Biol. anim. Biochim. Biophys.* **19**, 597–605. 597–605.

Fournier-Delpech, S., Colas, G. & Courot, M. (1981a) Observations sur les premiers clivages des oeufs intratubaires de Brebis après fécondation avec des spermatozoïdes épididymaires ou éjaculés. *C. r. hebd. Séanc. Acad. Sci, Paris* [III] **292**, 515–517.

Fournier-Delpech, S., Pisselet, C., Garnier, D.-H., Dubois, M. & Courot, M. (1981b) Mise en évidence chez le Bélier d'une préalbumine sécrétée par l'épididyme sous l'action de la Testostérone. *C. r. hebd. Séanc. Acad. Sci, Paris* [III] **293**, 589– 594.

Fournier-Delpech, S., Hamamah, S., Colas, G. & Courot, M. (1983) Acquisition of zona binding structures by ram spermatozoa during epididymal passage. In *The Sperm Cell*, pp. 103–106. Ed. J. André. Martinus Nijhoff, The Hague.

Fournier-Delpech, S., Hamamah, S., Courot, M. & Kuntz, C. (1984) Androgenic control of zona binding capacity of ram spermatozoa. In *The Male in Farm Animal Reproduction*, pp. 103–107. Ed. M. Courot. Martinus Nijhoff, The Hague.

Ganjam, V.K. & Amann, R.P. (1976) Steroids in fluids and sperm entering and leaving the bovine epididymis, epididymal tissue, and accessory sex gland secretions. *Endocrinology* **99**, 1618–1630.

Glover, T.D. & Nicander, L. (1971) Some aspects of structure and function in the mammalian epididymis. *J. Reprod. Fert., Suppl.* **13**, 39–50.

Goyal, H.O. (1985) Morphology of the bovine epididymis. *Am. J. Anat.* **172**, 155–172.

Goyal, H.O. & Vig, M.M. (1984) Histochemical activity of alkaline phosphatase and acid phosphatase in the epididymis of mature intact and androgen-deprived bulls. *Am. J. vet. Res.* **45**, 444–450.

Gunn, R.M.C. (1936) Fertility in sheep. Artificial production of seminal ejaculation and the characteristics of the spermatozoa contained therein. *Bull. Comm. Sci. Industr. Res. Austr.* **94**, 1–116.

Hamilton, D.W. (1972) The mammalian epididymis. In *Reproductive Biology*, pp. 268–337. Eds H. Balin & S. Glasser. Excerpta Medica, Amsterdam.

Hamilton, D.W. (1975) Structure and function of the epithelium lining the ductuli efferentes, ductus epididymidis, and ductus deferens in the rat. In *Handbook of Physiology*, Section 7: *Endocrinology*, Vol. 5, *Male Reproductive System*, pp. 259–301. Eds R. O. Greep & E. B. Astwood. American Physiological Society, Washington, D.C.

Hammerstedt, R.H & Parks, J.E. (1987) Changes in sperm surfaces associated with epididymal transit. *J. Reprod. Fert., Suppl.* **34**, 133–149.

Hammerstedt, R.H. & Rowan, W.A. (1979) Phosphatidylcholine of blood lipoprotein is the precursor of glycerophosphorylcholine found in seminal plasma. *Biochim. Biophys. Acta* **710**, 370–376.

Hinton, B.T. (1985) Physiological aspects of the blood-epididymis barrier. In *Male Fertility and its Regulation*, pp. 371–382. Eds T. J. Lobl & E. S. E. Hafez. MTP Press, New York.

Hinton, B.T., Snoswell, A.M. & Setchell, B.P. (1979) The concentration of carnitine in the luminal fluid of the testis and epididymis of the rat and some other mammals. *J. Reprod. Fert.* **56**, 105–111.

Hinton, B.T., White, R.W. & Setchell, B.P. (1980) Concentrations of *myo*-inositol in the luminal fluid of the mammalian testis and epididymis. *J. Reprod. Fert.* **58**, 395–399.

Huang, H.F.S. & Johnson, A.D. (1975) Comparative study of protein pattern of epididymal plasma of mouse, rat, rabbit and sheep. *Comp. Biochem. Physiol.* **51B**, 337–341.

Jauhiainen, A. & Vanha-Perttula, T. (1985) Acid and neutral α-glucosidase in the reproductive organs and seminal plasma of the bull. *J. Reprod. Fert.* **74**, 669–680.

Jégou, B., Dacheux, J.L., Garnier, D.H., Terqui, M., Colas, G. & Courot, M. (1979) Biochemical and physiological studies of androgen-binding protein in the reproductive tract of the ram. *J. Reprod. Fert.* **57**, 311–318.

Jones, R., Fournier-Delpech, S. & Willadsen, S.A. (1982) Identification of androgen-dependent proteins synthesized in vitro by the ram epididymis. *Reprod. Nutr. Develop.* **22**, 495–504.

Killian, G.J. & Amann, R.P. (1973) Immunoelectrophoretic characterization of fluid and sperm entering and leaving the bovine epididymis. *Biol. Reprod.* **9**, 489–499.

Klinefelter, G.R. & Amann, R.P. (1980) Metabolism of testosterone by principal and basal cells isolated from the rat epididymal epithelium. *Biol. Reprod.* **22**, 1149–1154.

Lino, B.F. & Braden, A.W.H. (1972) The output of spermatozoa in rams. I. Relationship with testicular output of spermatozoa and the effect of ejaculations. *Aust. J. biol. Sci.* **25**, 351–358.

Nicander, L. (1958) Studies on the regional histology and cytochemistry of the ductus epididymidis in stallions, rams, and bulls. *Acta morph. neerl.-scand.* **1**, 337–362.

Nicander, L. (1979) Fine structure of principal cells in the initial segment of the epididymal duct in the ram. *Zentbl. VetMed. C, Anat. Histol Embryol.* **8**, 318–330.

Nilnophakoon, N. (1978) Histological studies on the regional postnatal differentiation of the epididymis in the ram. *Zentbl. VetMed. C, Anat. Histol. Embryol.* **7**, 253–272.

Orgebin-Crist, M.-C. & Jahad, N. (1978) The maturation of rabbit epididymal spermatozoa in organ culture: inhibition of antiandrogens and inhibitors of ribonucleic acid and protein synthesis. *Endocrinology* **103**, 46–53.

Orgebin-Crist, M.-C., Danzo, B.J. & Davies, J. (1975) Endocrine control of the development and maintenance of sperm fertilizing ability in the epididymis. In *Handbook of Physiology*, Section 7: *Endocrinology*, Vol. 5, *Male Reproductive System*, pp. 319–338. Eds R. O. Greep & E. B. Astwood. American Physiological Society, Washington, D.C.

Ortavant, R. (1953) Existence d'une phase critique dans la maturation épididymaire des spermatozoïdes de belier et de taureau. *C. r. Séanc. Soc. Biol.* **147**, 1552–1556.

Pabst, R. & Schick, P. (1979) Proliferation of epithelial cells of vas deferens and epididymis in young adult rabbits. *Archs Androl.* **2**, 183–186.

Robaire, B. & Hermo, L. (1987) Efferent ducts, epididymis and vas deferens: structure, function and their regulation. In *The Physiology of Reproduction*, (in press). Eds E. Knobil & J. D. Neil. Raven Press, New York.

Robaire, B., Scheer, H. & Hachey, C. (1981) Regulation of epididymal steroid metabolizing enzymes. In *Bioregulators of Reproduction*, pp. 487–498. Eds G. Jagiello & H. J. Vogel. Academic Press, New York.

Schleicher, G., Drews, U., Stumpf, W.E. & Sar, M. (1984) Differential distribution of dihydrotestosterone and estradiol binding sites in the epididymis of the mouse. An autoradiographic study. *Histochemistry* **81**, 139–147.

Sinowatz, F. (1981) Ultrastrukturelle und enzymhistochemische Untersuchungen am Ductus epididymidis des Rindes. *Zentbl. VetMed. C, Anat. Histol. Embryol.*, Suppl. **32**, 1–99.

Sun, E.L. & Flickinger, C.J. (1979) Development of cell types and of regional differences in the postnatal rat epididymis. *Am. J. Anat.* **154**, 27–55.

Tröger, U. (1969) Mikroskopische Untersuchungen zum Afubau des Nebenhodenkopfes beim Stier. *Zentbl. VetMed. A* **16**, 385–399.

Turner, T.T., Jones, C.E., Howards, S.S., Ewing, L.L., Zegeye, B. & Gunsalus, G.L. (1984) On the androgen microenvironment of maturing spermatozoa. *Endocrinology* **115**, 1925–1932.

Voglmayr, J.K. & Amann, R.P. (1973) The distribution of free myo-inositol in fluids, spermatozoa, and tissues of the bull genital tract and observations on its uptake by the rabbit epididymis. *Biol. Reprod.* **8**, 504–513.

Voglmayr, J.K., Musto, N.A., Saksena, S.K., Brown-Woodman, P.D.C., Marley, P.B. & White, I.G. (1977) Characteristics of semen collected from the cauda epididymidis of conscious rams. *J. Reprod. Fert.* **49**, 245–251.

von Lanz, T. (1924) Der Nebenhoden einiger Saugetiere als Samenspeicher. *Anat. Anz.* **58**, 106–109.

Wagley, L.M., Versluis, T.D., Brown, D.V. & Amann, R.P. (1984) Culture of principal cells from the ram epididymis. A comparison of the morphology of principal cells in culture and in situ. *J. Androl.* **5**, 389–408.

Waites, G.M.H. (1977) Fluid secretion. In *The Testis*, Vol. 4, pp. 91–123. Eds A. D. Johnson & W. R. Gomes. Academic Press, New York.

J. Reprod. Fert., Suppl. **34** (1987), 133–149

Changes in sperm surfaces associated with epididymal transit

R. H. Hammerstedt and J. E. Parks*

*Program in Biochemistry, The Pennsylvania State University, University Park, PA 16802, and
Department of Animal Science, Cornell University, Ithaca, NY 14853, U.S.A.

Introduction

The paucity of progress towards understanding the functions of the individual gamete membrane systems in terms of their molecular structures is highlighted by the limited data base currently available. This is due in part to those factors that limit all studies of membrane structure–function: our understanding of two-phase chemical systems is quite primitive and assays for specific functional features of membranes are difficult to devise and validate. Superimposed on these general limits are the unique features of the sperm–oocyte system. Currently, it appears that while gametes from various species share a common endpoint (fusion of the haploid cells) they may differ in the molecular details and/or timing of events leading to this fusion. Care must therefore be taken to avoid overgeneralizations with regard to the data obtained from the few systems that have been studied in depth.

With respect to the topic of this specific symposium, the literature review and synthesis of common features of epididymal sperm membrane changes in the domestic ruminants (bull and ram) could be terse, as very limited data are available. Since a truncated presentation limited to reported membrane changes would be of minimal value, we have chosen to supplement the presentation by: (1) expanding the topic to include a more general description of types of molecular questions that should be considered for more rapid progress in this research area; (2) providing examples of apparent conflicts among reported results to reinforce the fact that different methods used to study the same phenomenon may yield different interpretations (at least until the methods themselves are completely understood); and (3) creating an honorary ruminant (the pig) to illustrate the detailed studies that are possible in an animal of great economic interest. The literature included in this review encompasses the period of 1980–1986, with a few references to earlier data in order to assemble the desired model. The interested reader would greatly benefit from review articles prepared by others and listed in Table 1.

Alteration of the sperm surface certainly occurs at all points of cellular transit from the testis through to penetration of the oocyte. A complete description of the role of the efferent, epididymal and deferent ducts is, therefore, necessary. Ruminant models for these studies will be most useful because of the large numbers of cells available from a single animal and the ability to cannulate the various portions of the male reproductive tract (see comparisons in Hammerstedt, 1981). Studies utilizing the ruminant are less frequent than those of rodents and most current concepts of the requirements for important biological events (e.g. the acrosome reaction and capacitation) stem from studies using spermatozoa from the cauda epididymidis. Such studies neglect the role of the accessory sex gland fluids on sperm function.

A systematic comparison of the features of cauda epididymal and ejaculated spermatozoa has been made in this laboratory, demonstrating important differences in sperm motility (Inskeep *et al.*, 1985), metabolism (Hammerstedt, 1981; Inskeep & Hammerstedt, 1982; Inskeep *et al.*, 1985) and membrane features (Hammerstedt *et al.*, 1976, 1979a, b, 1982). Use of data from studies utilizing cauda epididymal spermatozoa to predict molecular mechanisms for true physiological events are viewed as suspect. Such considerations argue for an increased use of species (ruminant, pig or rabbit) from which ejaculates can be easily obtained for studies of those events related to processing within the female reproductive tract.

Table 1. Recent reviews of sperm surface changes as related to epididymal transit

Topic emphasized	Reference
Carbohydrate determinants	Ahuja (1985)
Immunochemical characterization	Eddy *et al.* (1985)
Post-testicular sperm maturation of the boar	Crabo (1985)
Domains of sperm membranes	Koehler (1985)
General sperm surface aspects	Austin (1985)
Sperm membrane antigens	O'Rand (1985)
Overview of membrane features	Fraser (1984)
Sperm membrane features, with an emphasis on development of function	Holt (1982)
General sperm surface changes	Olson & Orgebin-Crist (1982)
General discussion of the sperm surface, with emphasis on the mouse	Koehler (1982)
General discussion of the sperm surface	Nicolson (1982)
Ultrastructural and cytochemical analysis of sperm membranes	Fléchon (1981)

Molecular considerations in membrane modification

The surface of the sperm cell must be viewed from at least two perspectives in order to appreciate the complexity of the system. The first perspective is one based on microscopic considerations and the second is based on deductions from chemical and physical analyses. These considerations allow comment on interpretation of membrane changes and experimental design.

The microscopic perspective

Similarities in the general organization of the spermatozoon from various species (Fig. 1) are apparent, and functions can be related to the distinct zones of the cell, e.g. motility to the tail, flux of metabolic substrates and end products to the middle-piece, oocyte binding and penetration to the anterior head. Existing data further enhance the concept of an organized and specialized surface in that heterogeneous distributions of proteins (Eddy *et al.*, 1985) and lipids (Wolf & Voglmayr, 1984; Schlegel *et al.*, 1986) have been described.

Two experimental approaches are required to determine the significance of such heterogeneity. The first would seek to establish the relationship between chemical features of one of the unique surface regions and the function(s) of that region. This requires a carefully validated assay for the function of interest, plus the ability to isolate experimentally the membrane components and complete the necessary chemical analysis of the isolate. Results of several analyses of isolated membrane components from the surface region overlying the acrosome have recently been published (Parks & Hammerstedt, 1985; Nikolopoulou *et al.*, 1985) but these data have not yet provided an explanation for the unique function(s) of that surface region. The second type of experiment characterizes the membrane features that allow heterogeneous surfaces to be formed and persist during extended periods of sperm transit through and storage within the male tract (Peterson & Russell, 1985). While the existence of such macromolecular domains is evident, available data do not satisfactorily describe any mode of accomplishing this segregation. One possibility is establishment of anchors, i.e. restriction of movement via interaction of surface components with integral membrane components and/or cytoplasmic components that bind to internal membrane surface. The second would be via formation of collars—as yet unidentified membrane–membrane junctions that define zones of heterogeneity. Implicit in either of these or any other mechanism is the requirement for continuous energy input to maintain a state of non-equilibrium. However, conventional wisdom assigns the ATP-generating systems to the middle-piece (see Mann & Lutwak-Mann, 1981) and mechanisms to transfer ATP to distant regions of the cell, e.g. the anterior head, have yet to be proposed.

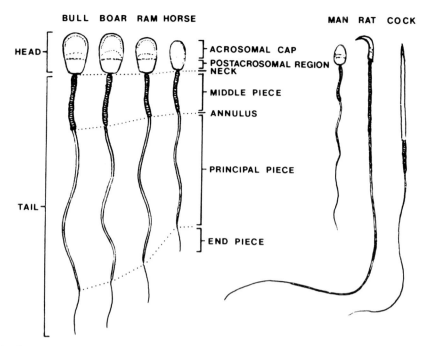

Fig. 1. Representative drawings of spermatozoa from various species, illustrating the general similarities with regard to the zones of the cell surface, i.e. anterior and posterior head, middle-piece and tail. Taken from p. 168 of Hafez (1980).

The submicroscopic perspective

The second perspective is one based on submicroscopic detail (Fig. 2). In this case studies are guided by rapidly evolving concepts of membrane structure as deduced from studies (reviewed in a series edited by Benga, 1985) of model membrane systems and selected cellular systems (e.g. the erythrocyte). Cytoplasmic domains, lipid bilayer integral components (both protein and lipid), lipid–water interfaces, protrusions of bilayer components into extracellular water and adsorbed components all interact to yield the functional unit of any surface zone.

Membrane modification is not trivial for any cell type, and the sperm cell probably has fewer options available than most cells because of its limited biosynthetic capacity for both lipids and proteins. To a first approximation this limits the possibilities to: (1) exchange of components between interior membrane components and the surface membranes; (2) lateral movement of components from one zone of the surface to another (within the restrictions imposed by anchors and collars); and (3) exchange of components between the surrounding milieu (epididymal lumen, seminal plasma, uterine and oviducal fluids) and the sperm surface. Addition or deletion of components from within the cell probably can occur via a minimally complex mechanism because the cytoplasm and inner surface of the membrane are contiguous. Transfer of lipid, for example, might only require a lipid transfer protein of the type found in many different somatic cells (reviewed by Wirtz, 1982; Helmkamp, 1983; Zilversmit, 1983). Transfer between spermatozoa and extracellular fluids requires an external approach to the bilayer surface and can be accomplished only after passage through a variety of extracellular components.

Any component destined to be moved to the cell and inserted as an integral membrane protein must display the solubilities, at the appropriate time, for both the aqueous fluids and the non-polar bilayer. Polar features are inherent in basic protein structure, but special post-translational modifications might be necessary for the non-polar aspects. Recent descriptions of events occurring in

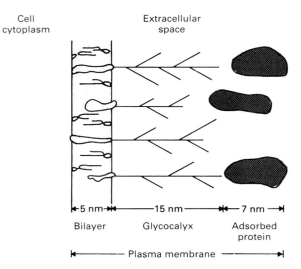

Fig. 2. Schematic representation of the plasma membrane, showing the scale of the bilayer and the types of molecules that could be associated with the exterior face of the membrane. It is important to note that the plasma membrane consists of much more than the bilayer.

somatic cell systems provide models for such complex post-translational modifications. One possibility is fatty acylation (Wold, 1986) whereby fatty acyl residues are added to the basic protein structure to yield a molecule with high affinity for the membrane. A second possibility is derivatization by transfer of phosphatidylinositol (Low *et al.*, 1986). This mechanism could be especially important in view of the changes in free inositol content of the epididymal lumen that have been described by others (reviewed by Amann, 1987), since the role of inositol in sperm maturation remains an enigma.

Interpretation of membrane changes

Empirical observations of the disappearance and appearance of membrane components are useful, but must ultimately be interpreted in terms of the membrane location relative to the bilayer (see Fig. 2). It is useful to remember that appearance may be due to addition to the most exterior aspect of the surface (adsorbed components) or unmasking of previously masked (to the test reagent) sites. Disappearance can be due to complete removal of the component or masking of the determinant of interest via addition of another moiety. To date no study has described mechanisms at this level.

Relationship to experimental design

Selected biological events may be uniquely dependent upon those aspects of the membrane displayed in Fig. 2. For example, acquisition by spermatozoa of the ability to bind to the oocyte may be related to an alteration of molecular features of the outermost aspects of the surface. In contrast, the fusion of plasma and outer acrosomal membrane during the acrosome reaction may be dependent on the molecular features of the inner surfaces of the plasma membrane network. Such considerations are necessary to the design of experiments to study these very different but related events in sperm–oocyte fusion.

Analysis of apparently contradictory results

Techniques usually are developed in model systems and then are applied to the study of more complex biological systems. Often, as information accumulates, it becomes apparent that the earlier interpretations must be carefully re-evaluated. Two examples of this, evaluation of membrane fluidity and analysis of lectin binding data, are presented below to illustrate the effect of choice of method on the perception of membrane characteristics.

Membrane fluidity

The difficulties of defining, much less measuring, membrane fluidity have been discussed in detail by others (see Lands, 1980). In a single phase and chemically defined medium the definition is simple and often is taken as the reciprocal of the measured viscosity; as such the viscosity can be evaluated by classical hydrodynamic techniques. For more complex systems, particularly biological membranes, greater care must be taken to define explicitly the desired feature. Current awareness of general membrane structure has introduced the consideration of motion in any of the three dimensions of the membrane surface; rotations perpendicular to the surface plane, lateral movement along the surface plane, and movement through the plane. Careful definition of the type of motion measured in each experiment must be given.

Another aspect that must be considered is the time scale of the measurement, since various spectroscopic techniques (nuclear magnetic resonance *vs* electron paramagnetic resonance) evaluated motional parameters on very different time scales. Usually, the latter methods involve adding trace amounts of spectroscopically detectable compounds to the cell, allowing them to bind to and segregate into the various lipid-containing domains and evaluating the distribution of the compound. Then conclusions can be reached, via comparisons to other systems that have been studied, with regard to the freedom of motion of the trace component in the membrane.

Results from three different studies of ram spermatozoa, as a function of epididymal transit, are compared to illustrate different types of conclusions that can be reached using different techniques.

Evaluation by relative packing density of a membrane probe. Schlegel *et al.* (1986) utilized the impermeable fluorescent membrane probe merocyanine 540 (MC540), which displays preferential binding to loosely packed phospholipids in model membranes, to assess changes in distribution and packing of lipids of the sperm surface as a function of epididymal maturation. Differential binding was noted, with the most intense fluorescence associated with the posterior portion of the sperm head and the middle-piece. This is consistent with the conclusion that these portions of the surface are loosely packed relative to other portions of the surface, i.e. these zones are more fluid. Since the probe cannot penetrate the bilayer, results gathered using this technique reflect the properties of the outer half of the plasma membrane layer. It is also important to note that this is a static method whereby dye partitioning is allowed to come to equilibrium and then the distribution of dye is subjectively evaluated by visual examination with the fluorescence microscope. No difference in dye distributions was detected for ram spermatozoa as a function of epididymal transit although differences were found for mouse spermatozoa. These results led to the conclusion that the lipid packing at the outer face of the bilayer (believed to be directly related to fluidity) differs across the various surface domains of the ram spermatozoa, but does not change during sperm maturation within the epididymis.

Evaluation by diffusion coefficient of a membrane probe. Wolf & Voglmayr (1984) used a different technique to evaluate motional properties in the membrane surface. An impermeable carbocyanine dye (C16diI) that partitions into the outer half of the bilayer of the plasma membrane was used and the partitioning of the dye was evaluated by visual comparison of the fluorescence patterns and also quantified by the technique of fluorescence recovery after photobleaching. This combination of measurements allows a comparison of the quantity of dye in any surface region (by the visual estimate) and the motional properties of the dye molecule in the plane of the membrane (by measurement of the diffusion constant using the photobleaching technique). A similar distribution pattern to that reported by Schlegel *et al.* (1986) was described. Comparison of testicular spermatozoa (those cells that have been released from the testis but have not been exposed to the epididymal environment) and ejaculated spermatozoa (complete epididymal maturation plus mixture with the accessory sex gland fluids) revealed differences in the estimated diffusion constant for C16diI. Specifically, all zones of the sperm surface except that of the middle-piece were altered to yield an environment where the dye had an increased diffusion constant; movements in the plane of the membrane surface occurred more rapidly. Wolf *et al.* (1986) have reported other results for the motional properties of membrane proteins in ram spermatozoa.

Evaluation by partition coefficient of a membrane probe. In contrast to the above techniques, which utilize intact spermatozoa and microscopic evaluation, selective membrane removal by cavitation treatment yields a preparation of membrane vesicles derived from the plasma membrane overlying the anterior portion of the sperm head. Such preparations have been analysed for changes in membrane composition (Parks & Hammerstedt, 1985) and now are available for analysis by appropriate spectroscopic techniques.

Electron paramagnetic resonance spectroscopy can be used to analyse the motional properties of nitroxyl spin probes in the membrane (see Hammerstedt *et al.*, 1976, for an overview of the application of the technique to sperm systems). Detailed evaluations of motional properties of such spin probes in a variety of membrane preparations have been presented and extensively analysed (Keith *et al.*, 1973; Marsh, 1981; Schwartz, 1982; Hemminga, 1984; Devaux & Seigneuret, 1985), but a detailed consideration (see Polnaszek *et al.*, 1978) has clearly outlined some of the pitfalls inherent in these interpretations. Results of a more restricted analysis are presented herein.

Systematic changes in the structure of the spin probe yields a series of compounds with different partition coefficients between the aqueous phase of a membrane vesicle suspension and the lipid phase of that system. This partitioning can be assessed by the spectroscopic features of the preparation, and the change in partitioning, as a function of temperature, can be readily assessed (Fig. 3). The spin probe 5-doxyldecane (Molecular Probes, Inc., Junction City, OR), at a ratio of

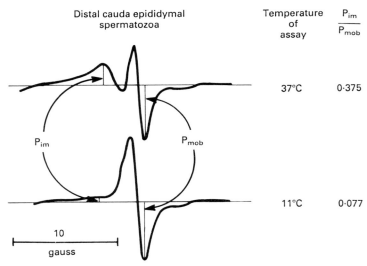

Fig. 3. Sample electron paramagnetic resonance spectra of the nitroxyl spin label 5-doxyldecane in a sperm membrane preparation. The high-field line of the spectrum for 5-doxyldecane in a ram sperm plasma membrane vesicle suspension at two temperatures is shown. The change in partitioning associated with temperature is reflected in the ratio of immobilized (P_{im}) and freely mobile (P_{mob}) aspects of the spectroscopic signal. These two peaks are proportional to the amount of 5-doxyldecane in the membrane lipid and in bulk water.

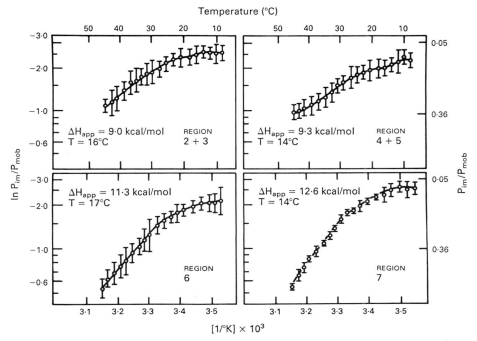

Fig. 4. Changes in the partitioning of the spin label 5-doxyldecane into plasma membrane vesicles of the anterior zone of ram spermatozoa. Sperm membrane preparations were isolated as described by Parks & Hammerstedt (1985) and the partitioning of 5-doxyldecane calculated as outlined in Fig. 3. Membrane vesicles were isolated from spermatozoa from the caput (Region 2 + 3), corpus (Region 4 + 5), proximal cauda (Region 6) and distal cauda plus vas deferens (Region 7). These correspond to the regions M2, M3a, F1 and F2 as defined by Amann (1987). The data for each region are presented in the form of an Arrhenius plot, with an estimate via computer analysis (Jones & Molitoris, 1984) of the temperature dependence of the partitioning (m) and apparent phase transition temperature (T). The values for the apparent enthalpy change (ΔH_{app}) (as estimated for the data between 20 and 45°C) increases with transit through the epididymis while no significant change was noted for the range of 8 to 20°C. The points represent the mean $\pm$ s.e.m. for 4 complete replicates.

20 nmol spin label per 150 nmol membrane phospholipid, yielded appropriate partitioning for analysis of ram sperm membrane vesicles. The spectral component labelled P_{im} relates to the relatively immobilized component (dissolved in the membrane) while the component P_{mob} relates to the spin label residing in bulk water. An Arrhenius plot [(ln P_{im}/P_{mob}) *vs* (1/K)] will yield a single straight line if temperature-induced changes in solubility of the spin label in lipid relative to that of water are not abruptly altered by changes in the lipid vesicle (i.e. by lipid phase changes). Changes in the slope of the line, as a function of temperature, reflect changes in the ability of the lipid phase to dissolve spin label and have generally been interpreted to mean changes in the bulk fluidity of the preparation. This refers to an average fluidity of total bilayer lipid and is not directly related to any specific surface distribution of the lipid.

Changes in this measure of fluidity were observed in vesicles isolated from ram spermatozoa (Fig. 4). The slight changes in apparent transition temperature (14–17°C) probably are not significant but the increased percentage of 5-doxyldecane that partitions into the membrane and the increased apparent enthalpy change (ΔH_{app}) are striking and reflect an abrupt change in membranes isolated from cells after transit into the proximal portion of the cauda epididymidis. To a

Fig. 5.

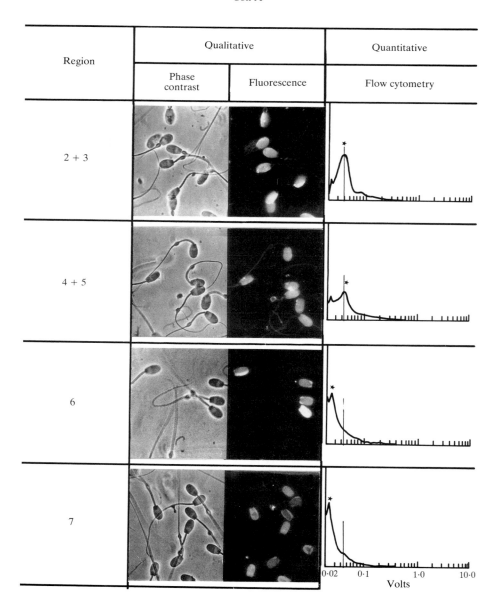

first approximation the plasma membrane overlying the acrosome becomes more fluid during epididymal transit, perhaps due to the large increase in the degree of unsaturation in the acyl groups of the phospholipids (Parks & Hammerstedt, 1985). Since the ability to fertilize ova is acquired at this same point in epididymal transit, a preliminary conclusion might be that the increase in fluidity is related to this aspect of membrane function.

A further extension of these data (not presented) to include comparison of membranes from cauda epididymal spermatozoa to those of ejaculated spermatozoa revealed the surprising result that the membranes of ejaculated spermatozoa changed in their fluidity features to resemble those of caput epididymal spermatozoa. This calls in to question the preliminary conclusion above that bulk membrane fluidity is uniquely related to some sperm membrane function, such as its ability to undergo an acrosome reaction. Interpretation of these latter changes awaits completion of the chemical analyses, but supports our earlier contention that extrapolation of data from cauda epididymal to ejaculated spermatozoa may be misleading. It is also important to remember that the effects of uterine and/or oviducal fluids on the membranes have yet to be evaluated.

Use of lectins to evaluate surface glycosyl residues

Lectins, a group of plant and invertebrate proteins that interact with carbohydrate components, can be used as probes for changes in the surface features of spermatozoa. Concanavalin A (con A) detects changes in ram spermatozoa in glycosyl residues (highest specificity for alpha mannosyl residues) associated with epididymal transit. With reference to Fig. 2, these reagents would be used to detect the appearance and disappearance of the determinant on all aspects of the exterior surface (glycolipid at the bilayer surface, glycoproteins of glycocalyx or adsorbed protein) that are accessible to the lectin.

Evaluation by lectin-induced agglutination. Hammerstedt *et al.* (1982) utilized the bivalent features of con A to induce a specific agglutination of cell suspensions. A careful titration was conducted to establish a lectin to sperm ratio that allowed specific agglutination (a head-to-tail pattern that was blocked by addition of an inhibitory sugar) to be evaluated for all sperm cell types tested. They concluded that caput spermatozoa agglutinated readily, corpus spermatozoa agglutinated to a lesser extent, and that the agglutination property was restored for cauda spermatozoa.

Evaluation by saturation analysis. Fournier-Delpech & Courot (1980) have studied lectin binding by caput, corpus and cauda epididymal ram spermatozoa and reached a different conclusion. Isotopically labelled con A and saturation or Scatchard plots were used to quantitate the number of binding sites per sperm cell; a decrease in sites per cell was detected. These reports further establish, using lectin affinity columns, that the molecular nature of the surface components changes during transit.

Fig. 5. Representative examples of FITC–con A binding to ram spermatozoa. Spermatozoa from regions of the epididymis (as defined in Fig. 4) were analysed for the distribution of lectin over the various zones of the sperm surface by microscopy (phase contrast and epifluorescence; left panel) and amount of lectin bound by the cell analyser (right panel). The photographs were taken under conditions where the optimum image for each field was recorded. This results in a valid comparison of intensities within each field (i.e. zones of the spermatozoa in that field) but does not yield valid information across fields (i.e. comparison of spermatozoa at different stages of maturation). Such quantitative comparisons are evident in the right panel where the gain of the amplifier was adjusted so that unstained cells do not appear on the output. The entire distribution (presented on a log scale) therefore illustrates the range of staining intensities but not the percentage of cells that were stained. This representative trace (>20 000 cells evaluated per sample) illustrates the decrease in staining intensity for spermatozoa isolated from the various regions of the same epididymis. An equivalent trend was seen in 5 other replicates.

The apparent contrast in conclusions from these studies and those of Hammerstedt *et al.* (1982) may be due to one of several factors. The first relates to the host of surface features that can influence the ability of a lectin to agglutinate cell suspensions. They include: (1) the total number of saccharide binding sites per cell; (2) binding constants of different sites for the same lectin; (3) the number of lectin–saccharide interactions that are needed for cell–cell agglutination; (4) mobility of lectin 'receptors' in the membrane; (5) interference of other cell-surface components; (6) net charge on the glycocalyx; and (7) possible influence of cytoskeletal elements. A second possibility resides in the fact that it has been very difficult to establish whether lectin binding follows the necessary quantitative features for Scatchard analysis. Susko-Parrish *et al.* (1985) carefully tested the binding features of isotopically labelled succinylated con A to ejaculated bull spermatozoa. This con A derivative is monovalent and allows the optimum opportunity to study binding without the complicating effects of agglutination. They concluded that the binding kinetics were too complex to allow any estimate of number of binding sites per cell for this cell type. If these conditions apply to ram epididymal spermatozoa, the apparent contradiction may be explained. The observance by Fournier-Delpech & Courot (1980) of differences in the molecular nature of the components isolated by lectin affinity columns still remains as strong evidence for alteration in the cell surface associated with epididymal transit.

Evaluation by flow cytometry. In an attempt to circumvent the above problems in quantitating the number of lectin binding sites per cell, we (S. Magargee, E. Kunze & R. H. Hammerstedt, unpublished observations) have adopted a different means of analysis. Fluorescently labelled con A distribution over the surface of ram spermatozoa was evaluated by microscopic analysis to assess the zones of the surface that were labelled (i.e. head *vs* tail). The amount of lectin bound to the cells was also quantified by passing the suspension through an Epics V flow cytometer (Coulter Electronics, Inc., Miami, FL, U.S.A.). A series of initial experiments tested a wide range of lectin to sperm ratios to establish conditions in which no cell type (caput, corpus or cauda) was saturated. This fixed ratio was used to test all cell types because discrimination in extent of binding is possible when all assays are tested under these subsaturating conditions.

Representative data (Fig. 5) for con A binding are presented. The left panel presents the distribution over the cell surface, where lectin is predominantly associated with the entire head but not the middle piece or tail of the spermatozoa at all stages of development. The right panel presents a quantitative evaluation of the amount of lectin per cell, as detected by the cell analyser. The population is somewhat heterogeneous, with a progressive decrease in fluorescence intensity noted as spermatozoa traverse the epididymis.

Recent studies in sperm surface changes associated with epididymal maturation

Results of recent efforts to examine changes on the sperm surface as a function of epididymal maturation have been assimilated, by species, into Tables 2–4. These reports are primarily descriptive in nature and have been categorized according to the type of membrane component(s) that were modified during epididymal transit. Changes in proteins therefore include modifications in the quantity and distribution of integral membrane proteins, adsorption or modification of proteins from testicular and epididymal fluids and changes in the specific activity of membrane associated enzymes. Carbohydrate changes refer to modifications in the carbohydrate moieties of glycoproteins and glycolipids initially part of the sperm surface as well as changes mediated by interaction with epididymal fluid components. Reports of modifications in sperm membrane lipids include changes in composition of the major classes of sperm plasma membrane lipids (phospholipids, sterols and phospholipid-bound fatty acyl chains) as well as the resulting physical changes in the sperm surface.

It is apparent that information on surface modifications to bull spermatozoa during epididymal maturation is very limited (Table 2). The ram may prove to be a very useful ruminant model for the

Table 2. Recent reports of sperm surface changes of bulls as related to epididymal transit and ejaculation*

PROTEINS

†1. Identification of ^{125}I-labelled surface proteins of M_r 16 000–100 000 on proximal caput spermatozoa by SDS-PAGE; loss of components of M_r 15 000–18 000 in distal cauda with decrease in components of M_r 90 000–100 000 and increase in those of M_r 42 000–47 000 (Vierula & Rajaniemi, 1981, 1982).

†2. Protein methylesterase activity 20-fold higher in caput than cauda spermatozoa (Gagnon *et al.*, 1984).

†3. Adenylate cyclase activity 4-fold greater in caput than cauda spermatozoa (Casillas *et al.*, 1980).

†4. Forward motility protein binds to caput spermatozoa (Acott & Hoskins, 1981).

‡5. Identification of antigenic determinants on washed, ejaculated spermatozoa by indirect immunofluorescence of monoclonal antibodies, not present on testicular or epididymal spermatozoa (Chakraborty, 1983).

CARBOHYDRATES

‡1. No difference in con A binding in acrosomal or post-acrosomal region between caput and cauda spermatozoa using colloidal gold technique; decreased con A binding on flagellum of cauda spermatozoa; no change in WGA binding except for increased intensity in distal, post-acrosomal region (Sinowatz & Friess, 1983; Friess & Sinowatz, 1984).

‡2. Decreased binding of rhodamine-conjugates of con A and WGA over acrosomal and post-acrosomal regions between caput and cauda; decrease in WGA but not con A binding to flagellum; increase in PNA binding in acrosomal region between caput and cauda (Arya & Vanha-Perttula, 1985).

†3. Epididymal fluid β-*N*-acetylglucosaminidase activity increased between caput and cauda; bound to epididymal spermatozoa (Jauhiainen & Vanha-Perttula, 1986).

LIPIDS

1. No reports

*Report describes an alteration that directly or indirectly affects the indicated component. Carbohydrate refers to changes in glycoprotein or glycolipid.
†Alteration detected via chemical or physical analysis.
‡Alteration detected via microscopic analysis.

Table 3. Recent reports of sperm surface changes of rams as related to epididymal transit and ejaculation*

PROTEINS

†1. Identification of ^{125}I-labelled surface proteins of M_r 78 000–132 000 on testicular spermatozoa by 2-dimensional SDS agarose–PAGE; lost by entry into proximal caput where components of M_r 17 000–65 000 appear; little change from distal caput to proximal cauda where components of M_r 4000, 65 000 and 97 000 become prominent and protein of M_r 24 000 appears; no direct relationship established between changes in fluid proteins and sperm surface proteins, although testicular spermatozoa adsorbed protein of M_r 24 000 from cauda fluid; many components identified as glycoproteins; some radioactivity from both integral and adsorbed proteins inaccessible to trypsin, suggesting internalization (Voglmayr *et al.*, 1980, 1982; Dacheux & Voglmayr, 1983).

†2. Detection of surface protein of M_r 90 000 accessible to ^{125}I-labelling on proximal caput spermatozoa by SDS-PAGE; appearance of additional proteins of M_r 24 000, 33 000, 45 000 and 67 000 on distal cauda spermatozoa (Vierula & Rajaniemi, 1982).

†3. Surface proteins (from M_r 13 500–145 000) of testicular spermatozoa ^{125}I-labelled and analysed by SDS-PAGE; proteins of M_r 72 000–88 000 lost in proximal caput, protein of M_r 17 500 lost in proximal corpus and proteins of M_r 108 000–115 000 lost in distal corpus; component of M_r 145 000 appeared in caput and components of M_r 13 500 and 66 000 appeared in corpus (Dacheux *et al.*, 1985).

‡4. Distribution of colloidal gold-labelled corpus fluid proteins found over anterior head of testicular spermatozoa, entire head of corpus spermatozoa, and posterior head except for the equatorial segment of cauda spermatozoa (Courtens *et al.*, 1982).

†‡5. Partial purification of a polymorphic antagglutinin from cauda fluid; reversed autoagglutination of corpus spermatozoa (Dacheux *et al.*, 1983).

†6. Activity of *myo*-inositol-1-phosphate synthase 3-fold higher in caput than corpus or cauda spermatozoa (Loewus *et al.*, 1983).

†7. Plasma membrane Ca^{2+}-ATPase activity 3-fold higher in ejaculated than cauda spermatozoa; 3-fold activation of enzyme in cauda spermatozoa by seminal protein calsemin (Bradley & Forrester, 1981).

CARBOHYDRATES

†1. Identification of surface glycoproteins of M_r 98 000 and 102 000 on testicular spermatozoa by galactose oxidase–$NaBH_4$ and 2-dimensional SDS agarose–PAGE; retained in cauda spermatozoa with addition of components of M_r 24 000, 35 000 and 350 000; loss of glycoproteins of M_r 24 000 and 35 000 from ejaculated spermatozoa with

continued

Table 3—*continued*

    component of M_r 350 000 very prominent; identification of surface sialoglycoproteins of M_r 79 000, 98 000 and 102 000 on testicular spermatozoa by $NaIO_4$–$NaBH_4$; not present on cauda or ejaculated spermatozoa where sialylated components of M_r 15 000 and 48 000 appear (Voglmayr *et al.*, 1983).

†2. Identification of surface glycoproteins of M_r 26 000–600 000 on testicular spermatozoa as above; only components of M_r 280 000–600 000 were detected upon transit into the proximal corpus; glycoproteins of M_r 350 000–600 000 very prominent in ejaculated spermatozoa where components of M_r 95 000 and 97 000 appear; similar pattern after neuraminidase treatment plus appearance of components of M_r 24 000–97 000 on cauda and ejaculated spermatozoa; surface sialoglycoproteins distributed similarly on testicular spermatozoa with fewer components of M_r 90 000–155 000; loss of components of M_r 280 000–600 000 by proximal corpus with appearance of sialoglycoproteins of M_r 24 000–36 000 on cauda and ejaculated spermatozoa (Voglmayr *et al.*, 1985).

†3. Identification of ^{125}I-labelled surface glycoproteins with affinity for con A by con A-agarose followed by acrylamide–agarose chromatography; major component of M_r 110 000 and minor components of M_r <20 000 on corpus and cauda but not testicular spermatozoa; ^{3}H-acetyl-con A binding sites on distal corpus spermatozoa only 25% of those on mid-corpus with no further decline on cauda spermatozoa (Fournier-Delpech & Courot, 1980).

†‡4. Identification of surface glycoproteins of M_r 71 000 and 117 000 with con A and WGA affinity as above; one distributed over acrosomal region of testicular spermatozoa but not present in corpus; one first appeared over post-acrosomal region in mid-corpus (Fournier-Delpech *et al.*, 1984).

‡5. Corpus fluid glycoprotein with affinity for con A bound to acrosomal region of testicular but not cauda spermatozoa, detected using a colloidal gold technique (Courtens *et al.*, 1982).

‡6. Head-to-head and head-to-tail agglutination of spermatozoa induced with RCA and UEA, head-to-tail with con A and WGA; degree of agglutination with RCA and UEA decreased between caput and corpus; con A agglutination was lower for corpus than caput or cauda spermatozoa; WGA agglutination increased between caput and cauda; FITC–con A distributed over entire sperm surface throughout the epididymis, but lectin bound per cell decreased based on cell fluorescence analysis (Hammerstedt *et al.*, 1982; S. Magargee, E. Kunze & R. H. Hammerstedt, unpublished results discussed in text).

†‡7. Negative surface charge assessed by quantitative analysis of colloidal iron hydroxide binding; binding much greater over surface of cauda spermatozoa; increase in negative charge almost entirely due to sialic acid residues, as assessed by neuraminidase treatment (Holt, 1980).

†8. Isoelectric focussing pH of 5·2–5·4 found for testicular and proximal caput spermatozoa, 5·0 for central caput, 4·7–5·0 for distal cauda spermatozoa and 4·8 for ejaculated spermatozoa (Hammerstedt, 1979; Hammerstedt *et al.*, 1979b, 1982).

‡9. Testicular spermatozoa did not bind cumulus-free rat oocytes; 50% of oocytes bound by mid-corpus spermatozoa; all oocytes bound by cauda spermatozoa with >25 spermatozoa/oocyte; spermatozoa intact and bound to unpenetrated zona by plasma membrane over acrosomal region (Fournier-Delpech *et al.*, 1983).

LIPIDS

†1. Comprehensive lipid analysis of plasma membrane overlying acrosome via HPLC and GLC; changes between caput and cauda spermatozoa included decrease in desmosterol to cholesterol ratio; increase in cholesterol to phospholipid ratio; high choline plasmalogen content with increase in choline to ethanolamine phosphatide ratio; increase in ratio of phospholipid-bound C22:6/C16:0, C22:6 was major acyl component of phosphatidylcholine (Parks & Hammerstedt, 1985).

†‡2. Diffusion coefficient of membrane lipid analogue C16diI determined for sperm plasma membrane by fluorescence recovery after photobleaching; coefficient highest in midpiece, intermediate in head and lowest in flagellum of testicular and ejaculated spermatozoa; coefficient was greater for ejaculated than testicular spermatozoa in all surface regions except midpiece, indicating increased membrane fluidity (Wolf & Voglmayr, 1984).

†3. Based on EPR measurements of surface directed spin labels, negative charge density increased between testicular and cauda spermatozoa and between cauda and ejaculated spermatozoa (Hammerstedt *et al.*, 1979b).

†4. Partitioning of lipid-directed spin labels into sperm plasma membrane vesicles was evaluated for membranes recovered from the anterior portion of the head, overlying the acrosome (Hall & R. H. Hammerstedt, unpublished results described in text).

†5. Partitioning of the impermeable membrane probe MC540 was evaluated for epididymal spermatozoa; no differences were detected over epididymal transit (Schlegel *et al.*, 1986).

†6. Analysis of goat spermatozoa; activities of sperm phospholipase(s) A and lysophospholipase were 4-fold higher in caput spermatozoa than cauda spermatozoa; phosphatidyl-ethanolamine, -choline, -inositol and phosphatidic acid were hydrolysable by sperm phospholipase; lysophospholipase activity was 10-fold higher than phospholipase in caput and cauda spermatozoa (Atreja & Anand, 1985).

---

*Report describes an alteration that directly or indirectly affects the indicated component. Carbohydrate refers to changes in glycoprotein or glycolipid.

†Alteration detected via chemical or physical analysis.

‡Alteration detected via microscopic analysis.

Table 4. Recent reports of sperm surface changes in boars as related to epididymal transit and ejaculation*

PROTEINS

†1. Comprehensive analysis of cauda epididymal and ejaculated sperm plasma membrane proteins and proteins of epididymal and individual accessory gland fluids by 2-dimensional SDS-PAGE with Coomassie blue and silver staining; no major differences in caput and cauda plasma membrane proteins; proteins of M_r 16 000 and 24 000 only major components common to epididymal fluid; acquisition and loss of several minor proteins between caput and cauda; major plasma membrane proteins of M_r 14–20 000, 110 000, 115 000 and 300 000 added by seminal plasma (Russell *et al.*, 1984).

†2. Identification of ^{125}I-labelled surface proteins of M_r 13 000–170 000 on testicular spermatozoa by 2-dimensional SDS-PAGE; loss of components of M_r 33 000 and 72–80 000 in proximal caput, in caput, and of 17 000 and 108 000 by distal corpus; appearance of proteins of M_r 13 500 and 145 000 in caput (Dacheux *et al.*, 1984).

†3. No distinct changes in ^{125}I-labelled surface proteins detected by SDS-PAGE during epididymal maturation (Vierula & Rajaniemi, 1982).

†4. Purified cauda fluid glycoprotein of M_r 133 000 first detected by immunofluorescence on elongating spermatids; over entire surface of testicular spermatozoa, most intense in post-acrosomal region; marked decrease in intensity by proximal corpus; comparable results with X-ray dispersion analysis using ferritin-conjugated antibody (Bostwick *et al.*, 1980).

†‡5. Integral plasma membrane proteins of M_r 44 000 and 47 000 common to caput and cauda spermatozoa; ELSIA endpoints at 50-fold greater dilution of monoclonal antibody using cauda *vs* caput spermatozoa as antigen; indirect immunofluorescence indicated increased distribution and intensity of fluorescence in cauda spermatozoa and cauda epithelial secretory cells; possibly secreted by epididymis and inserted into bilayer during maturation (Peterson *et al.*, 1985).

‡6. Dramatic shifts in localization of monoclonal antibodies against cauda sperm plasma membrane on surface of caput and cauda spermatozoa, based on indirect immunofluorescence; localization on flagellum and midpiece of caput spermatozoa, mid-piece on cauda spermatozoa and head on ejaculated spermatozoa; related to acquisition and loss of proteins from epididymal fluid and seminal plasma (Saxena *et al.*, 1984).

†‡7. Partial purification of a polymorphic antagglutinin from epididymal fluid; reversed autoagglutination of corpus spermatozoa; normal decrease in agglutination in distal corpus corresponded to increased binding to zona-free hamster oocytes (Dacheux *et al.*, 1983).

‡8. Freeze-fracture replicas demonstrated no distinct pattern of integral membrane proteins in proximal caput spermatozoa; domains with parallel arrays over acrosome and rectilinear arrays around neck and connecting piece in caput; hexagonal arrays around margin of head in distal corpus extend to post-acrosomal region in cauda, with most not detected in ejaculated spermatozoa (Suzuki, 1981).

†‡9. Sperm plasma membrane proteins of M_r 30 000, 45 000 and 70 000 with high affinity for dextran sulphate and zona pellucida; present in cauda but not caput spermatozoa; cauda but not caput spermatozoa proteins blocked sperm–zona binding; antibodies to cauda spermatozoa blocked sperm–zona binding, were absorbed by cauda but not caput membrane vesicles (Peterson *et al.*, 1986).

LIPIDS

†1. Comprehensive analysis of caput, corpus and cauda sperm plasma membrane via t.l.c. and g.l.c.; decrease in free fatty acids, major glycolipid, cholesterol and cholesterol to phospholipid ratio from caput to cauda spermatozoa, but increase in diacylglycerols, desmosterol and cholesterol sulphate; decrease in phosphatidyl-ethanolamine, -serine and -inositol; increase in phosphatidylcholine, sphingomyelin and phosphoinositides; high choline and ethanolamine plasmalogen content throughout; increase in C22:5 and decrease in C16:0 in phosphatidyl-choline and -ethanolamine acyl components (Nikopoulou *et al.*, 1985).

*Report describes an alteration that directly or indirectly affects the indicated component.
†Alteration detected via chemical or physical analysis.
‡Alteration detected via microscopic analysis.

study of epididymal sperm maturation because of its large testes and epididymides (and therefore sperm reserves) relative to body size. Changes in plasma membrane components of ram spermatozoa accessible to surface labels for proteins, glycoproteins, and sialoglycoproteins as well as lipid composition of plasma membrane from the region overlying the acrosome have been studied in detail (Table 3). Physical and compositional changes related to surface charge density and membrane fluidity also have been measured in epididymal ram spermatozoa. To date, none of these changes have been directly related to any aspect of sperm function.

The boar has been included as an honorary ruminant because of the elegant and detailed analyses (Table 4) of both the polypeptide and lipid composition which have been conducted on a defined and functionally important region of the plasma membrane, that overlying the acrosome. In addition, changes in specific polypeptides have been directly related to the acquisition of oocyte binding ability of this species.

Conclusion

This review has focussed on recent reports that describe changes in the sperm surface associated with epididymal transit. Since concepts of membrane structure are evolving rapidly, it is suggested that original interpretations of data should be carefully re-evaluated. This is especially true when different methods are used to evaluate the membrane system.

In the first case selected for study, changes in membrane fluidity associated with epididymal transit, the methods sample different aspects of the membrane surface and/or have different sensitivities for detection of changes. The apparently conflicting conclusions may, therefore, be resolved by careful definition of the membrane features actually sampled by the probes used in the experiments. Simply stated, different methods sampling different aspects of the bilayer may yield different observations. In the second case, changes in glycosyl residues that react with con A, alterations have been detected and quantitated by independent methods. Insensitive methods cannot detect slight changes, but application of recently developed techniques that quantitatively evaluate individual cells does provide the sensitivity and precision necessary to evaluate cells. In all cases, molecular mechanisms to account for these changes have yet to be proposed.

Finally, little progress in the study of sperm membrane structure–function relationships will be made without careful attention to the results of experiments utilizing membranes of other, more highly studied, biological systems. A further cataloguing of changes will be of little value, and design of experiments without detailed considerations of the interrelated features of the plasma membrane (Fig. 2) will only lead to ambiguity when the results are finally integrated into a general hypothesis regarding membrane function. It is especially important that new techniques be adopted or adapted for evaluating membrane function as related to sperm maturation. Progress in other membrane systems has been more rapid and models for the necessary general methods and overall experimental design are available (see Racker, 1985). Similar progress in this area will require: (1) in-vitro models for the events assumed essential (e.g. ion flux, specific vesiculation, membrane–membrane binding); (2) development of procedures for the isolation of specific membrane components; and (3) reconstitution of the system(s) for molecular analysis. Both the challenge and the required effort to accomplish these goals will be great, but so will be the satisfaction derived from finally gaining an understanding of any aspect of the molecular basis of gamete fusion.

Financial support was provided by NIH-NICHD-13099 (to R.H.H.) and NIH-NICHD-18628 (to J.E.P.).

References

Acott, T.S. & Hoskins, D.D. (1981) Bovine sperm forward motility protein: Binding to epididymal sperm. *Biol. Reprod.* **24,** 234–240.

Ahuja, K.K. (1985) Carbohydrate determinants involved in mammalian fertilization. *Am. J. Anat.* **174,** 207–223.

Amann, R.P. (1987) Function of the epididymis in bulls and rams. *J. Reprod. Fert., Suppl.* **34,** 115–131.

Arya, M. & Vanha-Perttula, T. (1985) Lectin-binding pattern of bull testis and epididymidis. *J. Androl.* **6,** 230–242.

Atreja, S.K. & Anand, S.R. (1985) Phospholipase and lysophospholipase activities of goat spermatozoa in transit from the caput to the cauda epididymidis. *J. Reprod. Fert.* **74,** 687–691.

Austin, C.R. (1985) Sperm maturation in the male and female genital tracts. In *Biology of Fertilization*, vol. 2, pp. 121–155. Eds C. B. Metz & A. Monroy. Academic Press, New York.

Benga, G. (1985) *Structure and Properties of Cell Membranes*, Vols I–III. CRC Press, Boca Raton.

Bostwick, E.F., Bentley, M.D., Hunter, A.G. & Hammer, R. (1980) Identification of a surface glycoprotein on porcine spermatozoa and its alteration during epididymal maturation. *Biol. Reprod.* 23, 161–169.

Bradley, M.P. & Forrester, I.T. (1981) Modulation of the flagellar plasma membrane calcium pump during mammalian sperm maturation. *Biol. Reprod.* 24 (Suppl. 1), 42A, Abstr.

Casillas, E.R., Elder, C.M. & Hoskins, D.D. (1980) Adenylate cyclase activity of bovine spermatozoa during maturation in the epididymis and the activation of sperm particulate adenylate cyclase by GTP and polyamines. *J. Reprod. Fert.* 59, 297–302.

Chakraborty, J. (1983) Monoclonal antibodies to bull sperm surface antigens. *J. Cell Biol.* 97, 9A, Abstr.

Courtens, J.L., Rozinek, J. & Fournier-Delpech, S. (1982) Binding of epididymal proteins to the spermatozoa of ram. *Andrologia* 14, 509–514.

Crabo, B.G. (1985) Post-testicular sperm maturation and its importance to deep freezing of boar sperm. In *Proc. 1st Int. Conf. Deep Freezing of Boar Semen*, pp. 17–36. Eds L. A. Johnson & K. Larsson. Swedish University of Agricultural Sciences, Uppsala.

Dacheux, J.L. & Voglmayr, J.K. (1983) Sequence of sperm cell surface differentiation and its relationship to exogenous fluid proteins in the ram epididymis. *Biol. Reprod.* 29, 1033–1046.

Dacheux, J.L., Paquignon, M. & Combarnous, Y. (1983) Head-to-head agglutination of ram and boar epididymal spermatozoa and evidence for an epididymal antagglutinin. *J. Reprod. Fert.* 67, 181–189.

Dacheux, J.L., Lanneau, M. & Paquignon, M. (1984) Epididymal surface change in boar spermatozoa. *Biol. Cell* 51, 23A, Abstr.

Dacheux, J.L., Paquignon, M. & Lanneau, M. (1985) Sequential analysis of the epididymal sperm maturation process in the boar. *Ann. N.Y. Acad. Sci.* 438, 526–529.

Devaux, P.F. & Seigneuret, M. (1985) Specificity of lipid-protein interactions as determined by spectroscopic techniques. *Biochim. Biophys. Acta* 822, 63–125.

Eddy, E.M., Vernon, R.B., Muller, C.H., Hahnel, A.C. & Fenderson, B.A. (1985) Immunodissection of sperm surface modifications during epididymal maturation. *Am. J. Anat.* 174, 225–237.

Fléchon, J.-E. (1981) Ultrastructural and cytochemical analysis of the plasma membrane of mammalian sperm during epididymal maturation. *Prog. Reprod. Biol.* 8, 90–99.

Fournier-Delpech, S. & Courot, M. (1980) Glycoproteins of ram sperm plasma membrane. Relationship of protein having affinity for Con A to epididymal maturation. *Biochem. Biophys. Res. Commun.* 96, 756–761.

Fournier-Delpech, S., Courtens, J.L., Pisselet, C.L., Delaleu, B. & Courot, M. (1983) Acquisition of zona binding by ram spermatozoa during epididymal passage, as revealed by interaction with rat oocytes. *Gamete Res.* 5, 403–408.

Fournier-Delpech, S., Hamamah, S., Courot, M. & Pisselet, C. (1984) Organization of glycoproteins of sperm membrane during the development of sperm-

egg recognition system into the epididymis. *Biol. Cell* 51, 49A, Abstr.

Fraser, L.R. (1984) Mechanisms controlling mammalian fertilization. *Oxford Rev. Reprod. Biol.* 6, 174–225.

Friess, A.E. & Sinowatz, F. (1984) Con A- and WGA-binding sites on bovine epididymal spermatozoa: TEM of specimens in toto. *Biol. Cell* 50, 279–284.

Gagnon, C., Harbour, D., deLamirande, E., Bardin, C.W. & Dacheux, J.-L. (1984) Sensitive assay detects protein methylesterase in spermatozoa: decrease in enzyme activity during epididymal maturation. *Biol. Reprod.* 30, 953–958.

Hafez, E.S.E. (1980) *Reproduction in Farm Animals*, 4th edn. Lea & Febiger, Philadelphia.

Hammerstedt, R.H. (1979) Characterization of sperm surfaces using physical techniques. In *The Spermatozoon*, pp. 205–216. Eds D. W. Fawcett & J. M. Bedford. Urban & Schwarzenberg, Baltimore.

Hammerstedt, R.H. (1981) Monitoring the metabolic rate of germ cells and sperm. In *Reproductive Processes and Contraception*, pp. 353–389. Ed. K. W. McKerns. Plenum Press, New York.

Hammerstedt, R.H., Amann, R.P., Rucinsky, T., Morse, P.D., II, Lepock, J., Snipes, W. & Keith, A.D. (1976) Use of spin labels and electron spin resonance spectroscopy to characterize membranes of bovine sperm: effect of butylated hydroxytoluene and cold shock. *Biol. Reprod.* 14, 381–397.

Hammerstedt, R.H., Keith, A.D., Boltz, R.C., Jr & Todd, P.W. (1979a) Use of amphiphilic spin labels and whole cell isoelectric focusing to assay charge characteristics of sperm surfaces. *Archs Biochem. Biophys.* 194, 565–580.

Hammerstedt, R.H., Keith, A.D., Hay, S., Deluca, N. & Amann, R.P. (1979b) Changes in ram sperm membranes during epididymal transit. *Archs Biochem. Biophys.* 196, 7–12.

Hammerstedt, R.H., Hay, S.R. & Amann, R.P. (1982) Modification of ram sperm membranes during epididymal transit. *Biol. Reprod.* 27, 745–754.

Helmkamp, G.M., Jr (1983) Phospholipid transfer proteins and membrane fluidity. In *Membrane Fluidity in Biology*, vol. 2, pp. 151–185. Ed. R. C. Aloia. Academic Press, New York.

Hemminga, M.A. (1984) Interpretation of ESR and saturation transfer ESR spectra of spin labelled lipids and membranes. *Chem. Phys. Lipids* 32, 323–383.

Holt, W.V. (1980) Surface bound sialic-acid on ram and bull spermatozoa: deposition during epididymal transit and stability during washing. *Biol. Reprod.* 23, 847–857.

Holt, W.V. (1982) Functional development of the mammalian sperm plasma membrane. *Oxford Rev. Reprod. Biol.* 4, 194–240.

Inskeep, P.B. & Hammerstedt, R.H. (1982) Changes in metabolism of ram sperm associated with epididymal transit or induced by exogenous carnitine. *Biol. Reprod.* 27, 735–743.

Inskeep, P.B., Magargee, S.F. & Hammerstedt, R.H. (1985) Alterations in motility and metabolism associated with sperm interaction with accessory sex gland fluids. *Archs Biochem. Biophys.* 241, 1–9.

Jauhiainen, A. & Vanha-Perttula, T. (1986) β-N-Acetylglucosaminidase in the reproductive organs and seminal plasma of the bull. *J. Reprod. Fert.* 76, 239–250.

Jones, R.H. & Molitoris, B.A. (1984) A statistical method for determining the breakpoint of two lines. *Analyt. Biochem.* **141**, 287–290.

Keith, A.D., Sharnoff, M. & Cohn, G.E. (1973) A summary and evaluation of spin labels used as probes for biological membrane structure. *Biochim. Biophys. Acta* **300**, 379–419.

Koehler, J.K. (1982) The mammalian sperm surface: An overview of structure with particular reference to mouse spermatozoa. In *Prospects For Sexing Mammalian Sperm*, pp. 23–42. Eds R. P. Amann & G. E. Seidel, Jr. Colorado Associated University Press, Boulder.

Koehler, J.K. (1985) Sperm membranes: segregated domains of structure and function. In *Proc. 1st Int. Conf. Deep Freezing of Boar Semen*, pp. 37–60. Eds L. A. Johnson & K. Larsson. Swedish University of Agricultural Sciences, Uppsala.

Lands, W.E.M. (1980) Fluidity of membrane lipids. In *Membrane Fluidity: Biophysical Techniques and Cellular Regulation*, pp. 69–73. Eds M. Kates & A. Kuskis. Humana Press, Clifton.

Loewus, M.W., Wright, R.W., Jr, Bondioli, K.R., Bedgar, D.L. & Karl, A. (1983) Activity of *myo*-inositol-1-phosphate synthase in the epididymal spermatozoa of rams. *J. Reprod. Fert.* **69**, 215–220.

Low, M.G., Ferguson, M.A.J., Futerman, A.H. & Silman, I. (1986) Covalently attached phosphatidylinositol as a hydrophobic anchor for membrane proteins. *Trends Biochem. Sci.* **11**, 212–215.

Mann, T. & Lutwak-Mann, C. (1981) *Male Reproductive Function and Semen*, pp. 195–268. Springer-Verlag, Berlin.

Marsh, D. (1981) Electron spin resonance: spin labels. In *Membrane Spectroscopy*, pp. 51–142. Ed. E. Grell. Springer-Verlag, New York.

Nicolson, G.L. (1982) Mammalian sperm plasma membrane. In *Prospects For Sexing Mammalian Sperm*, pp. 5–16. Eds R. P. Amann & G. E. Seidel, Jr. Colorado Associated University Press, Boulder.

Nikolopoulou, M., Soucek, D.A. & Vary, J.C. (1985) Changes in the lipid content of boar sperm plasma membranes during epididymal maturation. *Biochim. Biophys. Acta* **185**, 486–498.

Olson, G.E. & Orgebin-Crist, M.-C. (1982) Sperm surface changes during epididymal maturation. *Ann. N.Y. Acad. Sci.* **383**, 372–392.

O'Rand, M.G. (1985) Differentiation of mammalian sperm membranes. In *Biology of Fertilization*, vol. 2, pp. 103–120. Eds C. B. Metz & A. Monroy. Academic Press, New York.

Parks, J.E. & Hammerstedt, R.H. (1985) Developmental changes occurring in the lipids of ram epididymal spermatozoa plasma membrane. *Biol. Reprod.* **32**, 653–668.

Peterson, R.N. & Russell, L.D. (1985) The mammalian spermatozoon: a model for the study of regional specificity in plasma membrane organization and function. *Tissue & Cell* **17**, 769–791.

Peterson, R.N., Saxena, N.K., Saxena, N., Henry, L.H. & Russell, L.D. (1985) Organization of the plasma membrane of boar spermatozoa: the concentration of major proteins increases markedly during maturation in the epididymis. *J. Cell Biol.* **101**, 364A, Abstr.

Peterson, R.N., Hunt, W.P. & Henry, L.H. (1986) Inter-

action of boar spermatozoa with porcine oocytes: increase in proteins with high affinity for the zona pellucida during epididymal transit. *Gamete Res.* **14**, 57–64.

Polnaszek, C.F., Schreier, S., Butler, K.W. & Smith, I.C.P. (1978) Analysis of the factors determining the EPR spectra of spin probes that partition between aqueous and lipid phases. *J. Am. Chem. Soc.* **100**, 8223–8232.

Racker, E. (1985) *Reconstitutions of Transporters, Receptors, and Pathological States*. Academic Press, New York.

Russell, L.D., Peterson, R.N., Hunt, W. & Strack, L.E. (1984) Post testicular surface modifications and contributions of reproductive tract fluids to the surface polypeptide composition of boar spermatozoa. *Biol. Reprod.* **30**, 959–978.

Saxena, N., Russell, L.D., Saxena, N.K. & Peterson, R.N. (1984) Localization of boar sperm plasma membrane antigens at various stages of sperm maturation: do immunofluorescence results give a complete picture of antigen localization? *J. Cell Biol.* **99**, 258A, Abstr.

Schlegel, R.A., Hammerstedt, R.H., Cofer, G.P., Kozarsky, K., Freidus, D. & Williamson, P. (1986) Changes in the organization of the lipid bilayer of the plasma membrane during spermatogenesis and epididymal maturation. *Biol. Reprod.* **34**, 379–391.

Schwartz, H.M. (1982) Use of electron spin resonance to study complex biological membranes. In *Physical Methods on Biological Membranes and Their Model Systems*, pp. 39–53. Eds F. Conti, W. E. Blumberg, J. deGier & F. Pocchiari. Plenum Press, New York.

Sinowatz, F. & Friess, A.E. (1983) Localization of lectin receptors of bovine epididymal spermatozoa using a colloidal gold technique. *Histochemistry* **79**, 335–344.

Susko-Parrish, J.S., Hammerstedt, R.H. & Senger, P.L. (1985) Binding of 125 I-succinylated concanavalin A to bovine spermatozoa. *Biol. Reprod.* **32**, 129–136.

Suzuki, F. (1981) Changes in intramembranous particle distribution in epididymal spermatozoa of the boar. *Anat. Rec.* **199**, 361–376.

Vierula, M. & Rajaniemi, H. (1981) Changes in surface protein structure of bull spermatozoa during epididymal maturation. *Int. J. Androl.* **4**, 314–320.

Vierula, M. & Rajaniemi, H. (1982) Epididymal maturation of the surface protein structure of mammalian spermatozoa. *Med. Biol.* **60**, 323–327.

Voglmayr, J.K., Fairbanks, G., Jackowitz, M.A. & Colella, J.R. (1980) Post-testicular developmental changes in the ram sperm cell surface and their relationship to luminal fluid proteins of the reproductive tract. *Biol. Reprod.* **22**, 655–667.

Voglmayr, J.K., Fairbanks, G., Vespa, D.B. & Colella, J.R. (1982) Studies on mechanisms of surface modifications in ram spermatozoa during the final stages of differentiation. *Biol. Reprod.* **26**, 483–500.

Voglmayr, J.K., Fairbanks, G. & Lewis, R.G. (1983) Surface glycoprotein changes in ram spermatozoa during epididymal maturation. *Biol. Reprod.* **29**, 767–775.

Voglmayr, J.K., Sawyer, R.F., Jr & Dacheux, J.L. (1985) Glycoproteins: a variable factor in surface transformation of ram spermatozoa during epididymal transit. *Biol. Reprod.* **33**, 165–176.

Wirtz, K.W.A. (1982) Phospholipid transfer proteins. In *Lipid Protein Interactions*, vol. 1, pp. 151–231. Eds

P. C. Jost & O. H. Griffith. John Wiley and Sons, New York.

Wold, F. (1986) Fatty acylation of proteins (Keep fit with fat?). *Trends Biochem. Sci.* **11,** 58–59.

Wolf, D.E. & Voglmayr, J.K. (1984) Diffusion and regionalization in membranes of maturing ram spermatozoa. *J. Cell Biol.* **98,** 1678–1684.

Wolf, D.E., Hagopian, S.S., Lewis, R.G., Voglmayr, J.K. & Fairbanks, G. (1986) Lateral regionalization and diffusion of a maturation-dependent antigen in the ram sperm plasma membrane. *J. Cell Biol.* **102,** 1826–1831.

Zilversmit, D.B. (1983) Lipid transfer proteins: overview and applications. *Methods Enzymol.* **98,** 565–573.

J. Reprod. Fert., Suppl. **34** (1987), 151–165

In-vitro fertilization of ruminants

N. L. First and J. J. Parrish

University of Wisconsin, Department of Meat and Animal Science, Madison, Wisconsin 53706, U.S.A.

Introduction

The first ruminant offspring from in-vitro fertilization was a bull calf born in 1981 (Brackett *et al.*, 1982). Since then, fertilization has been accomplished *in vitro* for the three domestic ruminants—cattle, sheep and goats. These developments are largely the result of recent advances in our under-standing of oocyte maturation, sperm capacitation and embryonic development. It is the intent of this manuscript to address four questions concerning in-vitro fertilization with primary emphasis on cattle. The questions to be addressed are: (1) what are the uses or reasons for performing IVF, (2) can it be done, (3) what are the relative efficiencies of each step and which steps need further development and (4) what are the physiological mechanisms through which the normal in-vivo or in-vitro fertilization process is accomplished.

Potential use for in-vitro fertilization

The strongest motivation for development of systems for fertilization and development of embryos *in vitro* comes from the realization that development of the biotechnologies for producing trans-genic offspring or for multiplication *in vitro* of lines of superior offspring are dependent on volume production of precisely staged embryos and on the ability to develop these unique embryos to a stage compatible with transfer to recipient mothers. In commercial use it is not likely that cows of high genetic value will be subjected to slaughter or surgery in order to obtain pronuclear stage eggs for gene microinjection or tubal stage embryos for use in embryo multiplication. However, a pro-cedure based on in-vitro fertilization of laparoscopic recovered follicular oocytes would be a useful and much needed alternative. Similarly, commercial multiplication of identical embryos is expected to be dependent on the use of abattoir-obtained oocytes as recipients for transfer of nuclei from valuable multicellular stage embryos and on the ability to develop the resulting embryos to a trans-ferable stage. Important to the study of reproductive physiology is that fertilization *in vitro* of oocytes recovered in large numbers from abattoir ovaries could provide abundant numbers of embryos for research. The ability to mature and fertilize oocytes *in vitro* would also provide a means for production of a few calves from a dead or dying cow of great genetic value. Lastly, there is considerable hope that in-vitro fertilization tests will be useful for predicting fertility *in vivo*.

In-vitro fertilization, can it be done?

The production of calves from fertilization *in vitro* of oocytes matured *in vivo* (Brackett *et al.*, 1982, 1984; Sirard & Lambert, 1985, Sirard *et al.*, 1985) or *in vitro* (Critser *et al.*, 1986) and lambs from oocytes matured *in vitro* (Cheng *et al.*, 1986) attest to the fact that in-vitro fertilization can be successfully accomplished. The efficiency of producing offspring is less than achieved by natural processes (Tables 1 and 2). The overall efficiency is reduced by the sequential multiplicative effects of a series of seemingly efficient steps. This results in slightly more than 10% of recovered oocytes producing offspring (Table 3). The most limiting step at present appears to be the inefficiency of development of embryos *in vitro* from the pronuclear to the blastocyst stage (Table 3). For immature cow oocytes recovered from small follicles of 1–5 mm in diameter, an additional limitation is the failure to acquire competence for development from the pronuclear to blastocyst stage (Leibfried-Rutledge *et al.*, 1986a; Table 4). This deficiency by immature cow oocytes also occurs for immature oocytes of sheep after fertilization *in vivo* (Staigmiller & Moor, 1984). Embryo developmental competence can be acquired by immature oocytes through coculture with abundant numbers of preovulatory granulosa or cumulus cells which are receiving stimulation from FSH, LH and oestradiol (Staigmiller & Moor, 1984; Critser *et al.*, 1986). This developmental problem is but one manifestation of a need to understand the steps and mechanisms regulating oocyte maturation.

Table 1. Efficiency of in-vitro fertilization of in-vivo matured cow oocytes

	Brackett *et al.* (1982, 1984)	Lambert *et al.* (1986)	Leibfried-Rutledge *et al.* (1986a)*
Sperm capacitation method	High ionic strength medium	High ionic strength medium	Heparin
Oocyte maturation		67%*	43%* 78%†
Fertilization‡			
Follicular	14%	41%	57%
Oviducal	12%		32%
Embryo development§		15–20%	45%
Pregnancies	5/19 (21%)	6/13 (46%)	2/10 (20%)
Calves	3/19 (16%)	6/13 (46%)	0/10 (0%)

*Oocytes recovered from follicles with expanded cumulus.
†Oocytes recovered from oviducts with an evenly granulated cytoplasm and no evidence of fragmentation or degeneration.
‡Oocytes were recovered from follicles or oviducts. Oocytes were fertilized when two pronuclei, not polyspermic, or cleavage to the 2-cell stage occurred.
§Only oocytes obtained from follicles.

Table 2. Efficiency of in-vitro fertilization of in-vitro matured bovine oocytes

	% oocytes matured*	Type of spermatozoa	% Fertilized†
Ball *et al.* (1983)	>70	Epididymal	70
Iritani *et al.* (1984)	80	Ejaculated	44, 63
Hensleigh & Hunter (1985)	56	Ejaculated	15
Parrish *et al.* (1985a)	>70	Ejaculated	72
Fukui *et al.* (1983)	64	Ejaculated	28
Parrish *et al.* (1986b)	>70	Ejaculated, frozen–thawed	81
Leibfried-Rutledge *et al.* (1986a)	70‡	Ejaculated, frozen–thawed	94‡
	73	frozen–thawed	91

*Oocytes with expanded cumulus and/or at metaphase II of meiosis.
†No. of ova penetrated or cleaved to 2 cell/total no. oocytes.
‡During maturation, oocytes cocultured with granulosa cells.

Table 3. Efficiency of each step involved in successful
in-vitro fertilization of cow oocytes

	Efficiency (%)	% of original 100 oocytes
Oocyte maturation	80	80
Sperm penetration	80	64
2 Pronuclei	80	51
Sheep oviduct– recovery	80	41
1 Cell–blastocyst	55	23
Transfer pregnancy	60	14
Calves	80	11

Table 4. Development in sheep oviducts of in-vitro fertilized cow oocytes matured *in vitro* or *in vivo** (from Leibfried-Rutledge *et al.*, 1986a)

	Fertilization			Development*	
Maturation of oocytes	No. of ova penetrated/ total no. of ova (%)	No. of ova with 2 pronuclei/no. of penetrated ova (%)	No. of normally fertilized ova*/total no. of ova (%)	No. of cleaved embryos/no. recovered (%)	No. of morulae or blastocysts/no. recovered (%)
In vitro	158/207 (76)	115/158 (73)	81/207 (39)	15/33 (45)	0/33 –(0)
In vivo					
Follicular	22/ 24 (96)	18/ 22 (82)	19/ 24 (80)	66/80 (82)	36/80 (45)
Oviducal	24/ 28 (86)	20/ 24 (83)	15/ 24 (53)	38/56 (68)	14/56 (25)

*No. of ova with only two pronuclei, not polyspermic.

Mechanisms involved in fertilization

Fertilization *in vitro* requires the following: (1) a system for harvesting the oocyte which is efficient and does not damage the oocyte or donor, (2) nuclear, cytoplasmic and cumulus cell maturation of the oocyte/cumulus complex, (3) non-damaging culture systems, (4) a system for capacitating the spermatozoa and (5) efficient systems and conditions for accomplishing fertilization and development of the embryo from fertilization to a stage transferable to recipients. Accomplishment of each step is greatly aided by understanding the natural process as well as the mechanisms involved.

Harvesting of the oocyte

In-vivo matured oocytes are obtained from ovaries of superovulated cows by recovery from preovulatory follicles near the moment of ovulation or from the oviduct soon after ovulation. For research purposes, the ovaries or oviducts have usually been retrieved after slaughter of the donor. The cumulus-enclosed oocytes are then aspirated from the preovulatory follicles or flushed from the oviduct (Brackett *et al.*, 1982). Oocytes from both sources appear to have completed maturation and after fertilization, normal embryo development occurs (Brackett *et al.*, 1982; Leibfried-Rutledge *et al.*, 1986a; Table 4). However, the tubal oocytes must be recovered and placed in culture soon after ovulation. Bovine oocytes allowed to remain in excised oviducts for as little as 0·5–2 h will undergo a high frequency of parthenogenetic activation (L. Leibfried, unpublished). A more acceptable procedure for repeated oocyte recovery from cows of continued reproductive value is with the aid of a laparoscope. This procedure has been used most successfully by Sirard & Lambert (1985), Sirard *et al.* (1985) and Lambert *et al.* (1986). The efficiency of oocyte recovery by the laparoscopic method (Lambert *et al.*, 1986) appears to be equivalent to recovery from the excised ovary. Before cumulus expansion, oocytes remain firmly attached in small and medium-sized follicles and cannot be aspirated *in vivo* as they can from small follicles of excised ovaries (Ball *et al.*, 1983). Because a large proportion of the follicles in cows are in various stages of atresia (Choudary *et al.*, 1968; Leibfried & First, 1979) oocytes are commonly screened by selecting only those oocytes which have intact, expanded and complete cumulus cell enclosure and have uniformly granulated cytoplasm (Leibfried & First, 1979; Leibfried-Rutledge *et al.*, 1986a).

A major limitation to use of in-vivo matured oocytes is the number of oocytes obtained per female. Even when cows are stimulated with FSH, the usable number of oocytes/donor ranges from 5 to 10 (Lambert *et al.*, 1986; Leibfried-Rutledge *et al.*, 1986a). Whenever large numbers of embryos are required, in-vivo matured oocytes provide an expensive supply. Collection of immature oocytes from ovaries obtained at the abattoir would provide a much cheaper source.

Immature oocytes cannot be obtained by laparoscopy and must therefore come from ovaries at ovariectomy or at the abattoir. Commonly oocytes are aspirated from 1–5 mm follicles in these ovaries. Careful monitoring of temperature is important as oocytes recovered are incapable of maturing past metaphase I if cooled below 30°C (L. Leibfried-Rutledge, personal communication; A. G. Hunter, personal communication). Because of the large percentage of atretic follicles, only non-atretic follicles with intact and compact cumulus-enclosed oocytes are selected. This differs from procedures for in-vivo matured oocytes for which only cumulus–oocyte complexes responding to gonadotrophins *in vivo* with expanded cumulus are used. The other criteria for selection are the same as previously described (Leibfried & First, 1979; Leibfried-Rutledge *et al.*, 1986a).

Maturation of oocytes and completion of development competence

After spermatozoa are capacitated, successful fertilization *in vitro* is dependent on the correct maturity of the oocytes to be fertilized. The mechanisms regulating maturation of ruminant oocytes are discussed by Moor & Gandolfi (1987). Oocyte maturation is commonly considered to be the progression of nuclear maturation from the appearance of the germinal vesicle of the arrested dictyate stage, through meiosis I to metaphase II and extrusion of the first polar body. For successful fertilization and embryonic development, oocyte maturation must be considered in a much broader sense. Especially important are the development of proteins in the cytoplasm which regulate meiotic events, sperm decondensation, formation of the male pronuclear envelope, initiation of cleavage, completion of cleavage-stage embryo development and ability to produce a viable offspring (First & Haseltine, 1987; Moor & Gandolfi, 1987). The development of these regulatory signals can be affected by the stage of maturity of the follicle from which the oocyte is removed, by the duration of the in-vitro or in-vivo maturation period and by conditions of the in-vitro maturation culture.

Common defects resulting from fertilization of immature oocytes include failure of male pronuclear development (Thibault *et al.*, 1975; Thibault, 1977) and failure to progress to organized blastocyst-stage embryos (Staigmiller & Moor, 1984; Leibfried-Rutledge *et al.*, 1986a; Table 4). The latter has been corrected by co-culture *in vitro* of immature oocytes with granulosa or abundant cumulus cells under stimulation of gonadotrophins and oestrogen (sheep: Staigmiller & Moor, 1984; cow: Critser *et al.*, 1986). Some of these defects, such as failure of male pronuclear development, have been produced experimentally by shortening the maturation period such that a given regulatory protein is not yet produced. This was accomplished in the hamster to produce oocytes that were penetrated by spermatozoa but failed to form male pronuclei (Leibfried & Bavister, 1983).

An additional aspect of oocyte maturation is the full expansion of the cumulus cells surrounding the oocyte. While cumulus cells may impede the contact of spermatozoa with the egg and delay the time of fertilization (J. J. Parrish, unpublished), there is also evidence that, if expanded, their presence may enhance sperm capacitation and fertilization rate (Ball *et al.*, 1983). Additionally, a product of cumulus cells stimulated by FSH enhances motility of bovine spermatozoa (Bradley & Garbers, 1983). Bovine cumulus cells undergo expansion in response to FSH (Ball *et al.*, 1983) either *in vitro* or after a small FSH rise accompanying the preovulatory surge of LH *in vivo*. The principal secretory product of FSH-stimulated bovine cumulus cells is hyaluronic acid which forms a mucus matrix between and around cumulus cells (Ball *et al.*, 1983). In the hamster, hyaluronic acid will induce an acrosome reaction but only in capacitated epididymal spermatozoa. While hyaluronic acid is capable of promoting the acrosome reaction of epididymal bovine spermatozoa (Handrow *et al.*, 1982) it does not appear to do so in ejaculated bovine spermatozoa (Handrow *et al.*, 1986a). The principal function of hyaluronic acid in cattle may be similar to one of its functions in the hamster, which is to provide a large mass capable of being picked up and transported by the fimbria of the oviduct (Mahi-Brown & Yanagimachi, 1983). If so the transport function of the bovine cumulus mass is of short duration because the oviduct causes its rapid removal (Lorton & First, 1979).

Bovine oocytes cultured from immature follicles (1–5 mm; Leibfried & First, 1979) or ovine oocytes from non-atretic follicles (3–6 mm; Moor & Trounson, 1977; Staigmiller & Moor, 1984) require 24–27 h in culture to develop from the germinal vesicle stage to metaphase II. This progression in nuclear development is paralleled by a progression of cumulus expansion (Leibfried & First, 1979). While fertilization is highly temperature-dependent, nuclear maturation of bovine oocytes occurs over a temperature range of 35–41°C (Lenz *et al.*, 1983a; Katska & Smorag, 1985). Cooling oocytes below 30°C, however, can damage nuclear maturation ability by an apparent disruption of the spindle apparatus (Moor & Crosby, 1985).

Several different media have been used for successfully maturing cow or sheep oocytes *in vitro*. These include: Medium 199 (Moor & Trounson, 1977; Staigmiller & Moor, 1984; Leibfried-Rutledge *et al.*, 1986a; Katska & Smorag, 1985), Hams F-12 (Fukushima & Fukui, 1985) and TALP (Ball *et al.*, 1983; Lenz *et al.*, 1983a; Critser *et al.*, 1984; Leibfried-Rutledge *et al.*, 1986b).

There is as yet no satisfactory completely defined medium for ruminant oocyte cultures. For full development and subsequent fertilization the medium must contain a blood serum or serum extract. For the cow, whole sera such as fetal calf serum are superior to albumin both in supporting completion of nuclear maturation and viability of associated cumulus cells (Leibfried-Rutledge *et al.*, 1986b).

Capacitation and the acrosome reaction

Mammalian spermatozoa are not immediately capable of fertilizing oocytes, rather they must undergo a period of preparation, which normally occurs in the female reproductive tract, termed capacitation (Austin, 1951; Chang, 1951). Capacitation consists of at least two components, an initial sperm membrane alteration (Yanagimachi, 1981; O'Rand, 1982; Ahuja, 1984; Langlais & Roberts, 1985; Wolf *et al.*, 1986) which allows the spermatozoa to undergo the second phase, the fusion of the plasma membrane and outer acrosomal membrane (Yanagimachi & Usui, 1974). The first phase is now referred to as capacitation while the second phase is referred to as the acrosome reaction. The site of sperm capacitation, agents causing capacitation *in vivo* and mechanisms by which capacitation is manifest are not well understood for most mammalian species.

The site of capacitation within the female reproductive tract is uncertain. While sperm capacitation can take place within the uterus (Austin, 1951; Bedford, 1969, 1970; Barros, 1974) capacitation is accelerated by exposure of spermatozoa to the uterus and then the oviduct (Adams & Chang, 1962; Bedford, 1969; Hunter, 1969; Hunter & Hall, 1974). Events of sperm transport further complicate determination of the site of sperm capacitation. In the sheep (Hunter *et al.*, 1980, 1982; Hunter & Nichol, 1983) and the cow (Hunter & Wilmut, 1982, 1984; Wilmut & Hunter, 1984) adequate sperm numbers for fertilization require 6–12 h before being established in the lower isthmus of the oviduct, the sperm reservoir for these species. Spermatozoa involved in fertilization then move up to the ampullary–isthmic junction of the oviduct only at the time of ovulation. When spermatozoa are deposited in the female at the start of oestrus, they may reside in the isthmus for as long as 18–20 h in the cow (Hunter & Wilmut, 1984) and 17–18 h in the ewe (Hunter & Nichol, 1983) before fertilization. It is unknown how sperm transport and accumulation in the lower isthmus are affected by insemination later in oestrus. Similar situations occur in the rabbit (Overstreet *et al.*, 1978; Overstreet, 1982) and pig (Hunter, 1984) in which spermatozoa are held in the lower isthmus before ovulation and move up the oviduct at or near the time of ovulation. Whether spermatozoa are normally capacitated by the time they enter the oviduct is unknown. However, sperm fragility increases during capacitation (Hunter, 1980) and with the long time spermatozoa must reside in the lower isthmus before ovulation in the cow or ewe, it is most likely that capacitation is completed in the oviduct. For cattle, we have shown that spermatozoa undergo the acrosome reaction almost exclusively in the ampulla of the oviduct ipsilateral to the side of ovulation and only near and immediately after ovulation (Herz *et al.*, 1985). The results suggest that capacitation does occur in the oviduct but primarily in the oviduct ipsilateral to the side of ovulation. Additionally, the recovered contents of the oviducts of oestrous, but not luteal-phase, ewes caused bovine spermatozoa to be capacitated and undergo an acrosome reaction (Lee *et al.*, 1986). In the hamster, capacitation of spermatozoa *in vivo* is completed in the oviduct (Bedford, 1972) and the ability of the oviduct to promote capacitation is prevented by progesterone (Viriyapanich & Bedford, 1981). Oviduct fluid from the follicular stage of the oestrous cycle therefore appears capable of causing capacitation.

In vitro, glycosaminoglycans will induce an acrosome reaction or capacitate bovine epididymal spermatozoa (Lenz *et al.*, 1982, 1983b; Handrow *et al.*, 1982; Ball *et al.*, 1983; Parrish *et al.*, 1985a). Assay of the glycosaminoglycans in bovine (Lee & Ax, 1984) and ovine (Lee *et al.*, 1986) oviduct flushings revealed a high concentration of heparin-like material. *In vitro* heparin was shown to be the most potent glycosaminoglycan in its ability to induce the acrosome reaction in bovine epididymal spermatozoa (Handrow *et al.*, 1982) and to capacitate ejaculated bovine spermatozoa (Parrish *et al.*, 1985b). A characteristic of bovine sperm capacitation by heparin is that it is inhibited by 5 mM-glucose (Parrish *et al.*, 1985a; Susko-Parrish *et al.*, 1985). Capacitation of bovine spermatozoa by oviduct fluid is also inhibited by glucose (Parrish *et al.*, 1986a). In cows, glucose is present in very low levels in the oviduct, < 1 mM (Carlson *et al.*, 1970), levels which do not inhibit sperm capacitation by heparin. In an experiment to identify the capacitating agent of oviduct fluid in which the treatments were applied sequentially, the capacitating factor was found to be not sensitive to protease digestion, soluble in trichloroacetic acid, precipitated by ethanol and inacti-vated by nitrous acid, a reagent which degrades the heparin-like glycosaminoglycans, heparin or heparan sulphate (R. R. Handrow and J. J. Parrish, unpublished). This suggests a heparin-like glycosaminoglycan is responsible for the capacitating activity of oviduct fluid from oestrous cows but does not as yet identify it as heparin or heparan sulphate.

After capacitation, final preparation of spermatozoa for fertilization in cows and other mam-mals may occur at the zona pellucida as in the mouse (Florman & Storey, 1982; Bleil & Wassarman, 1983). There is support for the idea that mammalian spermatozoa may be capacitated by materials, perhaps glycosaminoglycans, in the fluids of the female reproductive tract or cumulus cells, but the fertilizing spermatozoa complete preparation for and undergo the acrosome reaction only at the zona pellucida. In cows, spermatozoa bound on the zona pellucida were all acrosome-reacted (Crozet, 1984). Furthermore, electron microscopy revealed that bovine spermatozoa recovered around and in the cumulus mass in the oviduct had acrosomes of a fluffy appearance but only spermatozoa bound to the zona pellucida were acrosome-reacted (Crozet, 1984). Collectively these findings suggest that bovine spermatozoa are capacitated in the oviduct and that fertilizing spermatozoa complete the acrosome reaction at or near the zona pellucida.

Precise molecular events in capacitation and the acrosome reaction are not well understood. It has been suggested that an alteration of membrane proteins occurs (Gordon *et al.*, 1975;

Koehler, 1978) and that subsequent sterol depletion may be a key event (Davis, 1981, 1982; Langlais & Roberts, 1985). A molecular membrane model describing changes in sperm membranes accompanying capacitation has been proposed by Langlais & Roberts (1985). By this model, capacitation is a reversible phenomenon which upon completion results in a modification of the sperm membrane surface, causing an efflux of membrane cholesterol which alters the membrane sterol/phospholipid ratio and results in an influx of Ca^{2+}. The Ca^{2+} activates phospholipase-A_2 which catalyses the synthesis and accumulation of fusogenic lysophospholipids such as lysophosphatidylcholine. In the presence of Ca^{2+} and reduced levels of cholesterol, such fusogenic compounds would induce the acrosome reaction. The critical component of this model is the sterol depletion of sperm membranes. The remaining changes are postulated to occur as a cascade from this event. The Langlais & Roberts (1985) model, however, does not address the role of a heparin-like glycosaminoglycan from oviduct fluid in capacitation of bovine spermatozoa. Under physiological conditions, a heparin-like glycosaminoglycan appears essential for efficient capacitation of ejaculated bovine spermatozoa (Parrish *et al.*, 1985a, b, 1986a, b). While the oviduct heparin-like glycosaminoglycan may be heparan sulphate or heparin, heparin can displace specifically bound heparan sulphate from cells (Kjellen *et al.*, 1980) and exert biological effects on cells similar to those of heparan sulphate (Laterra *et al.*, 1983). Based upon known effects of heparin on spermatozoa and other cell types we will propose several possible roles of an oviduct heparin-like glycosaminoglycan during capacitation.

There are two ways in which heparin-like glycosaminoglycans could modify the sperm plasma membrane resulting in the alterations suggested by Langlais & Roberts (1985) to be the first step in capacitation. The first method is by displacing a decapacitation protein (Oliphant *et al.*, 1985) from the sperm surface similar to heparin's ability to displace lipoproteinlipase from cells (Casu, 1985). The second method of altering the sperm plasma membrane could be a direct modification of membrane domains. Heparin-like glycosaminoglycans can bind to proteins, phospholipids and themselves through Ca^{2+} bridges (Srinivasan *et al.*, 1970). A network of heparin-like glycosaminoglycan molecules and membrane components could induce reorganization of the membrane by restricting or causing movement of membrane components. In support of a direct effect on membrane domain formation, heparin binding affinity to spermatozoa (Handrow *et al.*, 1984) and ability to capacitate spermatozoa is Ca^{2+}-dependent (J. J. Parrish, unpublished).

The second major event of capacitation proposed by Langlais & Roberts (1985) was an uptake of Ca^{2+} by spermatozoa. Heparin induces a linear increase in $^{45}Ca^{2+}$ uptake during capacitation of bovine spermatozoa that does not occur when spermatozoa are incubated under noncapacitating conditions (Handrow *et al.*, 1986b). It is, however, unknown whether the $^{45}Ca^{2+}$ uptake was a result of heparin acting as an ionophore, opening a Ca^{2+} channel in the membrane or the result of the membrane changes previously proposed. Heparin-like glycosaminoglycans may therefore have direct effects on Ca^{2+} uptake into spermatozoa during capacitation.

The third major step in capacitation proposed by Langlais & Roberts (1985) is the activation of phospholipase-A_2 by Ca^{2+} or acrosin. Heparin-like glycosaminoglycans may be able to activate phospholipase-A_2 through their ability to chelate Zn^{2+} (Casu, 1985), an inhibitor of phospholipase-A_2 (Thakkar *et al.*, 1984), or stimulate conversion of proacrosin to acrosin (Parrish *et al.*, 1980).

Our studies suggest an additional effect of heparin-like glycosaminoglycan on capacitation may be the activation of a cAMP-dependent protein kinase. We have shown that exogenously added 8-bromo-cAMP will reverse the inhibition by glucose on sperm capacitation with heparin (Susko-Parrish *et al.*, 1985). Potential elevation of cAMP levels in spermatozoa cannot be the only action of heparin on spermatozoa as 8-bromo-cAMP will not be itself capacitate bovine spermatozoa (Susko-Parrish *et al.*, 1985).

The complex potential effects of an oviducal heparin-like glycosaminoglycan on spermatozoa preclude assigning it a single role in capacitation. It is likely that further research will demonstrate multiple effects of heparin-like glycosaminoglycans regulating sperm capacitation and the acrosome reaction.

Methods to capacitate ruminant spermatozoa *in vitro* rely primarily on modification of techniques that have proven useful with rodent spermatozoa, since these species are the most studied in relation to sperm capacitation. In rodents, capacitation *in vitro* is easily accomplished by a variety of media manipulations. The functions of these manipulations have been to remove surface proteins that are inhibiting capacitation and/or the acrosome reaction (Wolf, 1979; Aonuma *et al.*, 1982; Fraser, 1983) and to stimulate sterol efflux from the sperm plasma membrane (Go & Wolf, 1985). Application of these techniques to ruminant spermatozoa has not always been successful. One difference is the use of ejaculated spermatozoa for in-vitro studies with ruminants and spermatozoa removed from the cauda epididymidis for rodents. It must be emphasized that ejaculated spermatozoa are responsible for in-vivo fertilization in all species. Differences in capacitating ability of ejaculated and epididymal spermatozoa do exist. Bovine epididymal spermatozoa can capacitate in simple salt solutions (Ball *et al.*, 1983; J. J. Parrish, unpublished) but ejaculated bovine spermatozoa do not capacitate well unless capacitating agents are added (Parrish *et al.*, 1985b, 1986a, b, c). Decapacitation factors (Chang, 1951) of seminal plasma are most probably responsible for these effects.

Despite the limitations of applying results from capacitation studies of rodent spermatozoa to ruminant spermatozoa, several procedures have been developed for capacitation of ruminant ejaculated spermatozoa. A medium of high ionic strength has been used to displace proteins from the surface of bovine spermatozoa which might inhibit sperm capacitation (Brackett *et al.*, 1982; Bousquet & Brackett, 1982; Lambert *et al.*, 1986). While this technique has been proved successful by the birth of calves, results have been variable among bulls (Brackett *et al.*, 1982; Sirard & Lambert, 1985). Long incubations of bovine and caprine spermatozoa, 18–24 h, which allows time for surface proteins to dissociate, have proved useful if spermatozoa remain viable for this period (Wheeler & Seidel, 1986; Song & Iritani, 1985). Sheep spermatozoa have also been successfully capacitated by incubation at elevated pH, 7·8–8·0 (Cheng *et al.*, 1986). This treatment presumably not only dissociates proteins from the sperm surface but also induces Ca^{2+} uptake in cells. In the goat, Ca^{2+} ionophore A23187 has been used to bypass capacitation and by directly increasing Ca^{2+} to induce the acrosome reaction (Shorgan, 1984). The last technique is the use of the glycosaminoglycan heparin with bovine spermatozoa (Parrish *et al.*, 1985a, 1986b). Capacitation of spermatozoa by heparin requires 4 h and can result in high rates of fertilization (Parrish *et al.*, 1985b, 1986b). An advantage to the use of heparin is that, while bull differences to a set dose exist (Parrish *et al.*, 1986b), the dose can be varied to increase or decrease the efficiency of capacitation (Parrish *et al.*, 1986c). Heparin may be effective at several points in proposed events of capacitation, as previously discussed, but the exact mechanism of its effect on spermatozoa remains to be elucidated. In earlier studies of in-vitro fertilization, particularly in cattle (Brackett *et al.*, 1982; Fukui *et al.*, 1983; Lambert *et al.*, 1986), techniques for sperm capacitation have been confounded by collection of oocytes in heparinized media. Although high-ionic strength medium or another treatment was intended to capacitate the spermatozoa, heparin carried into the gamete co-culture with cumulus–oocyte complexes may have actually been responsible for sperm capacitation.

Despite success in capacitating ruminant spermatozoa *in vitro*, procedures are not as efficient as *in vivo*. *In vivo*, fertilization most likely occurs when sperm and egg ratios are close to unity as in the hamster (Cummins & Yanagimachi, 1982). In support of this, Crozet (1984) found very few supernumerary spermatozoa present in cow eggs collected near the time of fertilization. Sperm–egg ratios *in vitro* for ruminants have generally been 10 000–200 000:1 (Brackett *et al.*, 1982; Parrish *et al.*, 1985a, 1986b; Lambert *et al.*, 1986). Using heparin to capacitate bovine spermatozoa (Parrish *et al.*, 1986b), we have been able to reduce the ratio for frozen–thawed bovine spermatozoa to sperm–egg ratios of 2000:1 and for some bulls to as low as 500:1 (J. J. Parrish, unpublished). Development of culture media able to maintain viability of spermatozoa at dilutions sufficient to give even a 10:1 sperm–egg ratio are lacking. This is an inherent difficulty in using sperm fertilizing ability to evaluate efficiency of capacitation. An alternative method has been developed using the fusogenic lipid lysophosphatidylcholine (Parrish *et al.*, 1985b). At 100 µg/ml, lysophosphatidylcholine induces an acrosome reaction only in capacitated bovine spermatozoa. This test has only been applied to the heparin capacitation system but should prove useful in comparing the efficiencies of the different capacitation systems to each other.

Fertilization

The success of fertilization *in vitro* relies upon completion of both oocyte maturation and sperm capacitation. Failure of either of these steps will result in failure of fertilization. In addition, the sperm concentration, time of sperm–egg interaction, medium utilized and temperature play a role in successful fertilization which will result in developmentally competent zygotes. Except when using fertilization as a means of detecting sperm capacitation or perhaps in-vitro fertilization as a predictor of in-vivo fertility, the intended outcome of in-vitro fertilization is to produce a viable embryo. In an in-vitro fertilization system one of the prime reasons for failure of the resulting zygotes to develop is the occurrence of polyspermy. *In vivo*, an oocyte encounters very few spermatozoa at the time of fertilization and blocks to polyspermy have time to be expressed. *In vitro*, large numbers of spermatozoa are placed with oocytes. If sperm capacitation is not efficient, few spermatozoa will fertilize oocytes and polyspermy is not a problem. However, as capacitation systems become more efficient, larger numbers of spermatozoa surrounding the oocyte are capable of fertilizing that oocyte and the probability of many spermatozoa penetrating the zona pellucida and vitelline membrane increases. The time that spermatozoa and oocytes are allowed to interact becomes important because early blocks to polyspermy in mammalian oocytes may not occur as rapidly as in invertebrates (reviewed by Wolf, 1981; Shapiro, 1981) and the block to polyspermy is reduced in aged oocytes (Hunter, 1967). Additionally, in-vitro matured oocytes may not develop the block to polyspermy to the same effectiveness as in-vivo matured oocytes (Leibfried-Rutledge *et al.*, 1986a). Sperm concentrations and time of sperm–oocyte interaction should therefore both be controlled to minimize polyspermy.

Table 5. In-vitro fertilization in goats

	Sperm capacitation treatment	Oocyte source	Fertilization (%)
Shorgan (1984)	A23187	In-vivo matured	49
Song & Iritani (1985)	18 h at 20°C	In-vitro matured	57

Medium composition can influence fertilization success. We have demonstrated that glucose blocks bovine sperm capacitation (Parrish *et al.*, 1985a; Susko-Parrish *et al.*, 1985). In other ruminants, unique compounds might also be important in inhibiting or aiding fertilization.

The effect of temperature cannot be overly emphasized. *In vivo*, fertilization occurs at core body temperature which is 38–39°C for cattle, 39°C for sheep and 38–39°C for goats. In cattle, fertilization frequencies *in vitro* are very dependent on temperature and highest frequencies occur at 39°C (Lenz *et al.*, 1983a). Temperature not only controls efficiency of capacitation but probably also the ability to undergo a physiological acrosome reaction in response to interaction with the zona pellucida.

While optimization of fertilization conditions for cattle have been reported (Brackett *et al.*, 1982; Sirard *et al.*, 1985; Parrish *et al.*, 1986b) differences between bulls still persist. Optimal conditions may need to be established for each male and possibly each ejaculate. Ruminants have a distinct advantage over other species in that their semen can be successfully frozen (Mann, 1964). Only a small portion of an ejaculate need then be utilized to determine fertilization conditions for each ejaculate or male. In-vitro matured oocytes are ideal for testing the fertilization ability of bull ejaculates. These oocytes appear to yield fertilization frequencies comparable to those with in-vivo matured oocytes (Leibfried-Rutledge *et al.*, 1986a) and large numbers can be obtained from ovaries collected at the abattoir.

Comparative results from in-vitro fertilization studies published since the review of Wright & Bondioli (1981) are shown for cattle in Tables 1 and 2. For sheep, 80% of in-vivo matured oocytes mature fully *in vitro* and 80% of these are penetrated by spermatozoa; 7 of 16 (44%) recipients of such fertilized eggs became pregnant (Cheng *et al.*, 1986). Fertilization results for goat gametes have also been good (Table 5). Overall, while fertilization frequencies have differed, rates of fertilization, >80%, with low rates of polyspermy (<15%) have been reported.

Development of embryos to a transferable stage

The production of offspring from in-vitro fertilization is considerably restricted by the absence of suitable in-vitro culture systems capable of efficiently supporting development to the morula or blastocyst stage which can be transferred non-surgically. Indeed, for mammals, suitable in-vitro culture systems are available only for embryos from a few strains of mice (Goddard & Pratt, 1983) and from women (Purdy, 1982). Cattle and sheep embryos cultured *in vitro* from the 1–4-cell stages rarely cleave beyond 8–16 cells, whereas embryos cultured from the 8–16-cell stage frequently develop into morulae or blastocysts (Thibault, 1966; Newcomb, 1982; Eyestone & First, 1986; reviewed by Wright & Bondioli, 1981). These observations suggest the existence of a block to in-vitro development at the 8–16-cell stage for cattle (Thibault, 1966; Camous *et al.*, 1984; Eyestone & First, 1986) and sheep (Bondioli & Wright, 1980) embryos. These observations also suggest that (1) the oviduct but not culture medium contains factors or conditions conducive to early embryonic development and (2) that certain developmental events occurring between the 1- and 16-cell stage require specific environmental factors or conditions normally provided by the oviduct. The time of blocked development occurs at a time of prolonged cell cycle, DNA synthesis and a transition from maternal to zygotic control of development for the mouse (Goddard & Pratt, 1983), cow (King *et al.*, 1985), sheep (Calarco & McLaren, 1976) and pig (Norberg, 1973). It is known that cells of the embryo remain alive during the period of blocked development (W. H. Eyestone, unpublished) and that they cannot be rescued once the block is initiated (Eyestone & First, 1986). In mice this period of transition from maternal to zygotic control of development is accompanied by the production of a class of proteins known as heat-shock proteins. More specifically, heat-shock proteins 68 and 70 have been shown to be the first products of the mouse embryonic genome (Flach *et al.*, 1982; Bensaude *et al.*, 1983; Bolton *et al.*, 1984). In mouse embryos which pass through this critical 2-cell period *in vitro*, heat-shock protein 70 returns to a low concentration by the 4-cell stage (Barnes *et al.*, 1987). Production of heat-shock proteins 68 and 70 at this transition period and under in-vitro culture conditions may reflect the inadequacy of the culture. Whether the culture medium fails to provide critical substrates for synthesis of essential proteins, cell cycle controlling signals, or a proper environment for transcription or translation is unknown. In spite of the absence of an adequate in-vitro embryo culture system, cattle and sheep embryos resulting from in-vitro fertilization have been developed to morula and blastocyst stages. This has been accomplished by direct surgical transfer of zygotes or early embryos to the oviduct of the cow (Brackett *et al.*, 1982, 1984). The oviducts of sheep *in situ* (Eyestone *et al.*, 1985) and rabbits (Sirard *et al.*, 1985) have also been used as temporary surrogate in-vivo incubators for development of cattle zygotes or early embryos to a transferable morula or blastocyst stage. A start to the development of an adequate in-vitro embryo culture system has been provided by the co-culture of embryos on a feeder layer of oviduct epithelial cells (Rexroad & Powell, 1986; F. Gandolfi, T. A. L. Brevini & R. M. Moor, unpublished). Development was limited for the experiments of Rexroad & Powell (1986) by culturing 1-cell sheep embryos for 24 h before transfer to recipient ewes. However, 43·8% of the sheep embryos reached the blastocyst stage after a 6-day culture period (F. Gandolfi, T. A. L. Brevini &

R. M. Moor, unpublished). The successful culture of cattle embryos on a layer of feeder cells has not been reported, but there is evidence that bovine trophoblast cells are competent to carry bovine oocytes through this period of blocked development (Camous *et al.*, 1984). It is also unknown whether cell types from other tissues or from tissues of animals not in oestrus can support embryo development. Since species differ in the length of time during which the oviduct will support development of an embryo (i.e. rabbit 3–4 days, sheep 5–6 days; Boland, 1984) it would be interesting to know whether epithelial cell cultures from these respective species also support embryo development for the same species specific times. It is hoped that well designed hypotheses and experiments utilizing these co-culture systems will identify the requirements for culture of embryos from fertilization to the blastocyst stage such that totally in-vitro systems for embryo development can be accomplished.

Use of in-vitro fertilization to predict fertility

As fertilization is one of the key events which determines in-vivo fertility of a male, it has been hoped that development of in-vitro fertilization will result in a procedure to predict fertility *in vivo*. However, fertility is complex and composed of multiple events such as sperm transport, capacitation, oocyte maturation, ovulation, female endocrine status, normality of fertilization and embryo development. The frequency of sperm penetration of zona-free hamster oocytes by bovine spermatozoa has been shown to be related to bull fertility (Bousquet & Brackett, 1982). Graham & Foote (1984) also found that penetration of zona-free hamster oocytes by bovine spermatozoa treated with acrosome reaction-inducing liposomes was correlated with bull fertility when the amount of liposome required to achieve maximum penetration of oocytes was used. However, in-vivo fertility has yet to be completely predicted by in-vitro fertilization tests. Because of the complexity of in-vivo fertility this is not unexpected. There is as yet no report on the correlation of in-vivo fertility of ruminants with results of homologous in-vitro fertilization tests. However, bull effects on embryonic development that were not related to sperm penetration frequency or normality of fertilization have been reported (Leibfried-Rutledge *et al.*, 1986a). A better prediction of fertility should therefore be obtained by including homologous fertilization data in a multiple regression model along with measures of semen quality and ability of embryos to develop.

Conclusions

It is apparent that our understanding of fertilization in ruminants has progressed considerably in recent years and that this understanding has resulted in workable systems for fertilizing oocytes of cattle, sheep and goats *in vitro*. In cattle in-vitro fertilization has resulted in birth of live offspring. There is also some evidence that in-vitro fertilization may provide a useful adjunct to present methods for evaluating the fertility of bulls. It is also apparent there is much yet to be learned about the processes of sperm capacitation and the acrosome reaction, the development of maturational competence by oocytes and the specific signals within the oocyte imparting the ability to complete maturation, to be fertilized, to initiate and complete maturation and, ultimately, result in normal embryo development.

This research was supported by the College of Agricultural and Life Sciences, University of Wisconsin-Madison; by a grant from the W. R. Grace and Company, NY; and by grant No. HD-18345-02 from the N.I.C.H.D.

References

Adams, C.E. & Chang, M.C. (1962) Capacitation of rabbit spermatozoa in the fallopian tube and in the uterus. *J. exp. Zool.* **151**, 159–165.

Ahuja, K.K. (1984) Lectin-coated agarose beads in the investigation of sperm capacitation in the hamster. *Devl Biol.* **104**, 131–142.

Aonuma, S., Okabe, M., Kishi, Y., Kawaguchi, M. & Yamada, H. (1982) Capacitation inducing activity of serum albumin in fertilization of mouse ova in vitro. *J. Pharm. Dyn.* **5**, 980–987.

Austin, C.R. (1951) Observations on the penetration of the sperm into the mammalian egg. *Aust. J. Sci. Res. B* **4**, 581–596.

Ball, G.D., Leibfried, M.L., Lenz, R.W., Ax, R.L., Bavister, B.D. & First, N.L. (1983) Factors affecting successful in vitro fertilization of bovine follicular oocytes. *Biol. Reprod.* **28**, 717–725.

Barnes, F.L., Robl, J.M. & First, N.L. (1987) Nuclear transplantation in mouse embryos: Assessment of nuclear function. *Biol. Reprod.* (in press).

Barros, C. (1974) Capacitation of mammalian spermatozoa. In *Physiology and Genetics of Reproduction*, pp. 3–24. Plenum Press, New York.

Bedford, J.M. (1969) Limitations of the uterus in the development of the fertilizing ability (capacitation) of spermatozoa. *J. Reprod. Fert., Suppl.* **8**, 19–26.

Bedford, J.M. (1970) Sperm capacitation and fertilization in mammals. *Biol. Reprod.* (Suppl. 2), 128–158.

Bedford, J.M. (1972) Sperm transport, capacitation and fertilization. In *Reproductive Biology*, pp. 338–392. Eds H. Balin & S. Glasser. Excerpta Medica, Amsterdam.

Bensaude, O., Bakinet, C., Morange, M. & Jacob, F. (1983) Heat shock proteins, first major products of zygotic gene activity in mouse embryo. *Nature, Lond.* **305**, 331–333.

Bleil, J.D. & Wassarman, P.M. (1983) Sperm-egg interactions in the mouse. Sequence of events and induction of the acrosome reaction by a zona pellucida glycoprotein. *Devl Biol.* **95**, 317–324.

Boland, M.P. (1984) Use of the rabbit oviduct as a screening tool for the viability of mammalian eggs. *Theriogenology* **21**, 126–137.

Bolton, V.N., Oades, P.J. & Johnson, M.H. (1984) The relationship between cleavage, DNA replication, and gene expression in the mouse 2-cell embryo. *J. Embryol. exp. Morph.* **79**, 139–163.

Bondioli, K.R. & Wright, R.W., Jr (1980) Influence of culture media on *in vitro* fertilization of ovine tubal oocytes. *J. Anim. Sci.* **51**, 660–667.

Bousquet, D. & Brackett, B.G. (1982) Penetration of zona-free hamster ova as a test to assess fertilizing ability of bull sperm after frozen storage. *Theriogenology* **17**, 199–213.

Brackett, B.G., Bousquet, D., Boice, M.L., Donawick, W.J., Evans, J.F. & Dressel, M.A. (1982) Normal development following in vitro fertilization in the cow. *Biol. Reprod.* **27**, 147–158.

Brackett, B.G., Keefer, C.L., Troop, C.G., Donawick, W.J. & Bennett, K.A. (1984) Bovine twins resulting from in vitro fertilization. *Theriogenology* **21**, 224, Abstr.

Bradley, M.P. & Garbers, D.L. (1983) The stimulation of bovine caudal epididymal sperm forward motility by bovine cumulus-egg complexes in vitro. *Biochem. Biophys. Res. Commun.* **115**, 777–787.

Calarco, P.G. & McLaren, A. (1976). Ultrastructural observations of preimplantation stages of the sheep. *J. Embryol. exp. Morph.* **36**, 609–622.

Camous, S., Heyman, Y., Meziou, W. & Menezo, Y. (1984) Cleavage beyond the block stage and survival after transfer of early bovine embryos co-cultured with trophoblastic vesicles. *J. Reprod. Fert.* **72**, 479–485.

Carlson, D., Black, D.L. & Howe, G.R. (1970) Oviduct secretion in the cow. *J. Reprod. Fert.* **22**, 549–552.

Casu, B. (1985) Structure and biological activity of heparin. *Advances in Carbohydrate Chem. Biochem.* **43**, 51–134.

Chang, M.C. (1951) Fertilizing capacity of spermatozoa deposited into the Fallopian tubes. *Nature, Lond.* **168**, 697–698.

Cheng, W.T.K., Moor, R.M. & Polge, C. (1986) In vitro fertilization of pig and sheep oocytes matured in vivo and in vitro. *Theriogenology* **25**, 146.

Choudary, J.B., Gier, H.T. & Marion, G.B. (1968) Cyclic changes in bovine vesicular follicles. *J. Anim. Sci.* **27**, 468–478.

Critser, E.S., Leibfried, M.L. & First, N.L. (1984) The effect of semen extension, cAMP and caffeine on in vitro fertilization of bovine oocytes. *Theriogenology* **21**, 625–631.

Critser, E.S., Leibfried-Rutledge, M.L., Eyestone, W.H., Northey, D.L., & First, N.L. (1986) Acquisition of developmental competence during maturation in vitro. *Theriogenology* **25**, 150, Abstr.

Crozet, N. (1984) Ultrastructural aspects of in vivo fertilization in the cow. *Gamete Res.* **10**, 241–251.

Cummins, J.M. & Yanagimachi, R. (1982) Sperm-egg ratios and the site of the acrosome reaction during in vivo fertilization in the hamster. *Gamete Res.* **5**, 239–256.

Davis, B.K. (1981) Timing of fertilization in mammals; sperm cholesterol/phospholipid ratio as a determinant of the capacitation interval. *Proc. natn. Acad. Sci. U.S.A.* **78**, 7560–7564.

Davis, B.K. (1982) Uterine fluid proteins bind sperm cholesterol during capacitation in the rabbit. *Experientia* **38**, 1063–1064.

Eyestone, W.H. & First, .N.L. (1986) A study of the 8- to 16-cell developmental block in bovine embryos cultured in vitro. *Theriogenology* **25**, 152, Abstr.

Eyestone, W.H., Northey, D.L. & Leibfried-Rutledge, M.L. (1985) Culture of 1-cell bovine embryos in the sheep oviduct. *Biol. Reprod.* **32** (Suppl. 1), 100, Abstr.

First, N.L. & Haseltine, F.P. (1987) Development of oocyte maturation and developmental competence and assessment of oocyte maturation. *J. Contraception* (in press).

Flach, G., Johnson, M.H., Braude, P.R., Taylor, R.A.S. & Bolton, V.N. (1982) The transition from maternal to embryonic control in the 2-cell mouse embryo. *EMBO Jl* **1**, 681–686.

Florman, H.M. & Storey, B.T. (1982) Mouse gamete interactions. The zona pellucida is the site of the acrosome reaction leading to fertilization in vitro. *Devl Biol.* **91**, 121–130.

Fraser, L.R. (1983) Mouse sperm capacitation assessed by kinetics and morphology of fertilization *in vitro*. *J. Reprod. Fert.* **69**, 419–428.

Fukui, Y., Fukushima, M. & Ono, H. (1983) Fertilization in vitro of bovine oocytes after various sperm procedures. *Theriogenology* **20**, 651–660.

Fukushima, M. & Fukui, Y. (1985) Effects of gonadotropin and steroids on the subsequent fertilizability of extrafollicular bovine oocytes cultured in vitro. *Anim. Reprod. Sci.* **9**, 323–332.

Go, K.J. & Wolf, D.P. (1985) Albumin-mediated changes in sperm sterol content during capacitation. *Biol. Reprod.* **32**, 145–153.

Goddard, M.J. & Pratt, H.P.M. (1983) Control of events during early cleavage of the mouse embryo: an analysis of the '2-cell block'. *J. Embryol. exp. Morph.* **73**, 111–133.

Gordon, M., Dandekar, P.V. & Bartoszewich, W. (1975) The surface coat of epididymal, ejaculated and capacitated sperm. *J. Ultrastruct. Res.* **50**, 199–207.

Graham, J.K. & Foote, R.H. (1984). In vitro fertilization of zona-free hamster ova by liposome-treated bovine sperm: a fertility assay. *Biol. Reprod.* **30** (Suppl. 1), 112, Abstr.

Handrow, R.R., Lenz, R.W. & Ax, R.L. (1982) Structural comparisons among glycosaminoglycans to promote an acrosome reaction in bovine spermatozoa. *Biochem. Biophys. Res. Commun.* **107**, 1326–1332.

Handrow, R.R., Boehm, S.K., Lenz, R.W., Robinson, J.A. & Ax, R.L. (1984) Specific binding of the glycosaminoglycan ³H-Heparin to bovine, monkey and rabbit spermatozoa in vitro. *J. Androl.* **5**, 51–63.

Handrow, R.R., Parrish, J.J. & Susko-Parrish, J.L. (1986a) Effect of glycosaminoglycans on capacitation and the acrosome reaction of bovine and hamster sperm. *Biol. Reprod.* **34** (Suppl. 1), 93, Abstr.

Handrow, R.R., Parrish, J.J. & First, N.L. (1986b) Heparin stimulates calcium uptake by bovine sperm in vitro. *J. Androl.* **7**, 23, Abstr.

Hensleigh, H.C. & Hunter, A.G. (1985) In vitro maturation of bovine cumulus enclosed primary oocytes and their subsequent in vitro fertilization and cleavage. *J. Dairy Sci.* **68**, 1456–1562.

Herz, Z., Northey, D., Lawyer, M. & First, N.L. (1985) Acrosome reaction of bovine spermatozoa in vivo, sites and effects of stages of the estrous cycle. *Biol. Reprod.* **32**, 1163–1168.

Hunter, R.H.F. (1967) The effects of delayed insemination on fertilization and early cleavage in the pig. *J. Reprod. Fert.* **13**, 133–147.

Hunter, R.H.F. (1969) Capacitation in the golden hamster, with special reference to the influence of the uterine environment. *J. Reprod. Fert.* **20**, 223–237.

Hunter, R.H.F. (1980) *Physiology and Technology of Reproduction in Female Domestic Animals.* Academic Press, London.

Hunter, R.H.F. (1984) Pre-ovulatory arrest and peri-ovulatory redistribution of competent spermatozoa in the isthmus of the pig oviduct. *J. Reprod. Fert.* **72**, 203–211.

Hunter, R.H.F. & Hall, J.P. (1974) Capacitation of boar spermatozoa, synergism between uterine and tubal environments. *J. exp. Zool.* **188**, 203–214.

Hunter, R.H.F. & Nichol, R. (1983) Transport of spermatozoa in the sheep oviduct; preovulatory sequestering of cells in the caudal isthmus. *J. exp. Zool.* **228**, 121–128.

Hunter, R.H.F. & Wilmut, I. (1982) The rate of functional sperm transport into the oviducts of mated cows. *Anim. Reprod. Sci.* **5**, 167–173.

Hunter, R.H.F. & Wilmut, I. (1984) Sperm transport in the cow. Peri-ovulatory redistribution of viable cells within the oviduct. *Reprod. Nutr. Dévelop.* **24**, 597–608.

Hunter, R.H.F., Nichol, R. & Crabtree, S.M. (1980) Transport of spermatozoa in the ewe; timing of the establishment of a functional population in the oviduct. *Reprod. Nutr. Dévelop.* **20**, 1869–1875.

Hunter, R.H.F., Barwise, L. & King, R. (1982) Sperm storage and release in the sheep oviduct in relation to the time of ovulation. *Br. vet. J.* **138**, 225–232.

Iritani, A., Kasai, M., Niwa, K. & Song, H.B. (1984) Fertilization *in vitro* of cattle follicular oocytes with ejaculated spermatozoa capacitated in a chemically defined medium. *J. Reprod. Fert.* **70**, 487–492.

Katska, L. & Smorag, Z. (1985) The influence of culture temperature on in vitro maturation of bovine oocytes. *Anim. Reprod. Sci.* **9**, 205–212.

King, W.A., Niar, A. & Betteridge, K.J. (1985) The nucleolus organizer regions of early bovine embryos. *J. Dairy Sci.* **68** (Suppl. 2), 249, Abstr.

Kjellen, L., Oldberg, A. & Hook, M. (1980) Cell surface heparan sulfate: mechanisms of proteoglycan-cell association. *J. biol. Chem.* **255**, 10407–10413.

Koehler, J.K. (1978) Observations on the fine structure of vole spermatozoa with particular reference to cytoskeletal elements in the mature head. *Gamete Res.* **1**, 247–259.

Lambert, R.D., Sirard, M.A., Bernard, C., Beland, R., Rioux, J.E., Leclerc, P., Menard, D.P. & Bedoya, M. (1986) In vitro fertilization of bovine oocytes matured in vivo and collected at laparoscopy. *Theriogenology* **25**, 117–133.

Langlais, J. & Roberts, K.D. (1985) A molecular membrane model of sperm capacitation and the acrosome reaction of mammalian spermatozoa. *Gamete Res.* **12**, 183–224.

Laterra, J., Silbert, J.E. & Culp, L.A. (1983) Cell surface heparan sulfate mediates some adhesive responses to glycosaminoglycan-binding matrices, including fibronectin. *J. Cell Biol.* **96**, 112–123.

Lee, C.N. & Ax, R.L. (1984) Concentrations and composition of glycosaminoglycans in the female bovine reproductive tract. *J. Dairy Sci.* **67**, 2006–2009.

Lee, C.N., Clayton, M.K., Bushmeyer, S.M., First, N.L. & Ax, R.L. (1986) Glycosaminoglycans in ewe reproductive tracts and their influence on acrosome reactions in bovine spermatozoa in vitro. *J. Anim. Sci.* **63**, 861–867.

Leibfried, M.L. & Bavister, B.D. (1983) Fertilizability of in vitro matured oocytes from golden hamsters. *J. exp. Zool.* **226**, 481–485.

Leibfried, L. & First, N.L. (1979) Characterization of bovine follicular oocytes and their ability to mature in vitro. *J. Anim. Sci.* **48**, 76–86.

Leibfried-Rutledge, M.L., Critser, E.S., Eyestone, W.H., Northey, D.L. & First, N.L. (1986a) Developmental potential of bovine oocytes matured in vitro or in vivo. *Biol. Reprod.* (in press).

Leibfried-Rutledge, M.L., Critser, E.S. & First, N.L. (1986b) Fetal calf serum is the preferred supplement for in vitro maturation of cumulus-oocyte complexes. *Biol. Reprod.* **35**, 850–857.

Lenz, R.W., Ax, R.L., Grimek, H.J. & First, N.L. (1982) Proteoglycan from bovine follicular fluid enhances an acrosome reaction in bovine spermatozoa. *Biochem. Biophys. Res. Commun.* **106**, 1092–1098.

Lenz, R.W., Ball, G.D., Leibfried, M.L., Ax, R.L. & First, N.L. (1983a) In vitro maturation and fertilization of bovine ocytes are temperature-dependent processes. *Biol. Reprod.* **29**, 173–179.

Lenz, R.W., Ball, G.D., Lohse, J.K., First, N.L. & Ax, R.L. (1983b) Chondroitin sulfate facilitates an acrosome reaction in bovine spermatozoa as evidenced by light microscopy, electron microscopy and in vitro fertilization. *Biol. Reprod.* **28**, 683–690.

Lorton, S.P. & First, N.L. (1979) Hyaluronidase does not disperse the cumulus oophorus surrounding bovine ova. *Biol. Reprod.* **21**, 301–308.

Mahi-Brown, C.A. & Yanagimachi, R. (1983) Parameters influencing ovum pickup by oviductal fimbria in the golden hamster. *Gamete Res.* **8**, 1–10.

Mann, T. (1964) *The Biochemistry of Semen and of the Male Reproductive Tract.* Methuen, London.

Moor, R.M. & Crosby, I.M. (1985) Temperature induced abnormalities in sheep oocytes during maturation. *J. Reprod. Fert.* **75**, 467–473.

Moor, R.M. & Gandolfi, F. (1987) Molecular and cellular changes associated with maturation and early development of sheep eggs. *J. Reprod. Fert., Suppl.* **34**, 55–69.

Moor, R.M. & Trounson, A.O. (1977) Hormonal and follicular factors affecting maturation of sheep oocytes *in vitro* and their subsequent developmental capacity. *J. Reprod. Fert.* **49**, 101–109.

Newcomb, R. (1982) Egg recovery and transfer in cattle. In *Mammalian Egg Transfer*, pp. 81–118. Ed. C. R. Adams. CRC Press, Boca Raton.

Norberg, H. (1973) Ultrastructural aspects of the pre-attached pig embryo: cleavage and early blastocyst stages. *Z. Anat. EntwGesch.* **143**, 95–114.

Oliphant, G., Reynolds, A.B. & Thomas, T.S. (1985) Sperm surface components involved in the control of the acrosome reaction. *Am. J. Anat.* **174**, 269–283.

O'Rand, M.G. (1982) Modification of the sperm membrane during capacitation. *Ann. N. Y. Acad. Sci.* **383**, 392–402.

Overstreet, J.W. (1982) Transport of gametes in the reproductive tract of the female mammal. In *Mechanism and Control of Animal Fertilization*, pp. 499–543. Ed. J. F. Hartman. Academic Press, New York.

Overstreet, J.W., Cooper, G.W. & Katz, D.F. (1978) Sperm transport in the reproductive tract of the female rabbit. II. The sustained phase of transport. *Biol. Reprod.* **19**, 115–132.

Parrish, J.J., Susko-Parrish, J.L. & First, N.L. (1985a) Effect of heparin and chondroitin sulfate on the acrosome reaction and fertility of bovine sperm in vitro. *Theriogenology* **24**, 537–549.

Parrish, J.J., Susko-Parrish, J.L. & First, N.L. (1985b) Role of heparin in bovine sperm capacitation. *Biol. Reprod.* **32** (Suppl. 1), 211, Astr.

Parrish, J.J., Susko-Parrish, J.L. & First, N.L. (1986a) Capacitation of bovine sperm by oviduct fluid or heparin is inhibited by glucose. *J. Androl.* **7**, 22, Abstr.

Parrish, J.J., Susko-Parrish, J.L., Leibfried-Rutledge, M.L., Critser, E.S., Eyestone, W.H. & First, N.L. (1986b) Bovine in vitro fertilization with frozen-thawed semen. *Theriogenology* **25**, 591–600.

Parrish, J.J., Susko-Parrish, J.L., Critser, E., Leibfried-Rutledge, L., Barnes, F., Eyestone, W. & First, N.L. (1986c) Bovine in vitro fertilization. *Proc. 11th Tech. Conf. A.I. & Reproduction* p. 120, Abstr.

Parrish, R.F., Wincek, T.J. & Polakoski, K.L. (1980) Glycosaminoglycans stimulation of the in vitro conversion of boar proacrosin into acrosin. *J. Androl.* **1**, 89–95.

Purdy, J.M. (1982) Methods for fertilization and embryo culture in vitro. In *Human Conception In Vitro*, p. 135. Eds. R. G. Edwards & J. M. Purdy. Academic Press, London.

Rexroad, C.E. & Powell, A.M. (1986) Co-culture of sheep ova and cells from sheep oviduct. *Theriogenology* **25**, 187, Abstr.

Shapiro, B.M. (1981) Awakening of the invertebrate egg at fertilization. In *Fertilization and Embryonic Development In Vitro*, pp. 233–255. Eds L. Mastroianni & J. D. Biggers. Plenum Press, New York.

Shorgan, B. (1984) Fertilization of goat and ovine ova in vitro by ejaculated spermatozoa after treatment with ionophore A23187. *Bull. Nippon Vet. Zootech. College* **33**, 219–221.

Sirard, M.A. & Lambert, R.D. (1985) In vitro fertilization of bovine follicular oocytes obtained by laparoscopy. *Biol. Reprod.* **33**, 487–494.

Sirard, M.A., Lambert, R.D., Menard, D.P. & Bedoya, M. (1985) Pregnancies after in-vitro fertilization of cow follicular oocytes, their incubation in rabbit oviduct and their transfer to the cow uterus. *J. Reprod. Fert.* **75**, 551–556.

Song, H.B. & Iritani, A. (1985) In vitro fertilization of goat follicular oocytes with epididymal spermatozoa capacitated in a chemically defined medium. *Proc. 3rd AAAP Animal Sci. Congress* **1**, 463–465.

Srinivasan, S.R., Lopez, S.A., Radhakrishnamurthy, B. & Berenson, G.S. (1970) Complexing of serum pre-β and β-lipoproteins and acid mucopolysaccharides. *Atherosclerosis* **12**, 321–334.

Staigmiller, R.B. & Moor, R.M. (1984) Effect of follicle cells on the maturation and developmental competence of ovine oocytes matured outside the follicle. *Gamete Res.* **9**, 221–229.

Susko-Parrish, J.L., Parrish, J.J. & Handrow, R.R. (1985) Glucose inhibits the action of heparin by an indirect mechanism. *Biol. Reprod.* **32** (Suppl. 1), 80, Abstr.

Thakkar, J.K., East, J. & Frason, R.C. (1984) Modulation of phospholipase A_2 activity with human sperm membranes by divalent cations and calcium antagonists. *Biol. Reprod.* **30**, 679–686.

Thibault, C. (1966) La culture in vitro de l'oeuf de vache. *Annls Biol. anim. Biochim. Biophys.* **6**, 159–164.

Thibault, C. (1977) Are follicular maturation and oocyte maturation independent processes? *J. Reprod. Fert.* **51**, 1–15.

Thibault, C., Gerard, M. & Menezo, Y. (1975) Acquisition par l'oocyte de lapine et de veau du facteur de decondensation do noyau do spermatozoide fecondant (MPGH). *Annls Biol. anim. Biochim. Biophys.* **15**, 705–714.

Viriyapanich, P. & Bedford, J.M. (1981) The fertilization performance in vivo of rabbit spermatozoa capacitated in vitro. *J. exp. Zool.* **216**, 169–174.

Wheeler, M.B. & Seidel, G.E. (1986) Time course of in vitro capacitation of frozen and unfrozen bovine spermatozoa. *Theriogenology* **25**, 216, Abstr.

Wilmut, I. & Hunter, R.H.F. (1984) Sperm transport into the oviducts of heifers mated early in oestrus. *Reprod. Nutr. Dévelop.* **24**, 461–468.

Wolf, D.E., Hagopian, S.S. & Ishijima, S. (1986) Changes in sperm plasma membrane lipid diffusibility after hyperactivation during in vitro capacitation in the mouse. *J. Cell Biol.* **102**, 1372–1377.

Wolf, D.P. (1979) Mammalian fertilization. In *The Biology of the Fluids of the Female Genital Tract*, pp. 407–414. Eds F. K. Beller & G. F. B. Schumacher. Elsevier North Holland, New York.

Wolf, D.P. (1981) The mammalian egg's block to polyspermy. In *Fertilization and Embryonic Development In Vitro*, pp. 183–197. Eds L. Mastroianni & J. D. Biggers. Plenum Press, New York.

Wright, R.W., Jr & Bondioli, K.R. (1981) Aspects of in vitro fertilization and embryo culture in domestic animals. *J. Anim. Sci.* **53**, 702–729.

Yanagimachi, R. (1981) Mechanisms of fertilization in mammals. In *Fertilization and Embryonic Development In Vitro*, pp. 81–182. Eds L. Mastroianni & J. D. Biggers. Plenum Press, New York.

Yanagimachi, R. & Usui, N. (1974) Calcium dependence of the acrosome reaction and activation of guinea pig spermatozoa. *Expl Cell Res.* **89**, 161–174.

J. Reprod. Fert., Suppl. **34** (1987), 167–186

Endocrine regulation of puberty in cows and ewes

J. E. Kinder, M. L. Day* and R. J. Kittok

Department of Animal Science, University of Nebraska, Lincoln, NE 68583, U.S.A.

Summary. Sexual maturation in cows and ewes is modulated through changes in hypothalamic inhibition. This inhibition results in little or no stimulation of the release of gonadotrophins from the anterior pituitary. The ovary has a primary role in inhibiting gonadotrophin secretion during the prepubertal period and the responsiveness to the negative feedback effects of oestrogen decreases during the peripubertal period. There is also an increased secretion of ovarian progesterone during the peripubertal period but its role in the process of sexual maturation is not clear. Photoperiodic cues and dietary intake act upon the hypothalamus to modulate gonadotrophin secretion during sexual maturation and, in turn, influence the time when puberty occurs.

Introduction

Puberty in this article is defined as the first behavioural oestrus accompanied by the development of a corpus luteum that is maintained for a period characteristic of a particular species. The maturation process that culminates in puberty occurs in a gradual fashion. It is initiated before birth and continues throughout the prepubertal and peripubertal periods of developing females. Some components of the endocrine system of prepubertal females are functional long before puberty occurs. For example, the gonadotrophs of the pituitary respond to hypothalamic secretagogues and the ovaries respond to exogenous gonadotrophins during the early phases of sexual maturation, well before first ovulation. Likewise the ovaries are able to respond to exogenous gonadotrophins administered before puberty. However, there appears to be at least one component of the endocrine system of the prepubertal ewe lamb and heifer that is incapable of functioning in an adult fashion until at or near the time of puberty. An alternative view is that all components of the endocrine system of prepubertal females are mature but one or more specific components are inhibited from functioning in an adult fashion.

Pituitary and ovarian function

Hypophysial stalk transection resulted in complete cessation of pulsatile release of luteinizing hormone (LH) at 5 months of age in heifers (Anderson *et al.*, 1981). Pituitary responsiveness to luteinizing hormone-releasing hormone (LHRH) was evaluated in prepubertal heifers by administration of LHRH at monthly intervals from 1 month of age to puberty (Schams *et al.*, 1981). Pituitary secretion of LH in response to LHRH was observed at all ages but the magnitude of this response increased with age. Increased follicle-stimulating hormone (FSH) secretion only occurred during the first 5 months of the 9-month prepubertal period. LHRH was administered at 4 and 10 months of age to prepubertal heifers (McLeod *et al.*, 1984). The amount of LH and FSH released in response to LHRH was greater in the older heifers. Pituitary concentrations of FSH and LH were higher in

---

*Present address: Department of Animal Science, The Ohio State University, 2029 Fyffe Road, Columbus, Ohio 43210–1095, U.S.A.

prepubertal than postpubertal heifers, indicating that the pituitary is able to synthesize and store LH and FSH before puberty (Desjardins & Hafs, 1968).

Exogenous administration of gonadotrophins has resulted in the production of fertile ova in heifers at 1 month of age (Seidel *et al.*, 1971). The number of induced ovulations resulting from exogenous gonadotrophins was greater in 5-month-old than in 1-month-old heifers. Therefore, both the pituitary and ovary respond to exogenously administered hormones early in life and maturation of these organs appears to occur before puberty.

Positive feedback of oestradiol in inducing the preovulatory surge of gonadotrophins

Increasing levels of oestradiol during the follicular phase of the oestrous cycle are thought to act at the hypothalamo–pituitary axis to induce the preovulatory surge of LH and FSH secretion in cattle (Echternkamp & Hansel, 1973). Induction of the preovulatory gonadotrophin surge by oestradiol is a component of the endocrine system that is essential for puberty to occur. Surges in LH secretion similar to those present before ovulation in mature cows have been induced in prepubertal heifers by administration of oestradiol (Schillo *et al.*, 1982). This component of the endocrine system appears to become functional between 3 and 5 months of age in heifers and the amount of LH released in response to oestradiol is actually greater in pre- than postpubertal heifers (Staigmiller *et al.*, 1979).

Squires *et al.* (1972) initially demonstrated that oestradiol would induce surges of LH in prepubertal ewe lambs that were similar to the preovulatory surge in sexually mature ewes. Foster & Karsch (1975) found that oestradiol failed to elicit a surge in secretion of LH at 3 weeks of age in ewe lambs, but surges progressively increased in magnitude as lambs were treated with oestradiol at 7, 12, 20 and 27 weeks of age. By 27 weeks of age, the surge in LH secretion was similar to the response of ewes given a similar dose of oestradiol during the anoestrous season. In that flock, puberty is normally attained in ewe lambs between 30 and 50 weeks after birth. Exogenous oestradiol does not appear to be effective in inducing ovulation and/or subsequent development of a corpus luteum that is maintained for a complete oestrous cycle (Foster & Karsch, 1975; Tran *et al.*, 1979).

Secretion of gonadotrophins during sexual maturation

Secretion of LH during the prepubertal period

Circulating concentrations of LH have been evaluated from birth to puberty in heifers (Schams *et al.*, 1981); values increased from birth to 3 months of age, declined from 3 to 6 months of age and then increased up to the time of puberty at 10 months of age. Frequency of pulses of LH increased and the amplitude of LH pulses decreased from 1 month of age until puberty. Swanson *et al.* (1972) reported that mean concentrations of LH increased during the 110-day period preceding puberty in heifers. An increase in mean concentration of serum LH and frequency of episodic LH pulses occurs during the 126 days preceding puberty (Day *et al.*, 1984; Table 1). This increase during the last 4 months before puberty coincides with the second increase in LH secretion reported by Schams *et al.* (1981).

Others have reported that serum concentrations of LH did not increase during the period of sexual maturation (Gonzalez-Padilla *et al.*, 1975b; McLeod *et al.*, 1984). The frequency and duration of blood collection is important for evaluating the dynamic changes that occur in gonadotrophin secretion before puberty. In a recent study we collected from heifers blood samples at 20-min intervals for 24 h every 2 weeks during the 140-day period before puberty. Our objective for this intensive evaluation of prepubertal secretion of LH was to develop a regression equation of LH

Table 1. Mean concentration of LH in serum and LH pulse frequency before puberty in intact heifers (N = 6) (from Day *et al.*, 1984)

Day*	Mean LH conc.† (ng/ml)	LH pulse frequency‡ (pulses/h)
−126	0·85 ± 0·26	0·04 ± 0·04
−112	1·22 ± 0·29	0·06 ± 0·06
− 98	0·93 ± 0·20	0·13 ± 0·06
− 84	1·11 ± 0·25	0·15 ± 0·05
− 70	1·33 ± 0·30	0·19 ± 0·07
− 56	1·16 ± 0·23	0·19 ± 0·08
− 42	1·48 ± 0·29	0·25 ± 0·08
− 28	1·48 ± 0·24	0·31 ± 0·05
− 14	2·30 ± 0·20	0·48 ± 0·10

*Days relative to puberty.
†$y = 0.53 + 0.15x; P < 0.0001, R^2 = 0.28$.
‡$y = 0.36 + 0.38x; P < 0.0001, R^2 = 0.36$.

on days before puberty that could be used to predict the stage of sexual maturation in heifers that were killed before puberty. The heifers in this study reached puberty at a mean of 366 (s.e.m. = 4) days of age at a mean weight of 244 (s.e.m. = 3) kg. Mean concentrations of serum LH tended to increase in a gradual linear fashion during the 140-day period before puberty (Fig. 1). Amplitude of pulses of LH declined ($P < 0.05$) in a cubic manner as puberty approached (Fig. 1). The regression line for frequency of LH pulses during the 130-day period preceding puberty was cubic ($P < 0.01$; Fig. 2). Of the 3 measurements of LH secretion (mean LH concentration, frequency of LH pulses and amplitude of LH pulses), frequency of LH pulses was the best predictor of age at puberty. The correlation coefficient of the regression of frequency of LH pulses on days before puberty was 0·88. This compares to a correlation coefficient of 0·60 from an earlier study (Day *et al.*, 1984) in which blood was collected at 12-min intervals for 8 h. This may indicate that long sampling periods are necessary to obtain sufficient data to assess the characteristics of gonadotrophin secretion and to predict age at puberty based on these characteristics. From 130 to 46 days before puberty, frequency of LH pulses did not increase and there was considerable variation amongst animals. Prediction of the occurrence of puberty from LH data collected during this period was therefore difficult. However, during the last 46 days before puberty the prediction equation for age at puberty was more useful. Blood samples were taken over 24-h periods at 20-min intervals to determine the stage of sexual maturation at the time of slaughter in a group (N = 20) of contemporary heifers. Using the regression of pulse frequency on days before puberty developed with the 6 control heifers, it was estimated that the mean age at puberty in the 20 heifers that were slaughtered would have been 360 (s.e.m. = 7) days of age. In 10 heifers reared in conditions similar to those utilized to formulate the prediction equation and those that were slaughtered, the average age at puberty was 360 (s.e.m. = 6) days. The equation was surprisingly accurate in prediction of puberty in this group of animals. We believe the data from this study clarify some of the inconsistencies between studies regarding the change in secretion of LH in prepubertal heifers. A definite increase in frequency of LH pulses occurred during the 50 days before puberty but changes in frequency of LH pulses before the 50 days prior to puberty are less obvious in the heifer.

The onset of pulsatile secretion of LH begins at about 11 weeks after birth in the ewe lamb. At this time, concentrations of LH increase to levels similar to those detected at the time of puberty. Frequency of pulses in LH appear to vary dramatically as puberty (35 weeks of age) approaches in ewe lambs (Foster *et al.*, 1975). There is evidence of a marked increase in frequency but not amplitude of LH pulses shortly before first ovulation (Huffman & Goodman, 1985). Concentrations of

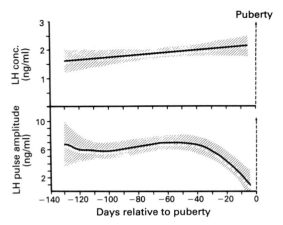

Fig. 1. Regression of mean concentration of LH and amplitude of LH pulses on days relative to puberty in heifers. Correlation coefficient for mean LH concentration; $r = 0.2$. Correlation coefficient for LH pulse amplitude, $r = 0.51$.

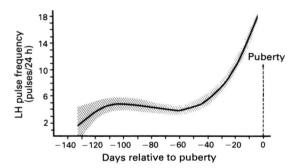

Fig. 2. Regression of frequency of LH pulses on days relative to puberty for 6 heifers that were bled at 20-min intervals for 24 h from 253 days of age until puberty at 366 days of age ($r = 0.88$).

LH were higher at puberty than 7 weeks before puberty in ewe lambs (Keisler *et al.*, 1985). In that study, the increased concentrations of LH during the luteal and follicular phases preceding first oestrus were attributed to increased amplitude of pulses since no increase in frequency of LH pulses was detected. Therefore, it appears that mean concentrations of LH increase before puberty in the ewe lamb. However, it is not clear whether increased frequency or increased amplitude of LH pulses is responsible for the increase in mean LH concentrations.

Administration of purified LH at a level of 15·5 μg/h in a pulsatile fashion over a 48-h period induced an LH surge and ovulation in 2 of 3 lambs (Foster *et al.*, 1984). By contrast, administration of a lower dose or 15·5 μg LH at 3-h intervals did not induce any surge in LH or ovulation. In another experiment (Foster *et al.*, 1984), the same dose of 15·5 μg of purified LH was administered in hourly pulses. Of 7 lambs, 4 had a sustained increase in oestradiol, a preovulatory surge of LH and ovulation. However, the luteal phase was normal in only one of the lambs and a short luteal phase of 6–11 days occurred in the other 3 lambs that ovulated. Keisler *et al.* (1985) conducted a study in which ovine LH was pulsed at 7·5 μg/h, 15 μg/h, 30 μg/2 h, 45 μg/3 h or was infused continuously at 15 μg/h into ewe lambs for 48 h (Table 2). Pulses of about 24 ng LH/ml serum were produced by pulsatile administration of LH at 2- or 3-h intervals but this did not consistently induce preovulatory-like surges of gonadotrophin. In contrast, pulses of 6–12 ng LH/ml serum

Table 2. Peak and basal concentrations of LH in plasma after infusion or pulsed delivery of ovine LH into prepubertal lambs and occurrence of surges of LH in response to exogenous LH (from Keisler *et al.*, 1985)

Treatment (dose)	Conc. (ng/ml) of LH achieved		Lambs responding with a surge of LH
	Peak	Basal	
Control (saline)		1·6 ± 0·2	1/6
Infused (15 μg LH/h)	6·3 ± 0·3	6·3 ± 0·2	3/6
Pulsed (7·5 μg LH/h)	6·0 ± 0·4	2·0 ± 0·2	5/6
Pulsed (15 μg LH/h)	11·7 ± 0·6	3·9 ± 0·3	4/6
Pulsed (30 μg LH/2 h)	23·9 ± 1·1	2·6 ± 0·2	1/5
Pulsed (45 μg LH/3 h)	31·9 ± 2·4	1·7 ± 0·2	1/6

Values are mean ± s.e.m.

were observed after hourly administration of LH and were effective in inducing preovulatory LH surges. Thus, hourly administration of the purified LH resulted in a frequency and amplitude of LH pulses similar to that detected during the follicular phase associated with the first oestrus as puberty is attained in the ewe lamb.

When 2 μg injections of LHRH were administered at 2-h intervals for 72 h in 5-month-old prepubertal heifers, mean circulating concentrations of LH increased (McLeod *et al.*, 1985). Preovulatory-like surges of LH occurred in 8 of 12 heifers between 17 and 59 h after the start of the treatment. The frequency of LH pulses resulting from administration of LHRH corresponds to that seen about 20 days before puberty.

It appears that administration of LH or a secretagogue that increases LH secretion at 1- or 2-h intervals will induce a preovulatory LH surge and ovulation in a majority of the ewe lambs or heifers that are treated in such a fashion. Therefore, it is likely that the putative hypothalamic pulse generator that regulates the pulsatile secretion of gonadotrophins is suppressed in the prepubertal ewe lamb and heifer. Very few of the prepubertal animals treated with exogenous hormones had a normal luteal phase or continued to exhibit oestrous cycles after the treatment regimen.

Secretion of FSH, during the prepubertal period

Schams *et al.* (1981) reported that FSH was secreted in a pulsatile fashion at a frequency of 3–6 pulses/24 h in heifers at 1, 2, 5 and 10 months of age. There is no evidence that an increased frequency of pulses of FSH occurs during that period. Likewise, Foster *et al.* (1975) reported that FSH concentrations increased from 3 to 11 weeks of age in the ewe lamb, but the concentration detected at 11 weeks of age was maintained until the lambs reached puberty at 35 weeks of age.

Compensatory ovarian hypertrophy occurs in prepubertal heifers after unilateral ovariectomy. There is a transient rise in plasma concentrations of FSH that peaks at 24 h and a return to baseline concentrations of FSH occurs by 36 h after ovariectomy (Johnson *et al.*, 1985). The administration of bovine follicular fluid devoid of steroids (removed by charcoal adsorption) prevented the transient rise in FSH and the compensatory ovarian hypertrophy that occurred after unilateral ovariectomy. Secretion of LH did not change following unilateral ovariectomy. In a study with prepubertal ewe lambs, unilateral ovariectomy resulted in an increase in the number of large follicles/ovary by 12 h after ovariectomy. A transient rise in circulating concentrations of FSH but not LH occurred coincident with the increase in follicle numbers (Smith *et al.*, 1984). The possible role of FSH in the pubertal process is not clear, but the results from studies with unilaterally ovariectomized animals indicate that nonsteroidal ovarian factors inhibit secretion of FSH in the prepubertal ewe lamb and heifer.

The role of the ovary in regulating gonadotrophin secretion during sexual maturation

Circulating concentrations of LH increased more than 3-fold between 2 and 8 weeks after ovariectomy at 19 weeks of age (Foster & Ryan, 1979). Basal concentrations of LH increased between 168 and 192 h after bilateral ovariectomy of prepubertal heifers (Kiser *et al.*, 1981). Secretion of LH increased earlier (72–80 h) after ovariectomy in postpubertal heifers. Odell *et al.* (1970) detected an increase in secretion of LH following ovariectomy as early as 1 month of age. An increase in secretion of LH has also been detected after ovariectomy of heifers at 3, 6 or 9 months of age (Anderson *et al.*, 1985).

Intrauterine ovariectomy of the fetal ewe lamb resulted in an increase in concentrations of LH and FSH above those of intact lambs by 5 weeks of age. The levels of gonadotrophins were maintained above those of intact animals throughout the first year of life (Bremner *et al.*, 1981).

Ovariectomy resulted in a marked increase in secretion of LH by 8 days after ovariectomy in the prepubertal heifer and subtle increases in secretion of LH continued over the next 130 days (Day *et al.*, 1984). The average age of puberty in intact control heifers was at about 130 days after ovariectomy in the treated heifers. It has not been determined whether these subtle changes in secretion of LH that occur after the acute increase in LH after ovariectomy have any physiological significance.

The ovary becomes involved in inhibiting gonadotrophin secretion very early in life. The inhibitory activity of the ovary is apparently maintained throughout the prepubertal period. Ovarian removal results in increased secretion of LH to levels that are equal to or higher than those detected during the follicular phase of the oestrous cycle of the adult cow and ewe. The increase in LH appears to be the result of increased frequency of LH pulses, with pulses being generated at a rate of about 1/h after ovariectomy.

The role of the ovarian steroids in sexual maturation

Oestradiol. Pulses of LH are maintained at a low level during the prepubertal state. High-frequency pulses of LH are one of the primary components absent in the prepubertal ewe lamb and heifer. However, the ovaries of these young animals are capable of responding to high frequency pulses of LH by developing a preovulatory follicle when this gonadotrophin is administered exogenously and the neuroendocrine system is capable of generating high-frequency pulses of LH if the ovary is removed.

The 'gonadostat hypothesis' emerged about 50 years ago and explains how the pubertal increase in secretion of LH might occur. According to this hypothesis, low concentrations of LH are maintained because of the responsiveness of the hypothalamo–pituitary axis to the inhibitory feedback action of oestradiol. As sexual maturation occurs, the responsiveness to steroid negative feedback decreases and secretion of LH increases to the point that follicular growth is stimulated, oestrogen secretion is enhanced and in turn the preovulatory surge of gonadotrophins is induced by the increasing oestrogen. The process culminates in ovulation. This hypothesis can be applied to prepubertal lambs and heifers since concentrations of oestradiol-17β that inhibit secretion of LH in prepubertal animals are no longer effective in inhibiting secretion of LH after puberty. Physiological concentrations of oestradiol-17β inhibit the increase in secretion of LH after ovariectomy in the ewe lamb. The inhibitory action of oestradiol on secretion of LH decreases coincident with the onset of puberty in contemporary intact ewe lambs (Foster & Ryan, 1979; Fig. 3).

The duration of suppression of secretion of LH after acute administration of oestradiol was longer in heifers that were 4 months of age than in heifers that were 8 and 12 months of age at the time of oestradiol administration (Schillo *et al.*, 1982). Oestradiol inhibited the postovariectomy rise in secretion of LH in prepubertal heifers that were 60 or 200 days of age at the time of ovariectomy (Staigmiller *et al.*, 1979). There was no difference in the degree of inhibition by oestradiol between the heifers in the two age groups. The ability of oestradiol to suppress secretion of LH decreased in ovariectomized heifers implanted with oestradiol and the change was coincident with

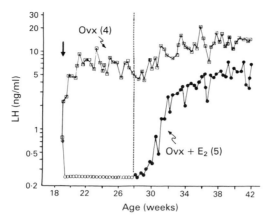

Fig. 3. Mean concentrations of circulating LH in ovariectomized (Ovx) lambs with or without chronic treatment with oestradiol (E_2). Arrow indicates the time of ovariectomy. Silastic capsules containing oestradiol-17β were inserted subcutaneously at the time of ovariectomy. Undetectable values for serum LH (0·25 ng/ml) are depicted by open symbols. (After Foster & Ryan, 1979.)

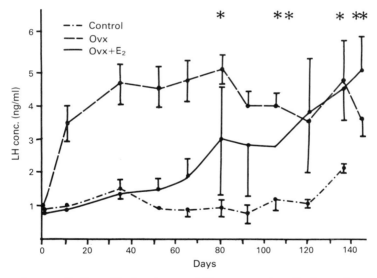

Fig. 4. Mean $\pm$ s.e.m. (indicated by bars) serum LH concentrations for intact (control, N = 6), ovariectomized (Ovx, N = 5) and ovariectomized, oestradiol-implanted (Ovx E_2, N = 5) heifers. * Indicates time of puberty for each control heifer. (After Day *et al.*, 1984.)

the time of puberty in contemporary intact heifers (Day *et al.*, 1984; Fig. 4). The increased mean concentrations of LH resulted from an increased frequency of LH pulses. Therefore, oestradiol can inhibit secretion of LH starting very early in life in heifers and ewe lambs and this inhibition is continued until the time when puberty occurs.

A decrease in concentration of cytosolic receptors for oestradiol in the anterior and medial basal hypothalamus and the anterior pituitary occurs during the period of sexual maturation in heifers (Fig. 5). This decline in receptors coincides with a decline in oestradiol negative feedback and an increased secretion of LH. The decline in receptor numbers may be responsible for the

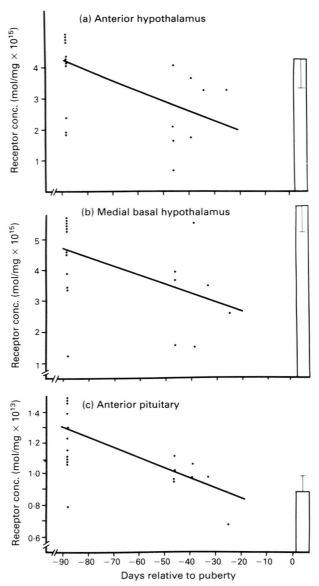

Fig. 5. Regression of concentration of receptors for oestradiol in cytosol of (a) the anterior hypothalamus, (b) the medial basal hypothalamus, and (c) the anterior pituitary of heifers on estimated days relative to puberty. Correlation coefficients for mean concentrations of receptors were $r = 0.48$ in (a), $r = 0.45$ in (b) and $r = 0.53$ in (c). The histograms give the mean ± s.e.m. values for 5 post-pubertal heifers.

reduction in negative feedback of oestradiol on secretion of LH during this time. As the concentration of oestradiol receptors decreases during sexual maturation, the corresponding regulatory effect on secretion of gonadotrophins also decreases. Changes in the concentration of receptors for oestradiol in the preoptic area and stalk median eminence did not occur during the process of sexual maturation in heifers.

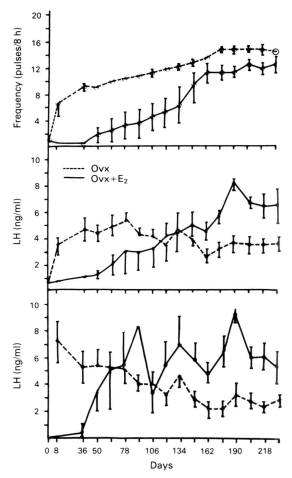

Fig. 6. Mean serum LH concentrations ± s.e.m. (indicated by bars), LH pulse frequency, and LH pulse amplitude for ovariectomized (Ovx) and ovariectomized–oestradiol implanted (Ovx + E_2) heifers. (After Day *et al.*, 1986a).

Chronic exposure of the prepubertal ewe lamb and heifer to oestradiol inhibits secretion of LH and the inhibitory effects of oestradiol are overcome near the time of puberty. An interesting phenomenon occurs in ovariectomized heifers that have had continuous exposure to oestradiol throughout the period of time that corresponds to the prepubertal, pubertal and postpubertal periods of contemporary intact heifers. The same dose of oestrogen that inhibited secretion of LH early in life enhances secretion of LH later in life (Day *et al.*, 1986b; Fig. 6). In our laboratory, we have seen this enhancement in secretion of LH in several studies in which cows that were ovariectomized after puberty were used. Heifers were ovariectomized after puberty and when oestradiol was administered at 3 different doses via implants, secretion of LH was increased above that detected in ovariectomized heifers without oestradiol (Kinder *et al.*, 1983; Table 3). Chronic implantation of ovariectomized cows that were sexually mature at the time of ovariectomy with oestradiol increases secretion of LH at the spring and autumn equinoxes and the summer and winter solstices above that of ovariectomized cows that did not receive oestradiol (unpublished data).

Although animals overcome the negative feedback effects of oestradiol at puberty, oestradiol can exert negative feedback on LH secretion in postpubertal animals in certain circumstances.

Table 3. Frequency and amplitude in pulses of LH and mean concentrations of oestradiol-17β and LH in serum of postpubertal heifers treated with different doses of oestradiol (taken from Kinder *et al.*, 1983)

Implant size (cm)	Oestradiol (pg/ml)	LH* (ng/ml)	Frequency of LH pulses (pulses/6 h)	Amplitude of LH pulses (ng)
0	2·1	2·35 ± 0·26	7·5 ± 0·5	0·98 ± 0·24
12·5	3·4	3·38 ± 0·17	6·0 ± 0·4	2·50 ± 0·41
27·0	4·7	3·60 ± 0·37	5·5 ± 0·7	3·24 ± 0·38
54·0	8·4	3·11 ± 0·42	4·0 ± 0·4	3·82 ± 0·58

Values are mean ± s.e.m. for 4 heifers/group.
*Concentrations of LH were evaluated at 30 days after ovariectomy and implantation of the oestradiol.

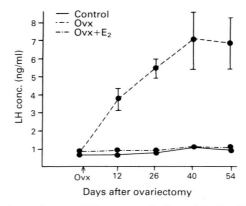

Fig. 7. Mean concentrations of serum LH ± s.e.m. (indicated by bars) in heifers. Heifers were fed diets restricted in energy until anoestrus resulted. At this time treatments were applied (Ovx, ovariectomized; Ovx + E₂, ovariectomized + oestradiol-17β) to the heifers and a diet adequate in energy was started. (After Imakawa *et al.*, 1986.)

Goodman & Karsch (1981) have shown that oestradiol suppresses secretion of LH in adult ovariectomized ewes during seasonal anoestrus. Oestradiol suppresses secretion of LH in cows during nutritionally-induced anoestrus (Imakawa *et al.*, 1986; Fig. 7). Oestradiol reduced frequency and amplitude of LH pulses during this physiological state. It would be useful to determine what happens to the concentration of oestradiol receptors at the hypothalamo–pituitary axis with the onset of anoestrus in animals of this type.

Oestrus without the initiation of oestrous cycles has been detected in prepubertal heifers and has been termed nonpubertal oestrus (Nelson *et al.*, 1985). The incidence of this phenomenon varied with breed of heifer during the 2-year study. This phenomenon could be related to photoperiod since, in the first year of the study, 64% of the heifers that exhibited behavioural oestrus before 1 January had a nonpubertal oestrus. In the 2nd year of the study, 80% of the heifers exhibiting oestrus before 1 January had a nonpubertal oestrus. The incidence of nonpubertal oestrus appears to decrease during the late winter, spring and summer months (Fig. 8). However, the possibility also exists that nonpubertal oestrus may be age related. In another study nonpubertal oestrus was exhibited in 62·8% of the heifers and there was a tendency for more of the heifers that were lighter in weight to exhibit nonpubertal oestrus than the heavier weight heifers (Rutter & Randel, 1986). More heifers that had a pubertal oestrus at the first observed behavioural oestrus also had a transient rise in serum concentrations of progesterone before their first oestrus

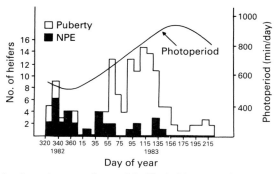

Fig. 8. Day of first behavioural oestrus for each heifer in Year 1. Puberty = oestrus accompanied by subsequent corpus luteum function; NPE = oestrus not accompanied by subsequent corpus luteum function. (After Nelson *et al.*, 1985.)

(64·3 *vs* 20%) and this transient elevation of serum progesterone was greater in heifers exhibiting the pubertal oestrus than in those showing the nonpubertal oestrus. On average, 63·5 days elapsed from the time that the nonpubertal oestrus was detected until the time of occurrence of pubertal oestrus. Therefore, situations appear to exist in which oestrogen secretion is occurring to the point that behavioural oestrus is induced but there is a failure of ovulation and/or subsequent luteal development to occur.

Progesterone. Two distinct elevations in progesterone concentration in blood serum were detected during the peripubertal period that immediately preceded puberty in the heifer (Gonzalez-Padilla *et al.*, 1975a). This increase in progesterone may have a role in the endocrine changes leading to the establishment of gonadotrophin and gonadal hormone secretions that are characteristic of the adult cow. The source of this progesterone is of ovarian origin in the heifer (Berardinelli *et al.*, 1979). When ovariectomy was performed 24–30 h after an increase in progesterone had been detected concentrations of progesterone returned to levels similar to those observed before the rise. Microscopic examinations of the ovaries revealed the presence of compact luteal tissue embedded within the ovary that could not be observed from the ovarian surface.

Fitzgerald & Butler (1982) have observed a transient increase in progesterone that lasts for 1 to 4 days before the time of the first ovulation in ewe lambs. Ovariectomy of ewe lambs within 8 h of the detected increases in progesterone concentration resulted in an immediate decline in serum progesterone concentration (Berardinelli *et al.*, 1980). It was determined that progesterone concentrations were significantly higher in the venous blood from the ovary containing luteal tissue. As in heifers, there were no ovulation papillae observed on the luteal structures of the ovaries that were removed at the time of the first rise in progesterone. Therefore, the source of the progesterone secreted in the short luteal phases preceding puberty in the ewe lamb and heifer is in the ovary but ovulation is not necessary for the formation of these structures.

The effect of progesterone during the short luteal phases upon subsequent endocrine and ovarian function in the prepubertal heifer and ewe lamb is unclear. Keisler *et al.* (1983) removed the ovary containing luteal tissue from 1 group of ewe lambs and removed the ovary without luteal tissue in another group of lambs within 24 h of the first increase of plasma progesterone concentrations. There was no difference in the time from the transient rise in progesterone concentrations to the initiation of the first luteal phase of normal duration in the ewe lambs of the two groups. Also, no difference in the duration of the normal luteal phase was detected (Table 4). The transient luteal structure that develops in the ewe lamb before development of the first normal corpus luteum therefore does not appear to be required for sexual maturation.

Quirke *et al.* (1985) found that the mean number of ovulations before the first behavioural oestrus averaged 1·6 in ewe lambs of several different breeds. It is therefore not uncommon for ewe lambs to ovulate more than once before the time when first behavioural oestrus is exhibited.

Table 4. Characteristics of profiles of progesterone exhibited before the first behavioural oestrus in ewe lambs (taken from Keisler *et al.*, 1983)

Treatment*	No. of lambs	Average maximum magnitude (ng/ml)†	Average duration (days)†	Days from transient rise to luteal phase preceding first oestrus†	Luteal phase preceding first oestrus	
					Average conc. of progesterone (ng/ml)†	Average duration (days)†
Sham-operation	5	0·50 ± 0·13	3 ± 1	4 ± 1	1·30 ± 0·25	12 ± 1
Nonluteal ovary removal	5	0·30 ± 0·02	3 ± 1	4 ± 1	1·20 ± 0·10	11 ± 1
Luteal ovary removal	6	0·32 ± 0·04	1	5 ± 1	1·23 ± 0·10	11 ± 1

*At the first rise in progesterone in the plasma.
†Determined from consecutive concentrations of progesterone that equalled or exceeded average baseline plus 3 s.e.m. concentrations.

The use of progestagen implants for induction of puberty in prepubertal heifers has been studied extensively (Gonzalez-Padilla *et al.*, 1975b; Short *et al.*, 1976; Rajamahendran *et al.*, 1981, 1982; Sheffield & Ellicott, 1982). Removal of the progestagen implant can induce puberty in a high percentage of heifers but the fertility at first oestrus increases with age of the heifer (Short *et al.*, 1976). The role that progesterone has in the process of sexual maturation is not well understood at the present time.

The influence of photoperiod on sexual maturation

The sheep is a seasonal breeder, and so it is not surprising that seasonal factors modify the time of puberty in the ewe lamb. The primary seasonal factor thought to be responsible for modification in age at puberty in the ewe lamb is daylength. Ewe lambs can attain puberty as early as 150 days of age, but the season dictates age at first ovulation. A delay in the onset of puberty occurs in ewe lambs born in the spring or autumn if they reach pubertal age during the nonbreeding season. Fitzgerald & Butler (1982) demonstrated that ewe lambs born in March were older at puberty than lambs born in July or August. Spring-born lambs were shown to begin reproductive cycles at 25–35 weeks of age (Foster *et al.*, 1986; Fig. 9). However, lambs born in the autumn attained the age at which puberty would have normally occurred (25–35 weeks) during the anoestrous season and thus remained anovulatory until the autumn when they were 48 to 50 weeks of age. Foster *et al.* (1986) have shown that a sequence of long days, followed by short days, was required to initiate and sustain oestrous cycles in ewe lambs. In addition, the age when long days were experienced was important, with later exposure to long days being more effective in producing oestrous cycles of normal duration than exposure to long days at an earlier age. There may therefore be a critical period during sexual maturation during which ewe lambs must be exposed to long days in order for normal sexual maturation to occur.

The re-establishment of oestradiol negative feedback on secretion of LH which accompanies seasonal anoestrus occurs about 3 weeks earlier in ewe lambs than in adult ewes. The timing of the increased negative feedback of oestradiol coincides with cessation of oestrous cycles in lambs and adult ewes (Foster & Ryan, 1981; Fig. 10). Ewe lambs may therefore maintain an enhanced response to oestradiol negative feedback throughout the first breeding season.

Even though reproductive activity in the cow is not limited to one season of the year, season appears to be able to modulate reproductive function. This action probably influences the hormonal secretions of the hypothalamo–pituitary–gonadal axis. Through this route, age at puberty in the

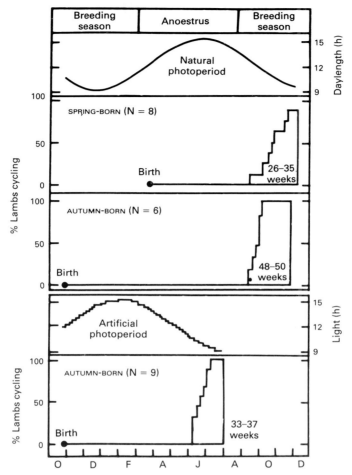

Fig. 9. Time of puberty in the spring- and autumn-born lamb. Growth and development occurs during the long days of the spring and summer anoestrous season, and reproductive cycles and matings begin during the decreasing daylengths of the autumn breeding season. (After Foster *et al.*, 1986.)

heifer could be influenced by photoperiod much as it is in the ewe lamb. Several studies have indicated that seasonal factors influence age at puberty in heifers. However, it is not possible to determine the time at which season influences sexual maturation because season of birth is confounded with the season when sexual maturation occurs (Petitclerc *et al.*, 1983; Hansen *et al.*, 1983). An intricate experiment was performed in a group of Angus–Holstein crossbred heifers that were born on 21 March 1978 or 23 September 1978 (Schillo *et al.*, 1983). All heifers were exposed to their natural environment during the first 6 months of life. However, during the second 6 months of life, heifers were exposed to a controlled environment in which temperature and photoperiod were both manipulated to give either a spring to autumn sequence of environmental events or an autumn to spring sequence of environmental events. Age at puberty was influenced by date of birth, indicating that the natural environmental conditions to which the heifers were exposed during the first 6 months of life influenced age at puberty. Heifers born in the autumn tended to reach puberty earlier in life than heifers born in the spring. March- and September-born heifers exposed to the spring to autumn climatic changes (controlled environment) during the second 6 months of life reached puberty earlier than the heifers

Table 5. Effects of month of birth and photoperiod chamber on age and body weight of heifers at puberty (from Schillo *et al.*, 1983)

Date of birth	Chamber* (after 6 months)	Age at puberty† (days)	Weight at puberty‡ (kg)
21 March	Sp–F	321 ± 29·5	281 ± 39·0
21 March	F–Sp	346 ± 50·1	318 ± 43·4
23 September	Sp–F	295 ± 12·6	268 ± 9·9
23 September	F–Sp	319 ± 32·5	306 ± 25·5
Date of birth means			
21 March		334	300
23 September		307	287
Chamber means			
Sp–F		308	274
F–Sp		333	312

Values are mean ± s.d. for 14 heifers born in March and 14 heifers born in September.

*Sp–F = spring to fall climatic changes; F–Sp = fall to spring climatic changes during second 6 months of life.

†There were effects of date of birth ($P < 0.06$) and chamber ($P < 0.08$).

‡There was an effect of chamber ($P < 0.01$).

exposed to the autumn to spring climatic changes. Therefore, exposure during the second 6 months of life to the short photoperiods and temperatures that would exist during the first 6 months of life in autumn-born heifers was associated with an earlier age at puberty (Table 5). The younger age at puberty in September-born heifers and in heifers exposed to the controlled environment that represented spring to autumn climatic changes suggests a mechanism by which heifers would tend to give birth to offspring in the spring and summer regardless of their own birth date. Heifers born in the autumn would tend to reach puberty at less than 1 year of age in the summer or early autumn after their birth. Spring-born heifers would tend to reach puberty later in life and give birth to offspring in the spring or early summer of the next year.

Season has been shown to influence secretion of LH in cows. Increased secretions of LH occur during the spring and suppression of secretion occurs during the autumn months in mature ovariectomized cows (Fig. 11). The fluctuation in concentration of LH with season results from changes in the amplitude of pulses (Day *et al.*, 1986a). The cow therefore has the ability to receive and integrate seasonal cues that modulate gonadotrophin secretion.

The influence of growth and nutrition on sexual maturation

It appears that growth-related cues influence the rate of sexual maturation and the onset of puberty in the ewe lamb and heifer. Low nutritional intake will delay the onset of puberty in the ewe lamb and heifer. Lambs remained anovulatory during the first breeding season when growth was retarded (Foster *et al.*, 1986). The frequency of pulses of LH was reduced and oestradiol continued to exert the negative feedback action on secretion of LH when growth rate was retarded in the ewe lamb. Therefore, the rate at which pulses were generated was much slower in ewe lambs that were given restricted amounts of food. When ewe lambs that had been fed diets restricted in quantity were switched to diets that were fed *ad libitum*, the sensitivity to oestradiol negative feedback on LH secretion was reduced and an increase in secretion of LH occurred (Foster & Olster, 1985; Fig. 12). Restricted food intake also resulted in a suppression of the preovulatory-like surge of LH secretion that resulted after oestradiol administration (Foster & Olster, 1985). Therefore, the

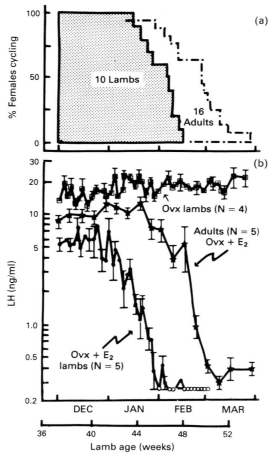

Fig. 10. Onset of anoestrus (1978) in (a) 10 of 11 March-born lambs and in 16 adult ewes and (b) seasonal change in mean ($\pm$ s.e.m.) concentrations of serum LH in 4 ovariectomized (Ovx) lambs, in 5 of 6 lambs and 5 adult ewes ovariectomized and treated with oestradiol (E_2). A female was considered to be 'cyclic' until the last ovulation, usually the last oestrus; failure of circulating progesterone concentration to increase to luteal-phase levels at the time of the next two expected ovulations was used as confirmation that the female was anoestrous. (After Foster & Ryan, 1981.)

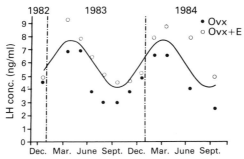

Fig. 11. Regression lines for mean concentration of LH on day of bleeding of cows ovariectomized (Ovx) and ovariectomized and treated with oestradiol-17β (Ovx + E). (After Day *et al.*, 1986a.)

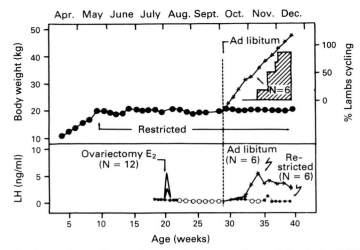

Fig. 12. Mean body weights and mean ($\pm$s.e.m.) concentrations of circulating LH in ovariecto-mized lambs (bottom panel) chronic treatment with low levels of oestradiol (Silastic implant). Results obtained during restricted feeding are presented as solid circles ($\bullet$) while those obtained during ad-libitum feeding are presented as solid stars ($\star$). Open circles ($\bigcirc$) designate undetectable (< 0.2 ng/ml) LH values. Histogram in top panel shows onset of reproductive cycles in lambs during ad-libitum feeding. (After Foster & Olster, 1985.)

pituitary reserves of LH to be released at the oestradiol-induced preovulatory LH surge were less, the amount of LHRH released was less or the response to LHRH was reduced in ewe lambs that received oestradiol and were fed restricted quantities of food.

Mean concentrations of LH were suppressed in prepubertal heifers fed restricted levels of dietary energy as compared to heifers fed higher energy diets during the 120 days preceding puberty (Day *et al.*, 1986c; Fig. 13). Both frequency and amplitude of pulses of LH were suppressed in the heifers fed restricted levels of dietary energy as compared to heifers that were fed a diet adequate in energy. Thus, as in the ewe lamb, the rate at which pulses of LH are generated was slower in heifers fed restricted levels of dietary energy during the period that sexual maturation would normally have occurred. The magnitude of the peak response of LH after administration of LHRH was less in the heifers fed the diet restricted in energy during the first 84 days following initiation of feed restriction as compared to heifers fed diets adequate in energy (Fig. 14). Therefore, the results of the studies in ewe lambs and heifers both indicate that pituitary responsiveness to exogenous secretagogues is decreased when quantities of food intake are restricted.

In a series of experiments in prepubertal heifers, increased concentrations of the volatile fatty acid propionate in the rumen resulted in an increased release of LH from the pituitary after administration of LHRH (Randel & Rhodes, 1980; Rutter *et al.*, 1983). In addition, an increase in propionate resulted in an increase in the amount of LH released during the preovulatory-like surge of gonadotrophin secretion after oestradiol administration in the prepubertal heifer (Randel *et al.*, 1982). It has also been reported that prepubertal heifers fed an additive that increases the propionate to acetate ratio in the rumen had an enhanced ovarian response with an increased ovarian weight, more corpora lutea and more follicular growth after administration of FSH and human chorionic gonadotrophin (Bushmich *et al.*, 1980). Therefore, increased concentrations of propionate result in an increased responsiveness of the pituitary to secretagogues, and thus an enhanced secretion of LH. In addition, ovarian responsiveness to the gonadotrophins is increased in heifers with a higher propionate to acetate ratio. The volatile fatty acid, propionate, might therefore be involved in modulating endocrine secretions in prepubertal heifers.

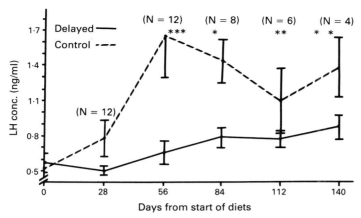

Fig. 13. Mean serum concentration of LH in heifers after start of feeding a growing diet (average daily gain = 0·79 kg/heifer/day; control) or energy-restricted diet (average daily gain = 0·21 kg/heifer/day; delayed). Day of puberty in individual control heifers is indicated by *; vertical bars indicate s.e. of each mean. Numbers in parentheses indicate the number of control heifers that were prepubertal at this period of blood collection. (After Day *et al.*, 1986c.)

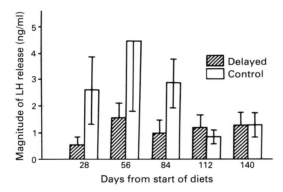

Fig. 14. Mean magnitude of the peak response of LH in heifers to 0·5 µg LHRH after start of feeding a growing diet (average daily gain = 0·79 kg/heifer/day; control) or energy-restricted diet (average daily gain = 0·21 kg/heifer/day; delayed). Vertical lines indicate s.e.m. (After Day *et al.*, 1986c.)

Puberty: a conceptual view

The low frequency endogenous pulses of LH generated in prepubertal sheep and cattle are not capable of driving follicle growth to the preovulatory stage. Ovarian oestradiol acts upon the hypothalamus to suppress the generation of pulses of LHRH which maintain the secretion of LH at low levels and in turn prevents the growth of ovarian follicles to the point that ovulation can occur.

Involvement of the opiates in modulating the responsiveness of the hypothalamus to oestradiol in female rats has been described (Bhanot & Wilkinson, 1983). The endogenous opiates sensitize the hypothalamus to oestradiol and thus inhibit generation of pulses of LH. The endogenous opiates may carry out this role by modulating the level of oestradiol receptors in the medial basal hypothalamus on which oestradiol can act in the inhibition of secretion of LH. Docke *et al.* (1984) suggested that the sensitivity of the medial basal hypothalamus to negative feedback of oestradiol

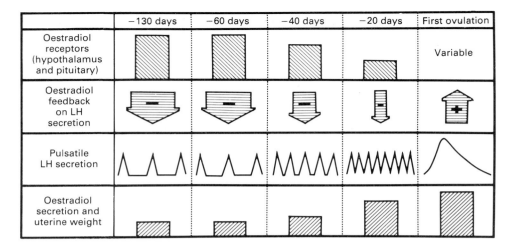

Fig. 15. Model for endocrine control of puberty in heifers.

decreased as puberty approached in the female rat. These data from the rat would be consistent with the data from heifers in which the population of oestradiol receptors in the medial basal hypothalamus declines as puberty approaches. As the concentration of oestradiol receptors in the medial basal hypothalamus declines, the frequency with which pulses of LH are generated increases. Follicles can develop to more advanced stages as frequency of pulses of LH increase and produce concentrations of oestrogen that stimulate uterine growth and development. The frequency at which pulses are generated reaches one pulse of LH per hour during the final stages of sexual maturation. At this point concentrations of LH are attained that drive ovarian follicle growth to the preovulatory stage, oestradiol induces a preovulatory surge of gonadotrophins and ovulation occurs as a result of gonadotrophins acting on the mature ovarian follicle (Fig. 15).

We thank our colleagues Kazuhiko Imakawa, Miguel Garcia-Winder, Douglas Zalesky, Patricia Wolfe, Todd Stumpf, Michael Wolfe and James Stotts for help in designing experiments, active participation in conducting the studies and help in preparation of this manuscript; Dr Jerry J. Reeves and Dr Norman Mason for supplying antisera; Dr Leo E. Reichert for providing purified LH for radioiodination; and Jane Ossenkop for her patience in preparation of this manuscript.

Published as Paper Number 8144, Journal Series, Nebraska Agricultural Research Division.

References

Anderson, L.L., Hard, D.L., Carpenter, L.S., Awotwi, E.K. & Diekman, M.A. (1981) Endocrine patterns associated with puberty in male and female cattle. *J. Reprod. Fert., Suppl.* **30**, 103–110.

Anderson, W.J., Forrest, D.W., Schulze, A.L., Kraemer, D.C., Bowen, M.J. & Harms, P.G. (1985) Ovarian inhibition of pulsatile luteinizing hormone secretion in prepuberal holstein heifers. *Dom. Anim. Endocr.* **2**, 85–91.

Berardinelli, J.G., Dailey, R.A., Butcher, R.L. & Inskeep, E.K. (1979) Source of progesterone prior to puberty in beef heifers. *J. Anim. Sci.* **49**, 1276–1280.

Berardinelli, J.G., Dailey, R.A., Butcher, R.L. & Inskeep, E.K. (1980) Source of circulating progesterone in prepubertal ewes. *Biol. Reprod.* **22**, 233–236.

Bhanot, R. & Wilkinson, M. (1983) Opiatergic control of gonadotropin secretion during puberty in the rat: a neurochemical basis for the hypothalamic 'gonadostat'. *Endocrinology* **113**, 596–603.

Bremner, W.J., Cumming, I.A., Williams, D.M., de Kretser, D.M. & Lee, V.W.K. (1981) The effect of intrauterine gonadectomy on fetal and neonatal gonadotrophin secretion in the lamb. *J. Reprod. Fert., Suppl.* **30**, 61–66.

Bushmich, S.L., Randel, R.D., McCartor, M.M. & Carroll, L.H. (1980) Effect of dietary monensin on ovarian response following gonadotrophin treatment in prepubertal heifers. *J. Anim. Sci.* **51**, 692–697.

Day, M.L., Imakawa, K., Garcia-Winder, M., Zalesky, D.D., Schanbacher, B.D., Kittok, R.J. & Kinder, J.E. (1984) Endocrine mechanisms of puberty in heifers. Estradiol negative feed-back regulation of luteinizing hormone secretion. *Biol. Reprod.,* **31**, 332–341.

Day, M.L., Imakawa, K., Pennel, P.L., Zalesky, D.D., Clutter, A.C., Kittok, R.J. & Kinder, J.E. (1986a) Influence of season and estradiol on secretion of LH in ovariectomized cows. *Biol. Reprod.* **62**, 1641–1648.

Day, M.L., Imakawa, K., Garcia-Winder, M., Kittok, R.J., Schanbacher, B.D. & Kinder, J.E. (1986b) Influence of prepubertal ovariectomy and estradiol replacement therapy on secretion of luteinizing hormone before and after pubertal age in heifers. *Dom. Anim. Endocr.* **3**, 17–25.

Day, M.L., Imakawa, K., Garcia-Winder, M., Zalesky, D.D., Kittok, R.J. & Kinder, J.E. (1986c) Effects of restriction of dietary energy intake during the prepubertal period on secretion of luteinizing hormone and responsiveness of the pituitary to luteinizing hormone-releasing hormone in heifers. *J. Anim. Sci.* **62**, 1641–1648.

Desjardins, C. & Hafs, H.D. (1968) Levels of pituitary FSH and LH in heifers from birth through puberty. *J. Anim. Sci.* **27**, 472–476.

Docke, F., Rohde, W., Stahl, F., Smollich, A. & Dorner, G. (1984) Serum levels of FSH, LH and estradiol-17β in female rats around the time of puberty onset. *Expl clin. Endocrinol.* **83**, 6–14.

Echternkamp, S.E. & Hansel, W. (1973) Concurrent changes in bovine plasma hormone levels prior to and during the first postpartum estrous cycle. *J. Anim. Sci.* **37**, 1362–1371.

Fitzgerald, J. & Butler, W.R. (1982) Seasonal effects and hormonal patterns related to puberty in ewe lambs. *Biol. Reprod.* **27**, 853–863.

Foster, D.L. & Karsch, F.J. (1975) Development of the mechanism regulating the preovulatory surge of luteinizing hormone in sheep. *Endocrinology* **97**, 1205–1209.

Foster, D.L. & Olster, D.H. (1985) Effect of restricted nutrition on puberty in the lamb: patterns of tonic luteinizing hormone (LH) secretion and competency of the LH surge system. *Endocrinology* **116**, 375–381.

Foster, D.L. & Ryan, K.D. (1979) Endocrine mechanisms governing transition into adulthood: a marked decrease in inhibitory feedback action of estradiol on tonic secretion of luteinizing hormone in the lamb during puberty. *Endocrinology* **105**, 896–904.

Foster, D.L. & Ryan, K.D. (1981) Endocrine mechanisms governing transition into adulthood in female sheep. *J. Reprod. Fert., Suppl.* **30**, 75–90.

Foster, D.L., Lemons, J.A., Jaffe, R.B. & Niswender, G.D. (1975) Sequential patterns of circulating luteinizing hormone and follicle-stimulating hormone in female sheep from early postnatal life through the first estrous cycles. *Endocrinology* **97**, 985–994.

Foster, D.L., Ryan, K.D. & Papkoff, H. (1984) Hourly administration of luteinizing hormone induces ovulation in prepubertal female sheep. *Endocrinology* **115**, 1179–1185.

Foster, D.L., Karsch, F.J., Olster, D.H., Ryan, D.R. &

Yellon, S.M. (1986) Determinants of puberty in a seasonal breeder. *Recent Progress in Hormone Research* **42**, 331–384.

Gonzalez-Padilla, E., Ruiz, R., Le Fever, D., Denham, A. & Wiltbank, J.N. (1975a) Puberty in beef heifers. III. Induction of fertile estrus. *J. Anim. Sci.* **40**, 1110–1118.

Gonzalez-Padilla, E., Wiltbank, J.N. & Niswender, G.D. (1975b) Puberty in beef heifers. I. The interrelationship between pituitary, hypothalamic and ovarian hormones. *J. Anim. Sci.* **40**, 1091–1104.

Goodman, R.L. & Karsch, F.J. (1981) A critique of the evidence on the importance of steroid feedback to seasonal changes in gonadotrophin secretion. *J. Reprod. Fert., Suppl.* **30**, 1–13.

Hansen, P.J., Kamwanja, L.A. & Hauser, E.R. (1983) Photoperiod influences age at puberty of heifers. *J. Anim. Sci.* **57**, 985–992.

Huffman, L.J. & Goodman, R.L. (1985) LH pulse patterns leading to puberty in the ewe lamb. *Biol. Reprod.* **32** (Suppl. 1), 208, Abstr.

Imakawa, K., Day, M.L., Garcia-Winder, M., Zalesky, D.D., Kittok, R.J., Schanbacher, B.D. & Kinder, J.E. (1986) Endocrine changes during restoration of estrous cycles following induction of anestrus by restricted nutrient intake in beef heifers. *J. Anim. Sci.* **63**, 565–571.

Johnson, S.K., Smith, M.F. & Elmore, R.G. (1985) Effect of unilateral ovariectomy and injection of bovine follicular fluid on gonadotropin secretion and compensatory ovarian hypertrophy in prepuberal heifers. *J. Anim. Sci.* **60**, 1055–1060.

Keisler, D.H., Inskeep, E.K. & Dailey, R.A. (1983) First luteal tissue in ewe lambs: influence on subsequent ovarian activity and response to hysterectomy. *J. Anim. Sci.* **57**, 150–156.

Keisler, D.H., Inskeep, E.K. & Dailey, R.A. (1985) Roles of pattern of secretion of luteinizing hormone and the ovary in attainment of puberty in ewe lambs. *Dom. Anim. Endocr.* **2**, 123–132.

Kinder, J.E., Garcia-Winder, M., Imakawa, K., Day, M.L., Zalesky, D.D., D'Occhio, M.L., Kittok, R.J. & Schanbacher, B.D. (1983) Influence of different estrogen doses on concentrations of serum LH in acute and chronic ovariectomized cow. *J. Anim. Sci.* **57**, (Suppl. 1), 350, Abstr.

Kiser, T.E., Kraeling, R.R., Rampacek, G.B., Landmeirer, B.J., Caudle, A.B. & Chapman, J.D. (1981) Luteinizing hormone secretion before and after ovariectomy in prepubertal and pubertal beef heifers. *J. Anim. Sci.* **53**, 1545–1553.

McLeod, B.J., Haresign, W., Peters, A.R. & Lamming, G.E. (1984) Plasma LH and FSH concentrations in prepubertal beef heifers before and in response to repeated injections of low doses of GnRH. *J. Reprod. Fert.* **70**, 1–8.

McLeod, B.J., Peters, A.R., Haresign, W. & Lamming, G.E. (1985) Plasma LH and FSH responses and ovarian activity in prepubertal heifers treated with repeated injections of low doses of GnRH for 72 h. *J. Reprod. Fert.* **74**, 589–596.

Nelson, T.C., Short, R.E., Phelps, D.A. & Staigmiller, R.B. (1985) Nonpuberal estrus and mature cow influences on growth and puberty in heifers. *J. Anim. Sci.* **61**, 470–473.

Odell, W.D., Hescox M.A. & Kiddy, C.A. (1970) Studies of hypothalamic-pituitary-gonadal interrelationships in prepubertal cattle. In *Gonadotropin and Ovarian Development*, pp. 371–385. Eds W. R. Butt, A. C. Crook & M. Ryle. Livingston, Edinburgh.

Petitclerc, D., Chapin, L.T., Emer, R.S. & Tucker, H.A. (1983) Body growth, growth hormone, prolactin and puberty response to photoperiod and plane of nutrition in Holstein heifers. *J. Anim. Sci.* **57,** 892–898.

Quirke, J.F., Stabenfeldt, G.H. & Bradford, G.E. (1985) Onset of puberty and duration of the breeding season in Suffolk, Rambouillet, Finnish Landrace, Dorset and Finn-Dorset ewe lambs. *J. Anim. Sci.* **60,** 1463–1471.

Rajamahendran, R., Lague, P.C. & Baker, R.D. (1981) Serum hormone levels and occurrence of oestrus following use of an intravaginal device containing progesterone and oestradiol-17β in heifers. *Anim. Reprod. Sci.* **3,** 271–277.

Rajamahendran, R., Lague, P.C. & Baker, R.D. (1982) Serum progesterone and initiation of ovarian activity in prepuberal heifers treated with progesterone. *Can. J. Anim. Sci.* **62,** 759–766.

Randel, R.D. & Rhodes, R.C., III (1980) The effect of dietary monensin on the luteinizing hormone response of prepuberal heifers given a multiple gonadotropin-releasing hormone challenge. *J. Anim. Sci.* **51,** 925–931.

Randel, R.D., Rutter, L.M. & Rhodes, R.C., III (1982) Effect of monensin on the estrogen-induced LH surge in prepuberal heifers. *J. Anim. Sci.* **54,** 806–810.

Rutter, L.M. & Randel, R.D. (1986) Nonpuberal estrus in beef heifers. *J. Anim. Sci.* **63,** 1049–1053.

Rutter, L.M., Randel, R.D., Schilling, G.T. & Forrest, D.W. (1983) Effect of abomasal infusion of propionate on the GnRH induced luteinizing hormone release in prepuberal heifers. *J. Anim. Sci.* **56,** 1167–1173.

Schams, D., Schallenberger, E., Gombe, S. & Karg, H. (1981) Endocrine patterns associated with puberty in male and female cattle. *J. Reprod. Fert., Suppl.* **30,** 103–110.

Schillo, K.K., Dierschke, D.J. & Hauser, E.R. (1982) Regulation of luteinizing hormone secretion in prepubertal heifers: increased threshold to negative feedback action of estradiol. *J. Anim. Sci.* **54,** 325–336.

Schillo, K.K., Hansen, P.J., Kamwanja, L.A., Dierschke, D.J. & Hansen, E.R. (1983) Influence of season on sexual development in heifers: age at puberty as related to growth and serum concentrations of gonadotropins, prolactin, thyroxine and progesterone. *Biol. Reprod.* **28,** 329–341.

Seidel, G.E., Jr, Larson, L.L. & Foote, R.H. (1971) Effects of age and gonadotropin treatment on superovulation in the calf. *J. Anim. Sci.* **33,** 617–622.

Sheffield, L.G. & Ellicott, A.R. (1982) Effect of low levels of exogenous progesterone on puberty in beef heifers. *Theriogenology* **18,** 177–183.

Short, R.E., Bellows, R.A., Carr, J.B. & Staigmiller, R.B. (1976) Induced or synchronized puberty in heifers. *J. Anim. Sci.* **43,** 1254–1258.

Smith, M.F., Agudo, L. Sp. & Schanbacher, B.D. (1984) Follicle stimulating hormone secretion and compensatory ovarian hypertrophy in prepubertal ewes. *Theriogenology* **21,** 969–979.

Squires, E.L., Scaramuzzi, R.J., Caldwell, B.V. & Inskeep, E.K. (1972) LH release and ovulation in the prepubertal lamb. *J. Anim. Sci.* **34,** 614–619.

Staigmiller, R.B., Short, R.E. & Bellows, R.A. (1979) Induction of LH surges with 17β-estradiol in prepubertal beef heifers: an age dependent response. *Theriogenology* **11,** 453–461.

Swanson, L.V., Hafs, H.D. & Morrow, D.A. (1972) Ovarian characteristics and serum LH, prolactin, progesterone and glucocorticoid from first estrus to breeding size in Holstein heifers. *J. Anim. Sci.* **34,** 284–293.

Tran, C. T., Edey, T.N. & Findlay, J.K. (1979) Pituitary response of prepuberal lambs to oestradiol-17β. *Aust. J. biol. Sci.* **32,** 463–467.

J. Reprod. Fert., Suppl. **34** (1987), 187–199

Printed in Great Britain
© 1987 Journals of Reproduction & Fertility Ltd

Photoperiodic control of the onset of breeding activity and fecundity in ewes

D. J. Kennaway, E. A. Dunstan* and L. D. Staples†

*Department of Obstetrics & Gynaecology, University of Adelaide, Adelaide, South Australia 5000;
*Department of Agriculture and Fisheries, Kybybolite Research Centre, Kybybolite, South Australia
5262; and †Department of Agriculture & Rural Affairs, Animal Research Institute, Werribee,
Victoria 3030, Australia*

The seasonal nature of fertility of sheep has been accepted for many years (Hafez, 1952). It has also placed major restraints on farmers wishing to maximize farm output. Because sheep have a 5-month pregnancy and a 3-month lactation period it should be possible to produce 3 sets of lambs every 2 years. A major restraint on such a programme is the very strong seasonal photoperiodic influence upon fertility. In the farming systems of Europe there is a need to maximize the efficiency of labour-intensive management while in Australia there is a need to minimize labour-extensive farming systems. Both management strategies require better control of the onset of breeding activity. In many areas of Australia the late autumn start of breeding of Suffolk, Romney, Border Leicester and Perindale ewes means that the lambs are often weaned at a time of deteriorating pasture quality in late spring/summer with subsequent poor weight gain. An earlier season would allow better growth of the lambs before the summer burn off.

This review covers recent studies on the photoperiodic control of reproduction in ewes. We concentrate on evidence for the role of light, the pineal gland and melatonin on seasonal breeding in the ewe and discuss the ways this information has been used to gain control of sheep fertility and fecundity.

Role of daylength in seasonal breeding

The first attempts to influence sheep breeding by manipulating daylength were performed by Yeates (1949). Ducker *et al.* (1970) showed that abrupt and gradual changes in daylength ranging from 18·5 h light to 4·5 h light per day resulted in earlier onset of oestrus (Table 1). These experiments were started around the time of the summer solstice. Ducker *et al.* (1970) suggested and later confirmed experimentally that the interval to first oestrus provoked by short daylength was longer when ewes were treated in spring compared to summer (Ducker & Bowman, 1970a). These same authors provided compelling evidence that long daylength terminates the breeding season of ewes (Ducker & Bowman, 1970b). Using this basic information it has been possible to devise practical (expensive) treatment regimens involving daylength manipulation (often supplemented with hormonal treatment) which allow 8-month breeding cycles (Ducker & Bowman, 1972; Vesely, 1975).

Seasonal hormonal changes

The question of how seasonal changes in daylength result in the onset and offset of breeding activity has been addressed in many different ways. Walton *et al.* (1977) were amongst the first to describe in detail LH, FSH and prolactin concentrations during a year. The most striking seasonal hormonal change occurred in plasma prolactin concentrations; highest prolactin secretion coincided with periods of long daylength and lowest secretion coincided with short daylength. Studies involving

Table 1. Time interval from start of various shortened daylength treatments to onset of first oestrus in Clun Forest ewes (taken from Ducker *et al.,* 1970, and Ducker & Bowman, 1970b)

Treatment	Mean interval to oestrus (days)
Natural daylength	66·4 ± 7·4
3·75 h reduction	
Summer	59·5 ± 3·3
Spring	87·0 ± 7·0
7·75 h reduction	44·8 ± 2·7
11·75 h reduction	33·6 ± 3·3

artificial daylength changes and more frequent blood sampling reinforced the concept that shortening daylength results in lower prolactin concentrations and onset of ovarian activity (Walton *et al.,* 1980). Walton *et al.* (1980) suggested that the seasonally high prolactin concentrations may be responsible for the impaired ovarian function. The importance of prolactin changes in ewes is, however, controversial. Many studies have demonstrated low prolactin concentrations coincident with the onset of ovarian activity (Thimonier *et al.,* 1978; Kennaway *et al.,* 1983; Fig. 1), but there are also reports of ovarian activity in the presence of high prolactin values induced by photoperiod manipulation (Worthy *et al.,* 1985). The hypothesis of Walton *et al.* (1977) is especially attractive since hyperprolactinaemic humans invariably experience amenorrhoea (Bohnet *et al.,* 1976). McNeilly & Baird (1983) have shown that stimulation of prolactin concentrations by repeated TRH injections impaired ovarian oestradiol secretion in ewes. Therefore, while a discrete function for prolactin in the non-pregnant ewe is yet to be unequivocally proven, photoperiodically driven changes in basal prolactin secretion represent one of the most robust seasonal hormonal events.

The pineal gland has been shown to have a central role in the control of reproductive activity in rodents and a similar involvement is apparent in ruminants (Kennaway, 1984). The pineal hormone, melatonin, is secreted only during the night in sheep (Rollag & Niswender, 1976; Kennaway *et al.,* 1977). Early studies indicated that melatonin was present in pinealectomized ewes (Kennaway *et al.,* 1977), but when more specific antibodies were used in the RIA the unique pineal origin of melatonin in the sheep was confirmed (Kennaway *et al.,* 1982/1983). Extensive investigations by Rollag *et al.* (1978) and Kennaway *et al.* (1983) have shown that maximum melatonin values do not vary either through the oestrous cycle or the year. With respect to season, when an adequate blood sampling regimen is used, the only consistent change observed is a lengthening of the period of secretion of the hormone during the extended nights of winter. An appreciation of the importance of this observation will become apparent later in this review.

Effects of manipulation of photoperiod on reproductive parameters

Perturbation of physiological systems is a valuable tool for gaining an understanding of the role played by different endocrine organs. The role of daylength in the seasonal onset of oestrus in ewes became apparent in the experiments of Ducker *et al.* (1970). Similarly, the importance of the pineal gland became apparent when pinealectomized ewes were studied. If left in a normal environment, pinealectomized ewes continue to exhibit normal seasonal hormonal and fertility cycles (Roche *et al.,* 1970; Kennaway *et al.,* 1984). When pinealectomized ewes are challenged with an artificial change in daylength, however, the alternate inhibitory and stimulatory photoperiods are ignored; there is in fact a tendency for the ewes to maintain their previous seasonal breeding pattern.

One of the most instructive endocrine manipulations employed in the study of seasonal reproduction in the ewe has involved ovariectomy with oestradiol replacement using Silastic capsules

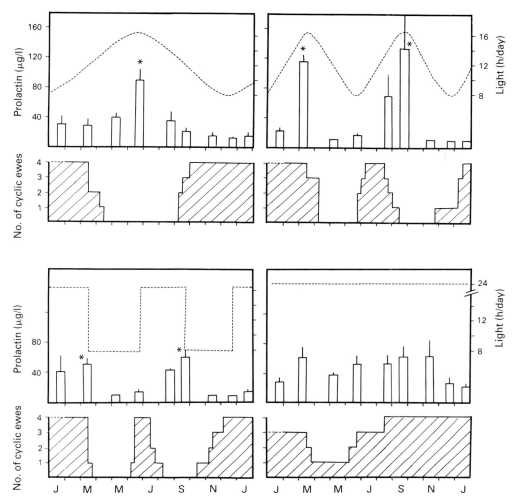

Fig. 1. Basal prolactin concentrations, number of ewes showing cyclic ovarian activity and the variations in daylength in four photoperiod regimens. The mean basal prolactin concentrations (N = 4) are indicated by the open bars with the s.e. represented by the vertical lines. The number of ewes cycling is shown by the hatched areas while the broken lines represent the hours of light per day in each room. *P 0·05 compared with December values. (From Kennaway *et al.*, 1983.)

(Legan *et al.*, 1977). Legan *et al.* (1977) advanced the hypothesis that anoestrus is a period when high sensitivity to oestrogen maintains low LH secretion. The breeding season of ewes is therefore characterized by lower oestradiol sensitivity at the hypothalamo/pituitary level and this permits the LH pulse frequency to increase during the periovulatory period, resulting in an LH surge and ovulation. Daylength is a major factor in the oestrogen sensitivity changes. If normal yearly swings in daylength are compressed into two 6-month periods there are 2 periods of breeding activity (Fig. 2) which is consistent with greater sensitivity to the negative feedback actions of oestrogen on LH secretion coincident with the periods of short daylength (Legan & Karsch, 1980; Fig. 3). Pineal involvement was implicated in studies of pinealectomized/ovariectomized/oestrogen-implanted ewes challenged with artificial daylength changes. Short-term pinealectomized ewes were unable to respond to the imposed daylengths with appropriate alterations in LH secretion (Bittman *et al.*,

1983; Fig. 3). Other studies have elegantly demonstrated that the duration of secretion of the pineal hormone melatonin is responsible for the changes in oestradiol feedback (Bittman *et al.*, 1983; Karsch *et al.*, 1984).

It was originally thought that ovariectomized ewes which were not treated with oestrogen had an unvarying LH pulse frequency during the year. Closer examination of these animals, however, indicated that the LH pulse frequency was lower during anoestrus than during the breeding season (Goodman *et al.*, 1982). This observation is described as a direct photoperiodic non-steroidal drive to the LHRH pulse-generating system. It is to be expected that this mechanism complements the steroidal feedback influences.

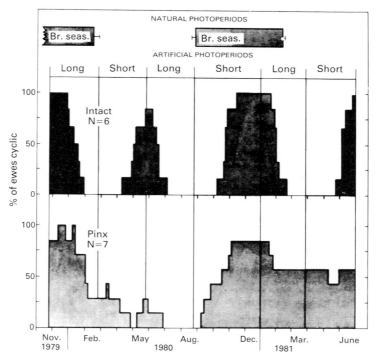

Fig. 2. Effects of artificial photoperiods on ovarian cyclicity in intact and pinealectomized (Pinx) ewes. Histograms indicate the percentage of ewes showing regular oestrous cycles. Photoperiodic treatments were 16 h light:8 h dark (long days) and 8 h light:16 h dark (short days). The normal breeding season was defined by 14 ewes maintained outdoors for 5 years. (From Bittman *et al.*, 1983.)

We have recently reported that long-term (>6 years) pinealectomized/ovariectomized/ oestrogen-treated ewes maintained in the field with pineal-intact/ovary-intact ewes show no synchronized seasonal swing in LH secretion (Kennaway *et al.*, 1984). The LH concentrations were low, indicating high sensitivity to negative feedback. Pinealectomized/ovary-intact ewes, however, continued to show ovarian cyclicity in phase with pineal-intact ewes. We interpreted these results as indicating that the pineal drive to the oestrogen negative feedback system could be overridden by pheromonal influences of the other ewes and rams in the flock or by the so-called direct photoperiodic drive to the LHRH pulse generator. If the latter were true then the direct non-steroidal drive occurs through a non-pineal retino/hypothalamic projection.

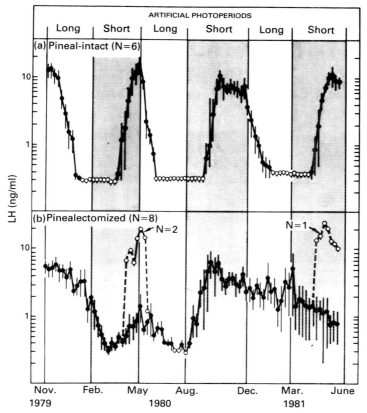

Fig. 3. Mean serum LH ($\pm$s.e.) in ovariectomized/oestradiol-implanted ewes. (a) Means of 2-weekly LH determinations in sham-pinealectomized ewes maintained in artificial photoperiod. (b) Alternating long (16L:8D) and short (8L:16D) daylength. (From Bittman *et al.*, 1983.)

Use of melatonin treatments to control the onset of breeding activity

Many groups have provided evidence supporting the role of light, the pineal gland and melatonin in the timing of the breeding season of ewes. While the mechanisms involved in the recognition by the hypothalamus of changes in duration of melatonin secretion are unknown, there is no doubt that appropriate administration of melatonin can alter the endocrine physiology of ewes.

In one of the first such experiments melatonin was chronically administered 8 h before darkness to anoestrous Border Leicester × Merino ewes maintained in long daylength (Kennaway *et al.*, 1982a). The melatonin was administered in food each day (when administered in this manner the melatonin is slowly absorbed and remains elevated for up to 8 h (Kennaway & Seamark, 1980)). The melatonin-treated ewes had an earlier onset of ovarian activity than did controls, with a lag time between the start of treatment and first oestrus of about 77 days, which is similar to the 'reaction time' observed in short daylength experiments. A decrease in prolactin concentrations after 30 days was a further indication that the treatment was mimicking the normal seasonal transition. Other groups have also confirmed the efficacy of melatonin in advancing the breeding season and promoting hormonal changes (Nett & Niswender, 1982; Arendt *et al.*, 1983; Howland *et al.*, 1984; Williams, 1984; Fig. 4).

More recently, continuous-release implants have been used in physiological studies and the hormonal changes observed are similar to those found after timed oral administration (Kennaway

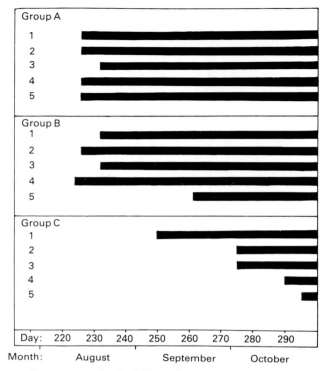

Fig. 4. Occurrence of oestrous cycles (solid bars) in 3 groups of 5 Suffolk cross ewes treated as follows. Group A: kept in 16 h light: 8 h darkness and fed 3 mg melatonin daily 8 h before darkness; Group B: kept in 8L:16D; Group C: kept in natural light. The experiment began on 15 June 1981. (From Arendt *et al.*, 1983.)

et al., 1982/1983, 1982b). When melatonin is continuously available during long daylengths, plasma prolactin concentrations are depressed within 14 days. Practical exploitation of this finding will be discussed below.

Use of light treatments and melatonin administration to gain control of fertility and fecundity

In this review much emphasis has been placed upon seasonal control of the onset of breeding activity and nothing about the quality of the fertility during the season. Clearly there are advantages in having control of when lambs are born and how many are born. The latter is normally genetically determined with some breeds being exceptionally fecund while in others twinning is rare. Seasonal influences upon fecundity in sheep have been recognized for many years (Land *et al.*, 1973; Scaramuzzi & Radford, 1983). Ovulation rate is highest in the late summer and autumn (the start of natural breeding activity) and declines progressively towards anoestrus. Hendy & Bowman (1974) have shown that ewes which had the earliest spontaneous ovulations had a higher ovulation rate than late breeding ewes of the same flock (Fig. 5). By the third cycle of the season, however, the ovulation rates were identical in early intermediate and late starters. Because daylength is the major environmental factor responsible for seasonality of sheep reproduction we have performed extensive studies on the effect of photoperiod on ovulation rate at first oestrus.

In the first such study, Dunstan (1977) showed that moving Border Leicester × Merino ewes into a darkened shed 4 h before sunset each day for 3–6 weeks in late spring, before the introduction of a ram, resulted in a 22–38% increase in lambs born. In subsequent studies we have shown that the

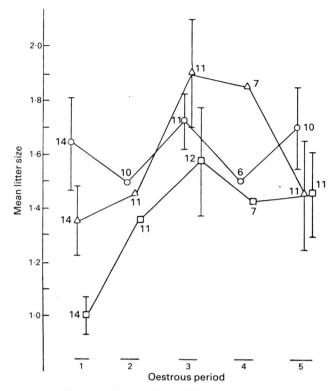

Fig. 5. Litter sizes in early (●), mid- (△) and late (□) mating phases in the first to fifth oestrous periods of the breeding season. Values are mean ± s.e. for the number of ewes indicated. (From Hendy & Bowman, 1974.)

artificial short daylength treatment can be successful if a vasectomized ram is used during the first 3 weeks and a fertile ram for the following 3 weeks (Figs 6 & 7; E. A. Dunstan, D. J. Kennaway & J. M. Obst, unpublished results). The advantage of delayed ram introduction, however, is a reasonably synchronized mating (21 days after introduction of rams) and a shorter lambing period, both being of practical importance. Most importantly, the phenomenon is highly reproducible; in one experiment on the same flock over a 6-year period a treatment involving 3 weeks short daylength, ram introduction and a further 6 weeks short daylength resulted in a mean 27% increase in lamb production each year (Fig. 8; E. A. Dunstan, D. J. Kennaway & J. M. Obst, unpublished results).

The hormone responsible for the short daylength-induced increases in fecundity was suspected to be melatonin because pinealectomized ewes had poor responses to the altered daylength regimen (Kennaway *et al.*, 1982/1983; E. A. Dunstan, D. J. Kennaway & J. M. Obst, unpublished results). When we compared a group of ewes on the short daylength treatment with ewes fed melatonin pellets (group-fed, average dose per ewe 2 mg) the number of lambs born was significantly increased in both groups (Fig. 9; D. J. Kennaway & E. A. Dunstan, unpublished results).

Further confirmation that melatonin was responsible for the response was obtained in trials on farms using much larger numbers of ewes. On 3 properties studied over 2 years, lamb production was increased by 10–20% using the 9-week treatment regimen. All the initial light and melatonin experiments had been performed on Border Leicester × Merino ewes in late anoestrus; under these conditions no effect on breeding season onset was observed because the rams induced early breeding in the control and treated animals. When melatonin was fed in January to a late breeding strain,

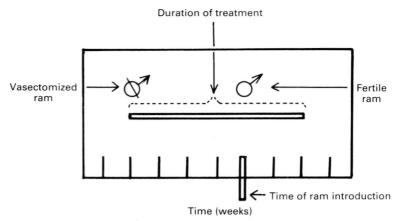

Fig. 6. Schematic representation of treatment designs used in short daylength, melatonin-feeding and melatonin-implant experiments with sheep (see Figs 7, 8, 9 and 12).

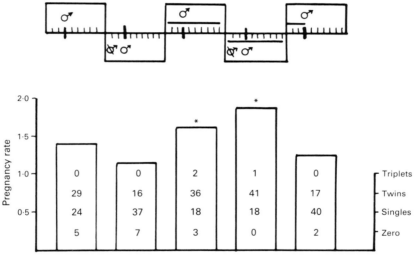

Fig. 7. Pregnancy rate (no. of lambs/ewe in group) in Border Leicester × Merino ewes given 4 h extra darkness each day for 3 weeks before fertile ram introduction (see Fig. 6 for treatment details). The actual numbers of singles, twins and triplets are also represented. The asterisk indicates the treatments resulting in significantly more lambs being born compared to the control groups ($P < 0.05$).

the Perindale (normal spontaneous ovarian activity and conceptions occur in March/April), we observed (a) an earlier onset of breeding activity in treated ewes versus control fed ewes in the same flock and (b) a 42% increase in lamb production (R. F. Seamark & D. J. Kennaway, unpublished results).

Daily light deprivation or melatonin feeding may not be very practical procedures for farm use except in intensive farming systems. We had, however, demonstrated hormonal effects (decreased prolactin concentrations) in ewes treated with Silastic envelopes of melatonin (Kennaway *et al.*, 1982/83) and proposed that continuous melatonin administration would result in earlier breeding and increased fecundity. The development of a continuous release/biodegradable formulation of

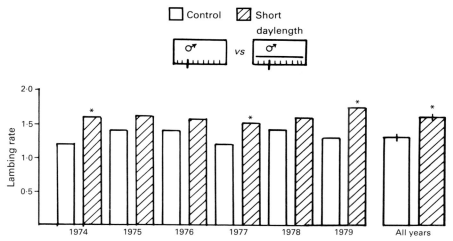

Fig. 8. Lambing rate (no. of lambs born/ewe in group) in Border Leicester × Merino ewes given 4 h extra darkness each day for 3 weeks before fertile ram introduction (see Fig. 6 for treatment details). The same flock was treated at the same time of the year for 6 years. The asterisks indicate the treatments resulting in significantly more lambs being born compared to the control groups in 3 of the 6 years ($P < 0.05$). When analysed over the entire 6-year period the difference was also significant ($P < 0.05$).

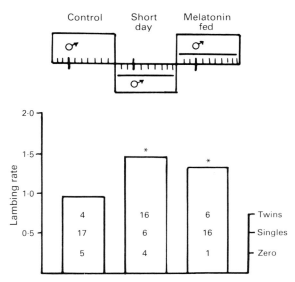

Fig. 9. Lambing rate of Border Leicester × Merino ewes given 4 h extra darkness each afternoon or fed melatonin (2 mg) daily 4 h before sunset for 3 weeks before ram introduction (see Fig. 6 for treatment details). Also shown are the actual numbers of single and twin lambs born.

melatonin by Gene Link (Australia) Pty Ltd in conjunction with the Institute of Drug Technology, Victoria, The Victorian Department of Agriculture and Rural Affairs and the University of Adelaide, has facilitated research in this area. When melatonin is continuously delivered to produce circulating concentrations of melatonin in excess of 1000 pmol/l the breeding season can be advanced and an increase in fecundity observed. Two experiments will be described which illustrate these effects.

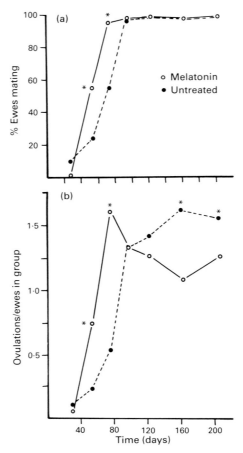

Fig. 10. Cumulative percentages of Corriedale ewes (100/group) (a) mating to vasectomized rams and (b) ovulating after treatment with melatonin implants or left untreated. Melatonin implants were inserted on Days 0 and 28 to give blood melatonin concentrations > 1000 pmol/l for ~ 70 days. *$P < 0.001$ compared with control values.

In the first experiment 200 Corriedale ewes were treated with melatonin implants or placebo on 24 October 1985 and again on 21 November 1985 (i.e. late spring) to determine whether the increased lamb production was due to increased ovulation rate or lower fetal wastage. Vasectomized harnessed rams were run with the ewes throughout the experiment and daily observations on oestrous activity were made. The ewes were subjected to endoscopy monthly to determine ovulation rates. Figure 10 shows that the melatonin implants caused a significant increase in the proportion of ewes ovulating at 53 and 74 days after the start of treatment. The treatment also significantly increased the number of ovulations per ewe ovulating on Days 53 and 74 and therefore increased the number of ovulations per ewe in the group (Fig. 10; L. D. Staples, A. Williams, S. McPhee & B. Ayton, unpublished results).

In the second experiment (L. D. Staples, S. McPhee, B. Ayton, J. Reeve & A. Williams, unpublished results) we tested the interrelationship between onset of breeding activity and increased pregnancy rate. We used 3 breeds of sheep, the Merino, Border Leicester × Merino, and Romney to represent 'non-seasonal', 'moderately seasonal' and 'highly seasonal' breeding patterns, respectively. The experiments were conducted such that the times of introduction of rams were conservatively 'early' for each breed, i.e. 8 November 1985, 18 November 1985 and 28 January

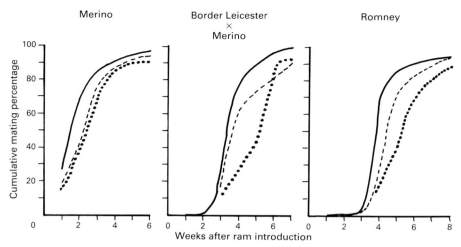

Fig. 11. Cumulative percentages of Merino, Border Leicester × Merino and Romney ewes mating to fertile rams after treatment with melatonin implants (——), left untreated but run with the melatonin group (– – –) or left untreated in an isolated part of the property (· · · · ·). Day 0 is the time of ram introduction.

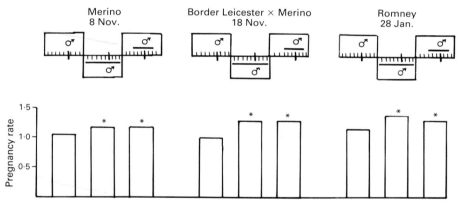

Fig. 12. Pregnancy rate at a Day-70 ultrasound scan in Merino, Border Leicester × Merino and Romney ewes (N = 93–140) given 5- or 9-week melatonin treatments (see Fig. 6 for treatment details). *$P < 0.001$ compared to control group value.

1986 respectively. Two treatments were used: the first involved the insertion of melatonin implants 6 weeks before ram introduction and giving a 9-week melatonin coverage, and the second treatment involved melatonin treatment 3 weeks before ram introduction with a total of 5 weeks melatonin coverage. The melatonin- and placebo-implanted ewes were run as a single flock and a third untreated control group was maintained in an isolated part of the property. Figure 11 shows the cumulative mating profiles for the 6-week pretreatment groups only. There was no effect on breeding of Merinos because many ewes in the flock were already cyclic at the start of the experiment. However, the Border Leicester × Merino and Romney ewes began to breed earlier than did placebo-implanted controls and the separate control flock. In all three breeds both treatment regimens resulted in significant increases in the number of fetuses at a Day-70 ultrasound scan (Fig. 12).

The chronically elevated melatonin concentrations achieved after implantation of the Gene Link (Aust.) formulation promotes early oestrus in several breeds of sheep. Moreover, the phenomenon first reported by Hendy & Bowman (1974) (the higher litter size of ewes mating earliest in the season) is likely to be the result of photoperiodic influences and can be mimicked by light or melatonin treatments. The increased ovulation rate does not simply follow on from the early onset of oestrus, because control ewes run with melatonin-treated ewes breed early due to a 'ewe effect' but do not have a higher ovulation response compared to the separate control group.

Conclusions

The seasonal nature of sheep breeding has been tolerated for centuries and management practices have evolved to compensate for it. With the desire to maximize sheep production numerous procedures have been used to induce out-of-season breeding (e.g. progesterone treatment, GnRH injections) but all have failed to account for the persisting inhibitory photoperiodic signals the sheep were receiving during anoestrus. A significant advance in our understanding of seasonality has come with the recognition of the vital roles the pineal gland and its hormone melatonin play in relaying photoperiodic information to the endocrine system. The development of biologically active, biodegradable, continuous release formulations of melatonin should allow producers to gain almost complete control of the time their animals can conceive and maximize their lamb production. From an endocrinologist's viewpoint, however, there are still many questions to be answered concerning how melatonin acts at the neuroendocrine level to promote cyclic ovarian activity and maximize the number of ova shed by the ewes.

We thank R. Male, G. Gleeson, R. F. Seamark, A. Gilmore, P. Royles, F. Carbone, H. Webb, A. Williams, S. McPhee, J. Reeve and B. Ayton. Original work cited in this review was generously supported by grants from the Australian Meat Research Committee, the National Health and Medical Research Council and Gene Link (Aust.) Pty Ltd.

References

Arendt, J., Symons, A.M., Laud, C.A. & Pryde, S.J. (1983) Melatonin can induce early onset of the breeding season in ewes. *J. Endocr.* 97, 395–400.

Bittman, E.L., Karsch, F.J. & Hopkins, J.W. (1983) Role of the pineal gland in ovine photoperiodism: regulation of seasonal breeding and negative feedback effects of estradiol upon luteinizing hormone secretion. *Endocrinology* 113, 329–336.

Bohnet, H.G., Dahlen, H.G., Wuttke, W. & Schneider, H.P.G. (1976) Hyperprolactinaemic anovulatory syndrome. *J. clin. Endocr. Metab.* 42, 132–143.

Ducker, M.J. & Bowman, J.C. (1970a) Photoperiodism in the ewe. 3. The effects of various patterns of increasing daylength on the onset of anoestrus in Clun Forest ewes. *Anim. Prod.* 12, 465–471.

Ducker, M.J. & Bowman, J.C. (1970b) Photoperiodism in the ewe. 4. A note on the effect on onset of oestrus in Clun Forest ewes of applying the same decrease in daylength at two different times of the year. *Anim. Prod.* 12, 513–516.

Ducker, M.J. & Bowman, J.C. (1972) Photoperiodism in the ewe. 5. An attempt to induce sheep of three breeds to lamb every eight months by artificial daylength changes in a non-light-proofed building. *Anim. Prod.* 14, 323–334.

Ducker, M.J., Thwaites, C.J. & Bowman, J.C. (1970) Photoperiodism in the ewe. 2. The effects of various patterns of decreasing daylength on the onset of oestrus in Clun Forest ewes. *Anim. Prod.* 12, 115–123.

Dunstan, E.A. (1977) Effects of changing daylength pattern around the mating period on the mating and lambing performance of Border Leicester–Merino cross ewes. *Aust. J. exp. Agric. Anim. Husb.* 17, 741–745.

Goodman, R.L., Bittman, E.L., Foster, D.L. & Karsch, F.J. (1982) Alterations in the control of luteinizing hormone-pulse frequency underlie the seasonal variation in estradiol negative feedback in the ewe. *Biol. Reprod.* 27, 580–589.

Hafez, E.S.E. (1952) Studies on the breeding season and reproduction of the ewe. *J. agric. Sci. Camb.* 42, 189–265.

Hendy, C.R.C. & Bowman, J.C. (1974) The association between variation in the seasonal onset of oestrus and litter size in the ewe. *J. Reprod. Fert.* 40, 105–112.

Howland, B.E., Palmer, W.M. & Vriend, J. (1984) Endocrine changes in ewes fed melatonin. *Proc. 10th Int. Congr. Anim. Prod. & A.I., Urbana-Champaign*, pp. 25–26, Abstr.

Karsch, F.J., Bittman, E.L., Foster, D.L., Goodman, R.L., Legan, S.J. & Robinson, J.E. (1984) Neuroendocrine basis of seasonal reproduction. *Recent Prog. Horm. Res.* **40**, 185–232.

Kennaway, D.J. (1984) Pineal function in ungulates. *Pineal Res. Rev.* **2**, 113–140.

Kennaway, D.J. & Seamark, R.F. (1980) Circulating levels of melatonin following its oral administration or subcutaneous injection in sheep and goats. *Aust. J. biol. Sci.* **33**, 349–353.

Kennaway, D.J., Frith, R.G., Phillipou, G., Matthews, C.D. & Seamark, R.F. (1977) A specific radioimmunoassay for melatonin in biological tissue and fluids and its validation by gas chromatography mass spectrometry. *Endocrinology* **101**, 119–127.

Kennaway, D.J., Gilmore, T.A. & Seamark, R.F. (1982a) Effect of melatonin feeding on serum prolactin and gonadotropin levels and the onset of seasonal estrous cyclicity in sheep. *Endocrinology* **110**, 1766–1772.

Kennaway, D.J., Gilmore, T.A. & Seamark, R.F. (1982b) Effects of melatonin implants on the circadian rhythm of plasma melatonin and prolactin in sheep. *Endocrinology* **110**, 2186–2188.

Kennaway, D.J., Dunstan, E.A., Gilmore, T.A. & Seamark, R.F. (1982/1983) Effects of shortened daylength and melatonin treatment on plasma prolactin and melatonin levels in pinealectomised and sham-operated ewes. *Anim. Reprod. Sci.* **5**, 287–294.

Kennaway, D.J., Sanford, L.M., Godfrey, B. & Friesen, H.G. (1983) Patterns of progesterone, melatonin and prolactin secretion in ewes maintained in four different photoperiods. *J. Endocr.* **97**, 229–242.

Kennaway, D.J., Dunstan, E.A., Gilmore, T.A. & Seamark, R.F. (1984) Effects of pinealectomy, oestradiol and melatonin on plasma prolactin and LH secretion in ovariectomized sheep. *J. Endocr.* **102**, 199–207.

Land, R.B., Pelletier, J., Thimonier, J. & Mauleon, P. (1973) A quantitative study of genetic differences in the incidence of oestrus, ovulation and plasma luteinizing hormone concentration in the sheep. *J. Endocr.* **58**, 305–315.

Legan, S.J. & Karsch, F.J. (1980) Photoperiodic control of seasonal breeding in ewes: modulation of the negative feedback action of estradiol. *Biol. Reprod.* **23**, 1061–1068.

Legan, S.J., Karsch, F.J. & Foster, D.L. (1977) The endocrine control of seasonal reproductive function in the ewe: a marked change in the response to the negative feedback action of estradiol on luteinizing hormone secretion. *Endocrinology* **101**, 818–824.

McNeilly, A.S. & Baird, D.T. (1983) Direct effect of prolactin, induced by TRH injection, on ovarian oestradiol secretion in the ewe. *J. Reprod. Fert.* **69**, 559–568.

Nett, T.M. & Niswender, G.D. (1982) Influence of exogenous melatonin on seasonality of reproduction in sheep. *Theriogenology* **17**, 645–652.

Roche, J.F., Karsch, F.J., Foster, D.L., Takagi, S. & Dziuk, P.J. (1970) Effect of pinealectomy on estrus, ovulation and luteinizing hormone in ewes. *Biol. Reprod.* **2**, 251–294.

Rollag, M.D. & Niswender, G.D. (1976) Radioimmunoassay of serum concentrations of melatonin in sheep exposed to different lighting regimens. *Endocrinology* **98**, 482–489.

Rollag, M.D., O'Callaghan, P.L. & Niswender, G.D. (1978) Serum melatonin concentrations during different stages of the annual reproductive cycle in ewes. *Biol. Reprod.* **18**, 279–285.

Scaramuzzi, R.J. & Radford, H.M. (1983) Factors regulating ovulation rate in the ewe. *J. Reprod. Fert.* **69**, 353–367.

Thimonier, J., Ravault, J.P. & Ortavant, R. (1978) Plasma prolactin variations and cyclic ovarian activities in ewes submitted to different light regimens. *Annls Biol. anim. Biochem. Biophys.* **18**, 1229–1235.

Vesely, J.A. (1975) Induction of lambing every eight months in two breeds of sheep by light control with or without hormonal treatment. *Anim. Prod.* **21**, 165–174.

Walton, J.S., McNeilly, J.R., McNeilly, A.S. & Cunningham, F.J. (1977) Changes in concentrations of follicle stimulating hormone, luteinizing hormone, prolactin and progesterone in the plasma of ewes during the transition from anoestrus to breeding activity. *J. Endocr.* **75**, 127–136.

Walton, J.S., Evins, J.D., Fitzgerald, B.P. & Cunningham, F.J. (1980) Abrupt decrease in daylength and short-term changes in the plasma concentrations of FSH, LH and prolactin in anoestrous ewes. *J. Reprod. Fert.* **59**, 163–171.

Williams, H.L.L. (1984) The effects on the onset of the breeding season of sheep of feeding melatonin during the late summer. *Proc. 10th Int. Congr. Anim. Prod. & A.I., Urbana-Champaign,* pp. 31–32, Abstr.

Worthy, K., Haresign, W., Dodson, S., McLeod, B.J., Foxcroft, G.R. & Haynes, N.B. (1985) Evidence that the onset of the breeding season in the ewe may be independent of decreasing plasma prolactin concentrations. *J. Reprod. Fert.* **75**, 237–246.

Yeates, N.T.M. (1949) The breeding season of the sheep with particular reference to its modification by artificial means using light. *J. agric. Sci., Camb.* **39**, 1–43.

J. Reprod. Fert., Suppl. **34** (1987), 201–213

Printed in Great Britain
© 1987 Journals of Reproduction & Fertility Ltd

Function of the hypothalamic–hypophysial axis during the post-partum period in ewes and cows

T. M. Nett

Department of Physiology, Colorado State University, Fort Collins, Colorado 80523, U.S.A.

Summary. During pregnancy the hypothalamic–hypophysial axis is suppressed by the high concentrations of progesterone and oestradiol in the circulation. The high concentrations of these steroids appear to inhibit secretion of GnRH from the hypothalamus, resulting in inadequate stimulation of pituitary gonadotrophs to maintain synthesis of LH. This produces a depletion of LH in the anterior pituitary gland that must be restored after parturition before normal oestrous cycles can begin.

Introduction

Post-partum anoestrus refers to the absence of ovarian cycles after parturition. In domestic ruminants the factors known to affect the duration of post-partum anoestrus include pre- and post-partum level of nutrition, whether the young are allowed to suck, and season in which the young are born. Theoretically, the lack of ovarian cycles after parturition could be due to an inability of the ovaries to respond to gonadotrophins, or of the hypophysis to secrete a sufficient quantity of gonadotrophins to stimulate follicular growth and ovulation. Moreover, since the production and secretion of gonadotrophins is dependent on hypothalamic input, i.e. gonadotrophin-releasing hormone (GnRH), a lack of sufficient gonadotrophin may be due to inadequate production and/or secretion of GnRH. Early studies indicated that ovulation could be induced soon after parturition in cows by injection of urine from pregnant women and a hypophysial emulsion (Zawadowsky *et al.*, 1935) or by injecting pituitary gonadotrophins (Casida *et al.*, 1943). The ovary therefore appears to be capable of responding to gonadotrophic stimulation very early in the post-partum period. From these observations, one can infer that the lack of ovarian activity following parturition is due to the lack of gonadotrophic stimulation. The remainder of this review will address factors that are perhaps responsible for the suppression of gonadotrophin secretion during the post-partum period in cattle and sheep.

Circulating concentrations of luteinizing hormone (LH) are low during the early post-partum period in beef cows (Arije *et al.*, 1974), dairy cows (Edgerton & Hafs, 1973) and ewes (Restall & Starr, 1977), apparently due to a decreased frequency of pulses of LH. The secretion of LH must return to a level similar to that observed in cyclic animals before normal ovarian cycles are initiated. Although the circulating concentration of LH is lower in post-partum animals than in cyclic animals, the circulating concentrations of follicle-stimulating hormone (FSH) are not suppressed (Walters *et al.*, 1982a; Moss *et al.*, 1985). This may be due to the relative lack of follicular development during late gestation, and without the negative feedback of folliculostatin, secretion of FSH remains normal (Miller *et al.*, 1982).

What component of the hypothalamic–hypophysial axis causes a deficiency in the secretion of LH during the post-partum period?

There are several components of the hypothalamic–hypophysial axis that, if their function was suppressed, could lead to a reduction in the secretion of LH, probably manifested as a decrease in frequency and/or amplitude of LH pulses. These include the amount of GnRH synthesized and stored in the hypothalamus, the amount of GnRH secreted into the hypophysial portal circulation, the number of receptors for GnRH in the anterior pituitary gland and the amount of LH synthesized and stored in the anterior pituitary gland. If the quantity of GnRH synthesized in the hypothalamus and stored in the median eminence were reduced to very low levels, then it is possible that the amount of GnRH available for release would be insufficient to stimulate normal function of the anterior pituitary gland. Likewise, direct inhibition of the secretion of GnRH could occur without a concomitant reduction in the content of GnRH in the hypothalamus. If either of these scenarios occurred, one would anticipate a reduction in function of the gonadotrophs due to a lack of trophic support. It is also possible that the anterior pituitary gland could become insensitive to trophic stimulation due to an inadequate number of receptors for GnRH or to a post-receptor defect. If this occurred, then the synthesis and secretion of GnRH could be normal but the gonadotrophs could not respond appropriately. This, in turn, would result in insufficient secretion of LH to stimulate ovarian cyclicity. Finally, it is possible that each portion of the hypothalamic–hypophysial axis described above is functionally competent, but that the anterior pituitary gland contains too little LH to elevate circulating concentrations of LH into the normal range after GnRH stimulation.

Hypothalamic content of GnRH

To date, only a few investigators have examined the content of GnRH in the hypothalamus in the periparturient animal. Moss *et al.* (1980) did not observe a significant change in content of GnRH in the preoptic area, medial basal hypothalamus or median eminence of ewes from 1 day after parturition until the ewes began cycling, about Day 40 after delivery (Fig. 1). Likewise, there was no significant change in the content of GnRH in the same areas of the hypothalamus of beef cows between Days 5 and 30 after parturition (Moss *et al.*, 1985). Similar observations for beef heifers have been reported by Cermak *et al.* (1983), but the hypothalamic content of GnRH was less in cyclic cows (Braden *et al.*, 1983) than in post-partum cows, suggesting that release of GnRH may have been inhibited during the post-partum period leading to increased stores of GnRH. Carruthers *et al.* (1980) did not find a difference in the content of GnRH between suckling and non-suckling dairy cows. From these observations, one can conclude that during the post-partum period the hypothalamus contains sufficient quantities of GnRH to stimulate the anterior pituitary gland.

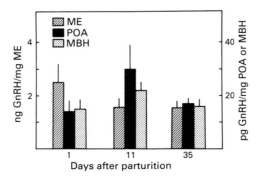

Fig. 1. Concentration of GnRH in the median eminence (ME), preoptic area (POA) and medial basal hypothalamus (MBH) of ewes at various times after parturition. Each of the ewes in the Day 35 group had ovulated and established a functional corpus luteum. Values represent mean ± s.e., *n* = 4–8 per group. (Adapted from Moss *et al.*, 1980.)

Secretion of GnRH

Although there is sufficient evidence to conclude that adequate amounts of GnRH are present in the hypothalamus during the early post-partum period, the question that must be answered is, 'Is the GnRH secreted in a manner that will stimulate reproductive activity?' Unfortunately, to date no one has examined the concentrations of GnRH in hypophysial portal or cavernous sinus blood during the post-partum period of domestic ruminants. However, by utilizing the relatively new technique for collecting portal blood developed by Clarke & Cummins (1982) or the push–pull technique for sampling secretions from the median eminence of sheep (Levine *et al.*, 1982), it should now be possible to obtain such information. Considering the reported insensitivity of the anterior pituitary to exogenous GnRH during the early post-partum period in cattle (Carter *et al.*, 1980) and sheep (Jenkin *et al.*, 1977), such information is essential for elucidating the mechanisms responsible for restoring sensitivity of the anterior pituitary to GnRH.

Sensitivity of the anterior pituitary: receptors for GnRH

The amount of LH released in response to GnRH decreases as gestation progresses and remains low during the early post-partum period in cattle (Carter *et al.*, 1980) and sheep (Rippel *et al.*, 1974; Chamley *et al.*, 1974; Jenkin *et al.*, 1977; Crowder *et al.*, 1982). This has led some investigators to conclude that the pituitary was 'insensitive' to the effects of GnRH (Jenkin & Heap, 1974; Chamley *et al.*, 1974). A reduction in sensitivity of the anterior pituitary to GnRH could be interpreted to mean that there was a decrease in the number of receptors for GnRH. With this in mind, we undertook two series of experiments to examine the sensitivity of the anterior pituitary to GnRH and to quantify the number of hypophysial receptors for GnRH during the post-partum period of the ewes. In the first series of experiments, ewes were induced to ovulate and were mated during anoestrus so that lambing would occur during the breeding season. The anterior pituitary glands of these ewes were collected at various times during the post-partum period and dissociated into single cells. The cells were cultured overnight, incubated with various doses of GnRH and the amount of LH released into the incubation medium was measured.

Release of LH from cultured cells increased progressively with increasing doses of GnRH. Further, there was a progressive increase in the amount of LH released from cells in response to all doses of GnRH as time from parturition increased from Day 1 to Day 35. However, the dose of GnRH (0.5×10^{-9} M) that caused half-maximal release of LH from the pituitary cells was similar at each of the times examined (Moss *et al.*, 1980). These results have been interpreted to mean that the anterior pituitary cells contain similar numbers of receptors for GnRH at each of the times after parturition. Moreover, based on these data, there does not appear to be a post-receptor defect in the mechanism responsible for secretion of LH.

In the second series of experiments, the number of receptors for GnRH was measured during late pregnancy and on Days 1, 11, 22 and 35 after parturition. The number of GnRH receptors was higher ($P < 0.05$) at Days 1 and 11 after parturition than during late gestation or on Days 22 and 35 after parturition (Fig. 2). However, at no time during late gestation or after parturition did the number of GnRH receptors fall below that seen in ovariectomized ewes (Crowder *et al.*, 1982) in which secretion of LH is always high. In a similar study in post-partum beef cows, we observed a transient increase in the number of hypophysial receptors for GnRH between Days 1 and 15 *post partum*. The number of receptors then returned to the lower level by Days 30 and 45 after parturition (Cermak *et al.*, 1983). Moss *et al.* (1985) quantified hypophysial GnRH receptors on Days 5, 10, 20 and 30 after parturition in beef cows. They also found the number of receptors for GnRH to be reduced at 20 and 30 days after parturition when compared to Days 5 and 10 after parturition.

Collectively, these data regarding the number of GnRH receptors and the in-vitro responsiveness of anterior pituitary cells to GnRH indicate that the sensitivity of the pituitary gland to GnRH is not reduced during the early post-partum period. Therefore, this portion of the regulatory system

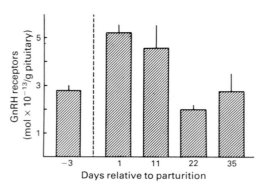

Fig. 2. Concentration of GnRH receptors during late pregnancy and after parturition in ewes. Values represent mean ± s.e., *n* = 3 or 4 per group. (Adapted from Crowder *et al.*, 1982.)

for secretion of LH does not appear to be responsible for the anoestrus that occurs after parturition.

Pituitary content of LH

The content of LH in the anterior pituitary glands of cows (Nalbandov & Casida, 1940) and ewes (Chamley *et al.*, 1976; Jenkin *et al.*, 1977) is reduced by as much as 95% during gestation. In both species there is a gradual increase in the pituitary content of LH after parturition. In suckling beef cows the content of LH in the anterior pituitary increased from very low levels at parturition to levels similar to those present in cyclic animals by Day 30 *post partum* (Saiddudin *et al.*, 1968; Cermak *et al.*, 1983; Moss *et al.*, 1985). Likewise, the pituitary content of LH in ewes gradually increased during the post-partum period, reaching levels similar to those observed in intact ewes between 3 and 10 weeks *post partum* (Fig. 3). The length of time for complete restoration of pituitary LH depends on whether lambing occurred during anoestrus or during the breeding season, and whether the ewes were suckling (Jenkin *et al.*, 1977; Moss *et al.*, 1980).

During the post-partum period in ewes, the pituitary content of LH was highly correlated (*r* > 0·9) with the amount of LH released in response to a half-maximal dose of GnRH (Jenkin *et al.*, 1977; Crowder *et al.*, 1982). Therefore, during the early post-partum period in cattle and sheep, it seems likely that if pulses of GnRH of normal amplitude were released into the hypophysial portal circulation, the resulting pulses of LH would be lower than those observed in cyclic animals. From this information, I suggest that a lack of stores of LH in the anterior pituitary gland, rather than reduced sensitivity to GnRH, is one of the initial limitations to the resumption of normal oestrous cycles in post-partum animals.

What causes depletion of LH in the anterior pituitary gland during gestation?

From the information presented above, it is clear that in cattle and sheep the amount of LH in the anterior pituitary gland dramatically decreases during gestation, and is slowly replenished during the post-partum period. It therefore appears that some factor(s) secreted during pregnancy might inhibit synthesis of LH, thereby resulting in depletion of LH in the anterior pituitary. Since progesterone and oestradiol are present in the circulation for most of gestation and are particularly high during the last third (Stabenfeldt *et al.*, 1972; Burd *et al.*, 1976; Carnegie & Robertson, 1978),

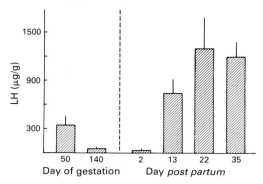

Fig. 3. Concentration of LH in the anterior pituitary glands removed from ewes during pregnancy and at various times after parturition. Values represent mean ± s.e., $n = 4$ or 5 per group. (Adapted from Wise *et al.*, 1986b.)

these steroids are likely candidates. Furthermore, receptors for oestradiol and progesterone are present in both the hypothalamus and anterior pituitary gland (Clarke *et al.*, 1981; Kamel & Krey, 1982; Cermak *et al.*, 1983; Wise *et al.*, 1986a). Accordingly, we suggested that these steroids, alone or in combination, might be responsible for the reduction in pituitary stores of LH. The basis for this hypothesis is strengthened by the fact that progesterone reduces the frequency of pulsatile LH release whereas oestradiol inhibits the amplitude of LH pulses in ovariectomized ewes (Goodman & Karsch, 1980). Since pulses of LH released into the peripheral circulation are the result of pulses of GnRH in the hypophysial–portal circulation (Clarke & Cummins, 1982; Levine *et al.*, 1982), then it is reasonable to presume that there is a reduction in frequency and/or amplitude of GnRH pulses when these steroids are present in the circulation. Given the fact that GnRH is required for synthesis of LH (Fraser *et al.*, 1975), if both the amplitude and frequency of GnRH pulses are inhibited by progesterone and oestradiol, then an adequate stimulus for synthesis of LH is probably absent during gestation. A reduction in synthesis of LH would thus explain the observed decrease in pituitary content of LH. To test this hypothesis, we initiated a series of experiments to determine the effects of chronically elevated concentrations of progesterone and/or oestradiol on serum and pituitary concentrations of LH in ovariectomized ewes.

Groups of ovariectomized ewes served as controls or were treated with progesterone, oestradiol or oestradiol plus progesterone for 3 weeks at doses that produced serum concentrations of these steroids approximating those observed during the 2-week period before parturition. As expected, each steroid, alone and in combination, produced a dramatic decrease in serum concentrations of LH. Although progesterone had no effect on the hypophysial concentration of LH, oestradiol alone, or in combination with progesterone, resulted in a dramatic (>95%) reduction in the pituitary concentration of LH (Fig. 4). As in the pregnant and post-partum animals, the reduction in pituitary content of LH was not due to a lack of GnRH in the hypothalamus, or to a deficiency in the number of receptors for GnRH in the anterior pituitary gland (Moss *et al.*, 1981).

These data are consistent with previous reports for the ewe indicating that, when the anterior pituitary gland is under the influence of progesterone, the release of LH induced by GnRH is not diminished, whereas when under the influence of oestradiol the GnRH-induced release of LH is dramatically diminished (Goodman & Karsch, 1980; Tamanini *et al.*, 1986). Since there was actually an increase ($P < 0.01$) in the number of GnRH receptors in the oestradiol-treated ewes in the study described above, it appears that oestradiol (1) directly inhibits synthesis of LH at the level of the pituitary, or (2) so completely inhibits secretion of GnRH that there is insufficient trophic stimulation for synthesis of LH.

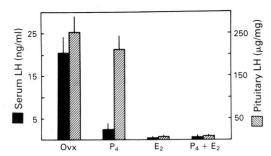

Fig. 4. Pituitary and serum concentrations of LH in ovariectomized (Ovx) ewes treated for 3 weeks with progesterone (P_4), oestradiol (E_2) or progesterone plus oestradiol. Values represent mean $\pm$ s.e., N = 5 per group. (Adapted from Moss *et al.*, 1981.)

How are the negative feedback effects of progesterone and oestradiol exerted?

Since progesterone inhibits secretion of LH by reducing the frequency of episodes of LH release without affecting the pituitary content of LH or pituitary responsiveness to GnRH, it has been postulated that progesterone suppresses secretion of GnRH from the hypothalamus (Goodman & Karsch, 1980). Indeed, progesterone will decrease the concentration of GnRH in pituitary stalk blood of ovariectomized rats (Sarkar & Fink, 1979). If progesterone decreases release of GnRH, then it is tempting to speculate that it may also suppress the synthesis of GnRH. This speculation is based on the fact that the content of GnRH in the hypothalamus does not change as a result of treatment with progesterone. If the only effect of progesterone was to inhibit secretion of LH (i.e. without affecting synthesis) then an increase in the hypophysial content of LH would be expected, unless there was an increase in degradation of LH by the gonadotroph. However, when ovariectomized ewes were treated with progesterone, there was no effect on the number of hypophysial receptors for GnRH (Moss *et al.*, 1981) nor was there a significant change in the content of mRNA for the subunits of LH (Hamernik & Nett, 1986). Therefore, there does not appear to be a direct effect of progesterone on the secretion or synthesis of LH by the anterior pituitary gland in domestic ruminants.

Since the pituitary content of LH does not decrease in ewes treated with progesterone, one can conclude that infrequent pulses of GnRH (one every 6–8 h) are sufficient to maintain pituitary stores of LH. A complete absence of GnRH pulses (in hypothalamic–pituitary-disconnected ewes) results in a rapid depletion (<48 h) in the content of mRNA for the subunits of LH and a more gradual decrease in the pituitary content of LH (Hamernik *et al.*, 1986). Therefore, it appears that, if the anterior pituitary gland is exposed to infrequent pulses of GnRH that are apparently of normal magnitude, then synthesis of LH by the anterior pituitary will be maintained. Considering this information, one would predict that administration of GnRH to pregnant animals might maintain pituitary content of LH.

The inhibitory effects of oestradiol on release of LH might be explained by decreased secretion of GnRH, similar to the actions of progesterone, but it is not clear how oestradiol acts to decrease synthesis of LH. One dramatic effect of oestradiol is to decrease the pituitary content of mRNAs for the subunits of LH (Fig. 5). In fact, after treating ovariectomized ewes with oestradiol for 3 weeks, the concentration of mRNA for alpha-subunit had decreased by approximately 85%, whereas the concentration of mRNA for the beta-subunit of LH had decreased by 98% (Nilson *et al.*, 1983). A similar decrease in the pituitary content of mRNAs for LH was observed during gestation (Wise *et al.*, 1985). Whether oestradiol directly inhibits synthesis of LH at the level of the anterior pituitary gland, or indirectly by inhibiting secretion of GnRH, remains to be determined.

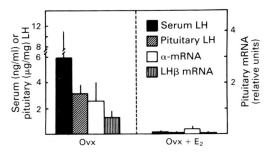

Fig. 5. Pituitary and serum concentrations of LH and pituitary concentrations of mRNA for alpha-subunit and LH-beta subunit of ovariectomized (Ovx) ewes and ovariectomized ewes treated with oestradiol (E_2) for 3 weeks. Values represent mean $\pm$ s.e., N = 4 per group. (Adapted from Nilson *et al.*, 1983.)

Since the oestradiol receptor has been shown to interact with the rat prolactin gene (Maurer, 1985), there is a precedent for a direct effect of oestradiol on the anterior pituitary gland. To date, similar studies have not been extended to gonadotrophin genes or to the prolactin gene in species other than rats. Likewise, oestradiol has direct effects on secretion of LH from ovine (Moss & Nett, 1980) and bovine (Padmanabhan *et al.*, 1978) pituitary cells cultured *in vitro*; however, the effects *in vitro* are not nearly as dramatic as those observed *in vivo*.

Oestradiol may also act at the level of the hypothalamus to decrease secretion of GnRH. Karsch *et al.* (1987) observed a nearly complete absence of GnRH pulses in the hypophysial–portal circulation of ovariectomized ewes treated with oestradiol during the anoestrous season. Since hypothalamic input is required for maintaining the pituitary content of mRNAs for the subunits of LH (Hamernik *et al.*, 1986), if oestradiol sufficiently reduces the magnitude and frequency of pulses of GnRH, this alone could account for the reduction in pituitary content of LH. Whether the inhibitory effect of oestradiol on synthesis of LH is mediated primarily at the hypothalamus or directly on the gonadotroph must be determined before a treatment regimen to prevent reduced synthesis of LH during pregnancy can be designed.

Not only is the pituitary content of LH decreased during gestation, but there are correlated changes in the morphological characteristics of gonadotrophs as well. On Day 1 *post partum* these changes include a 30% decrease in volume of individual gonadotrophs, a 50% decrease in the percentage of the gonadotroph volume occupied by secretory granules and an apparent 40–50% reduction in the percentage of gonadotrophs (identified immunocytochemically) in the anterior pituitary gland (Wise *et al.*, 1986b). The data on the percentage of gonadotrophs in the anterior pituitary gland at 1 day after parturition must be interpreted with caution since the content of LH in the pituitary decreases so dramatically. That is, the content of LH may have been so low in some gonadotrophs at the end of gestation that they could not be identified immunocytochemically. Whether an actual decrease in the number of gonadotrophs occurs during pregnancy therefore remains to be determined. This will probably require identification of gonadotrophs by a means other than immunocytochemical localization of LH. At present, it is not known whether the change in the morphology of gonadotrophs during pregnancy is due to a direct effect of oestradiol and/or progesterone on these cells or to a lack of trophic support from the hypothalamus, i.e. decreased secretion of GnRH during pregnancy.

Recovery of the anterior pituitary gland after parturition

Removal of the fetal–placental unit at parturition is accompanied by a dramatic decrease in the concentration of oestradiol and progesterone in the circulation (Burd *et al.*, 1976). This results in

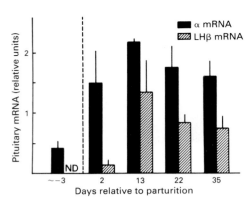

Fig. 6. Pituitary concentrations of mRNA for the alpha- and beta-subunits of LH in ewes during late pregnancy and at various times after parturition. Values represent mean ± s.e., N = 4 or 5 per group. (Adapted from Wise *et al.*, 1985.)

the removal of their negative feedback actions on the hypothalamic–hypophysial axis and permits a gradual recovery in gonadotroph function. In sheep, the first detectable change in the hypothalamic–hypophysial axis that we have observed during the post-partum period is an increase in the concentration of mRNAs for the subunits of LH (Wise *et al.*, 1985). Compared to late gestation, within 2 days after parturition there is an approximate 4-fold increase in the concentration of mRNA for the alpha-subunit of LH, and at least a 10-fold increase in the concentration of mRNA for the beta-subunit of LH (Fig. 6). The maximum concentration of mRNA for the subunits of LH in the anterior pituitary gland was noted on Day 13 *post partum* (mRNA for alpha-subunit was 5-fold greater than in late gestation, mRNA for LH beta-subunit was at least 50-fold greater than in late gestation). The concentrations of these mRNAs then decreased slightly, but remained high until normal oestrous cycles began about 35 days after parturition. A few days after the increase in mRNAs, there was an increase in the quantity of LH contained in pituitary cells (Fig. 3). As a result of these changes during the early post-partum period, there is an increase in the frequency of LH pulses observed in the peripheral circulation with time after parturition (Peters *et al.*, 1981; Walters *et al.*, 1982a).

The concentration of receptors for oestradiol in the hypothalamus and the anterior pituitary gland was very low at the end of gestation (Fig. 7). These concentrations began to increase at Day 22 *post partum* and were highest at Day 35 *post partum* (Wise *et al.*, 1986b). This was after the maximum concentration of LH was noted in the anterior pituitary. Perhaps the relatively low concentration of receptors for oestradiol renders the hypothalamic–pituitary axis less sensitive to the positive feedback effects of this hormone during the early post-partum period. If this were the case, then a higher concentration of oestradiol would probably be needed to induce an ovulatory surge of LH during the early post-partum period. Indeed, Wright *et al.* (1980) reported that 50–64% of ewes failed to show a positive feedback to a 40 µg challenge with oestradiol early in the post-partum period. This dose of oestradiol induced an LH surge in essentially 100% of anoestrous ewes (Beck & Reeves, 1973).

Concomitant with these biochemical changes, the morphology of the gonadotrophs also returns to a state similar to that observed in cycling ewes. Changes include an increase in the percentage of pituitary cells that can be identified immunocytochemically as gonadotrophs, an increase in the percentage of gonadotroph volume occupied by secretory granules and an increase in total volume of individual gonadotrophs (Table 1). Therefore, in ewes that were induced to ovulate and were mated during anoestrus so that lambing would occur during the breeding season, the function of the hypothalamic–hypophysial axis appears to have returned to normal by about 35 days *post*

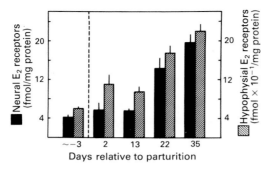

Fig. 7. Concentrations of receptors for oestradiol in the hypothalamus and anterior pituitary gland of ewes before and at various times after parturition. Hypothalamic receptors include those in the preoptic area, the medial basal hypothalamus and median eminence. Values represent mean ± s.e., N = 4 or 5 per group. (Adapted from Wise *et al.*, 1986a.)

Table 1. Changes in the characteristics of gonadotrophs of ewes at various times after parturition

Day post partum	Volume (μm³)	% Staining for LH	% Gonadotroph volume occupied by granules
2	225 ± 3[a]	5·6 ± 0·2[a]	5·5 ± 1·0[a]
13	277 ± 12[b]	6·4 ± 0·1[a]	6·7 ± 1·2[a]
22	303 ± 20[b]	9·5 ± 1·2[b]	12·4 ± 1·1[b]
35	327 ± 11[b]	10·6 ± 1·1[b]	13·5 ± 0·9[b]

Values are mean ± s.e.m. for 4 or 5 ewes/group.
Values with different superscripts differ ($P < 0.05$).

partum. This is consistent with the time of the first ovulation resulting in an oestrous cycle of normal length in our flock (Moss *et al.*, 1980; Wise *et al.*, 1986a).

Effect of suckling on gonadotrophin release

The changes described above represent a recovery of the hypothalamic–hypophysial axis from the prolonged negative feedback effects of progesterone and oestradiol to which the animal was exposed during pregnancy. From an endocrinological viewpoint, once this recovery phase is complete, the female should be ready to resume normal oestrous cycles. However, even though the function of the hypothalamic–hypophysial axis returns to normal within a relatively short period of time after parturition, frequent suckling will continue to suppress concentrations of LH in the peripheral circulation, particularly in cows (Short *et al.*, 1972; Randel *et al.*, 1976). That this effect is due to suckling, rather than lactation, is suggested by the fact that suckling will suppress pulsatile secretion of LH for a longer period after parturition than milking (Peters *et al.*, 1981), even when milking frequency is increased to 4 times per day (Carruthers & Hafs, 1980). In some cases the suckling-induced inhibition of reproductive function may persist for months. Recently, much speculation has centred on the endogenous opioid peptides as factors that might be involved in the suckling induced inhibition of LH secretion.

Evidence for this supposition is that administration of naloxone, an opioid antagonist, will stimulate secretion of LH in suckling beef cows (Gregg *et al.*, 1985), but this treatment is ineffective in non-suckling cows (Whisnant *et al.*, 1985). The fact that treatment with naloxone does not increase serum concentrations of LH in non-suckling cows provides indirect evidence that endogenous opioid peptides might be involved in the suckling-induced suppression of LH secretion.

Administration of naloxone to post-partum ewes also resulted in increased serum concentrations of LH. In contrast to the cow, however, the effect of naloxone in ewes was independent of suckling (Gregg *et al.*, 1985). The effect of suckling on length of the post-partum period in ewes is controversial. Fletcher (1973) and Moss *et al.* (1980) found no differences in time from birth to first post-partum ovulation between suckling and non-suckling ewes. Kann & Martinet (1975) reported that denervation of the udder was associated with a more rapid return to oestrus. At any rate, the effect of suckling on the duration of post-partum anoestrus appears to be less profound in the ewe than in the cow. The fact that there is an interaction between suckling and the effect of naloxone on secretion of LH in the cow (a species in which suckling reduces secretion of LH) but not in the ewe (a species in which suckling does not decrease secretion of LH), provides a further indication that suckling stimulates secretion of the endogenous opioid and this, in some way, suppresses secretion of LH.

If functional recovery of the hypothalamic–hypophysial axis is complete by 20–30 days after parturition, then by administering pulses of GnRH to suckling animals it should be possible to stimulate an increase in serum concentrations of LH and subsequent follicular development that would lead to ovulation. This appears to be the case in post-partum ewes (Wright *et al.*, 1983) and cows (Walters *et al.*, 1982b). Injections of GnRH (500 ng every 2 h for 4 days) beginning about 20 days after calving reduced the interval from parturition to first ovulation in suckling beef cows. Likewise, administration of pulses of GnRH (100 ng/h for 48 h) to ewes about 28 days after parturition induced an increase in basal circulating concentrations of LH which was followed by a surge of LH in about 85% of the animals. Inadequate secretion of GnRH therefore appears to limit the resumption of oestrous cycles during the post-partum period after the initial recovery of the hypothalamic–hypophysial axis from the negative effects of gestation.

Model for post-partum anoestrus

During pregnancy the high circulating concentrations of progesterone and oestradiol result in a prolonged negative feedback on the hypothalamic–hypophysial axis. This feedback results in an inhibition of the synthesis of LH by the anterior pituitary gland. Because synthesis of LH is inhibited for an extensive period of time, pituitary stores of this gonadotrophin become depleted and the basal release of LH is diminished. The mechanisms for releasing LH, however, appear to remain intact and functional throughout pregnancy.

During the post-partum period, a two-phase recovery of the hypothalamic–hypophysial–gonadal axis occurs. The first phase, lasting from 2 to 5 weeks after parturition, is presumed to be characterized by relatively infrequent discharges of GnRH into the hypothalamic–hypophysial portal circulation, i.e. one pulse every 4–8 h. This mode of GnRH secretion effectively stimulates the biosynthetic machinery in the gonadotroph and the rate of synthesis of LH increases. However, the pulses of GnRH are sufficiently spaced so that only a small portion of the newly synthesized LH is secreted. The increased rate of synthesis of LH coupled with the relatively slow rate of release creates a situation in which pituitary stores of LH are replenished. Since the magnitude of the LH pulse is dependent on the quantity of LH stored in the anterior pituitary gland, then during the early portion of this phase of recovery, the pulses of LH are of insufficient magnitude to induce follicular maturation.

Only after pituitary stores of LH have returned to their normal level are pulses of LH that are released into the circulation of sufficient amplitude to stimulate follicular growth. This marks the beginning of the second phase of the recovery process. During this phase the increased circulating concentrations of LH stimulate growth of ovarian follicles and the resultant secretion of oestradiol. I suggest that the first effect of oestradiol is to stimulate production of its own receptor in the hypothalamus and anterior pituitary gland, thus increasing the sensitivity of these tissues to the positive feedback effects of oestradiol. The positive feedback of oestradiol is the result of small increases in circulating concentrations for short periods in contrast to the prolonged, very high concentrations of oestradiol during late gestation that produce a powerful negative feedback effect. At this point, the frequency of discharges of GnRH increases, in turn, producing more frequent pulses of LH. These events lead to the final stages of follicular development and culminate in ovulation.

It is my supposition that the first phase of this recovery process (i.e. events leading to increased pituitary stores of LH) is relatively independent of the suckling stimulus and environmental stressors. The second phase of the recovery (i.e. events leading to an increased frequency of discharges of LH) appears to be tightly coupled to the suckling stimulus and environmental stressors in cows. Suckling and environmental stressors both appear to induce secretion of endogenous opioid peptides which, in turn, inhibit discharges of GnRH from the hypothalamus. The inhibitory effect(s) of the endogenous opioids appear to be short-lived since cows must suckle more than 4 times per day before the post-partum period is extended. Therefore, the inhibition of LH secretion (and proposed inhibition of GnRH secretion) is observed in cows whose calves are allowed to suck *ad libitum* but not in cows that are milked twice daily (Peters *et al.*, 1981). This inhibition in GnRH secretion results in a reduction in the frequency of LH pulses to such an extent that the final stages of follicular growth do not occur. Inhibition of this phase of recovery from post-partum anoestrus persists until the suckling stimulus and/or environmental stressors are reduced to the point where more frequent discharges of GnRH begin to occur. Only then are oestrous cycles initiated.

Future treatments designed to shorten the post-partum anoestrus interval should focus on two problems: (1) methods to increase the ability of the anterior pituitary to synthesize LH during late gestation, and (2) methods to prevent the suckling-induced inhibition of GnRH secretion during the post-partum period. It is possible that one treatment, namely chronic administration of GnRH, may solve each of these problems. In fact, such a treatment will induce ovulation in suckling beef cows when initiated about 20 days after parturition (Walters *et al.*, 1982b). In contrast, the ability of such a treatment to stimulate synthesis of LH during late gestation has not been examined. If the inhibitory effect of oestradiol on synthesis of LH is mediated at the anterior pituitary, then treatment with GnRH during late gestation may be without effect. If so, then one must consider other means of inhibiting the negative effects of oestradiol. Possibilities worthy of consideration include treatment with anti-oestrogens such as clomiphene and tamoxifen, or passive immunization to oestradiol. Clearly, more basic information is needed before a treatment that effectively shortens the post-partum anoestrus interval can be developed.

The research described herein was supported in part by NIH grant HD-07841, USDA grant 84-CRSR-2-2466 and a grant from the CSU Experiment Station. I thank Heywood Sawyer, Marsha Crowder, Deb Hamernik, Robyn Herring and David Gregg for help in preparing this manuscript.

References

Arije, G.R., Wiltbank, J.N. & Hopwood, M.L. (1974) Hormone levels in pre- and post-parturient beef cows. *J. Anim. Sci.* **39**, 338–346.

Beck, T.W. & Reeves, J.J. (1973) Serum luteinizing hormone (LH) in ewes treated with various dosages of 17β-estradiol at three stages of the anestrous season. *J. Anim. Sci.* **36**, 566–570.

Braden, T.D., Cermak, D.L., Manns, J., Niswender, G.D. & Nett, T.M. (1983) Hypothalamic GnRH, pituitary FSH and LH, and pituitary receptors for GnRH and estradiol in cycling beef cows. *Proc. West. Sect. Amer. Soc. Anim. Sci.* **34**, 215–218.

Burd, L.I., Lemons, J.A., Makowski, E.L., Meschia, G. & Niswender, G. (1976) Mammary blood flow and endocrine changes during parturition in the ewe. *Endocrinology* **98**, 748–754.

Carnegie, J.A. & Robertson, H.A. (1978) Conjugated and unconjugated estrogens in fetal and maternal fluids of the pregnant ewe: a possible role for estrone sulfate during pregnancy. *Biol. Reprod.* **19**, 202–211.

Carruthers, T.D. & Hafs, H.D. (1980) Suckling and four-times daily milking: Influence on ovulation, estrus and serum luteinizing hormone, glucocorticoids and prolactin in postpartum Holsteins. *J. Anim. Sci.* **50**, 919–925.

Carruthers, T.D., Convey, E.M., Kesner, J.S., Hafs, H.D. & Cheng, K.W. (1980) The hypothalamo-pituitary gonadotrophic axis of suckled and nonsuckled dairy cows postpartum. *J. Anim. Sci.* **51**, 949–957.

Carter, M.L., Dierschke, D.J., Rutledge, J.J. & Hauser, E.R. (1980) Effect of gonadotropin-releasing hormone and calf removal on pituitary-ovarian function and reproductive performance in postpartum beef cows. *J. Anim. Sci.* **51**, 903–910.

Casida, L.E., Meyer, R.K., McShan, W.H. & Wisnicky, W. (1943) Effects of pituitary gonadotrophins on the ovaries and induction of superfecundity in cattle. *Am. J. vet. Res.* **4**, 76–81.

Cermak, D.L., Braden, T., Manns, J., Niswender, G.D. & Nett, T.M. (1983) Contents of hypothalamic GnRH, pituitary FSH and LH, and pituitary receptors for GnRH and estradiol in postpartum suckled beef cows. *Proc. West. Sect. Amer. Soc. Anim. Sci.* **34**, 208–210.

Chamley, W.A., Findlay, J.K., Cumming, I.A., Buckmaster, J.M. & Goding, J.R. (1974) Effect of pregnancy on the LH response to synthetic gonadotropin-releasing hormone in the ewe. *Endocrinology* **94**, 291–293.

Chamley, W.A., Jonas, H.A. & Parr, R.A. (1976) Content of LH, FSH and growth hormone in the pituitaries of pregnant and anestrous sheep. *Endocrinology* **98**, 1535–1538.

Clarke, I.J. & Cummins, J.T. (1982) The temporal relationship between gonadotropin releasing hormone (GnRH) and luteinizing hormone (LH) secretion in ovariectomized ewes. *Endocrinology* **111**, 1737–1739.

Clarke, I.J., Burman, K., Funder, J.W. & Findlay, J.K. (1981) Estrogen receptors in the neuroendocrine tissues of the ewe in relation to breed, season, and stage of the estrous cycle. *Biol. Reprod.* **24**, 323–331.

Crowder, M.E., Gilles, P.A., Tamanini, C., Moss, G.E. & Nett, T.M. (1982) Pituitary content of gonadotropins and GnRH-receptors in pregnant, post-partum and steroid-treated ovx ewes. *J. Anim. Sci.* **54**, 1235–1242.

Edgerton, L.A. & Hafs, H.D. (1973) Serum luteinizing hormone, prolactin, glucocorticoid and progestin in dairy cows from calving to gestation. *J. Dairy Sci.* **56**, 451–458.

Fletcher, I.C. (1973) Effects of lactation, suckling and oxytocin on *post-partum* oestrus in ewes. *J. Reprod. Fert.* **33**, 293–298.

Fraser, H.M., Jeffcoate, S.L., Gunn, A. & Holland, D.T. (1975) Effect of active immunization to luteinizing

hormone releasing hormone on gonadotrophin levels in ovariectomized rats. *J. Endocr.* **64**, 191–192.

Goodman, R.L. & Karsch, F.J. (1980) Pulsatile secretion of luteinizing hormone: differential suppression by ovarian steroids. *Endocrinology* **107**, 1286–1290.

Gregg, D.W., Moss, G.E., Hudgens, R.E. & Malven, P.V. (1985) Endogenous opioid modulation of LH and PRL secretion in postpartum beef cows and ewes. *J. Anim. Sci.* **61** (Suppl. 1), 418, Abstr.

Hamernik, D.L. & Nett, T.M. (1986) Effect of progesterone (P) on hypothalamic and hypophyseal parameters which regulate synthesis and secretion of luteinizing hormone (LH) in ovariectomized ewes. *J. Anim. Sci.* **63** (Suppl. 1), 339–340, Abstr.

Hamernik, D.L., Crowder, M.E., Nilson, J.H. & Nett, T.M. (1986) Measurement of mRNA for luteinizing hormone β-subunit, α-subunit, growth hormone and prolactin following hypothalamic-pituitary disconnection in the ewe. *Endocrinology* **119**, 2704–2710.

Jenkin, G. & Heap, R.B. (1974) The lack of response of the sheep pituitary to luteinizing hormone releasing hormone stimulation in gestation and early lactation; the possible role of progesterone. *J. Endocr.* **61**, xii–xiii.

Jenkin, G., Heap, R.B. & Symons, D.B.A. (1977) Pituitary responsiveness to synthetic LH-RH and pituitary LH content at various reproductive stages in the sheep. *J. Reprod. Fert.* **49**, 207–214.

Kamel, F. & Krey, L.C. (1982) Intracellular receptors mediate gonadal steroid modulation of GnRH-induced LH release. *Molec. cell. Endocr.* **28**, 471–478.

Kann, G. & Martinet, J. (1975) Prolactin levels and duration of postpartum anoestrus in lactating ewes. *Nature, Lond.* **257**, 63–64.

Karsch, F.J., Cummins, J.T., Thomas, G.B. & Clarke, I.J. (1987) Steroid feedback inhibition of the pulsatile secretion of gonadotropin-releasing hormone in the ewe. *Biol. Reprod.* (in press).

Levine, J.E., Pau, K.-Y.F., Ramirez, V.D. & Jackson, G.L. (1982) Simultaneous measurement of luteinizing hormone-releasing hormone and luteinizing hormone release in unanesthetized, ovariectomized sheep. *Endocrinology* **111**, 1449–1455.

Maurer, R.A. (1985) Selective binding of the estradiol receptor to a region at least one kilobase upstream from the rat prolactin gene. *DNA* **4**, 1–9.

Miller, K.F., Crister, J.K. & Ginther, O.J. (1982) Inhibition and subsequent rebound of FSH secretion following treatment with bovine follicular fluid in the ewe. *Theriogenology* **18**, 45–53.

Moss, G.E. & Nett, T.M. (1980) GnRH interaction with anterior pituitary. IV. Effect of estradiol-17β on GnRH-mediated release of LH from ovine pituitary cells obtained during the breeding season, anestrous season and period of transition into or out of the breeding season. *Biol. Reprod.* **23**, 398–403.

Moss, G.E., Adams, T.E., Niswender, G.D. & Nett, T.M. (1980) Effects of parturition and suckling on concentrations of pituitary gonadotropins, hypothalamic GnRH and pituitary responsiveness to GnRH in ewes. *J. Anim. Sci.* **50**, 496–502.

Moss, G.E., Crowder, M.E. & Nett, T.M. (1981) GnRH-receptor interaction. VI. Effect of progesterone and estradiol on hypophyseal receptors for GnRH and serum and hypophyseal concentrations of gonadotropins in ovariectomized ewes. *Biol. Reprod.* **25**, 938–944.

Moss, G.E., Parfet, J.R., Marvin, C.R., Allrich, R.D. & Diekman, M.A. (1985) Pituitary concentrations of gonadotropins and receptors for GnRH in suckled beef cows at various intervals after calving. *J. Anim. Sci.* **60**, 285–293.

Nalbandov, A. & Casida, L.E. (1940) Gonadotropic action of pituitaries from pregnant cows. *Endocrinology* **27**, 559–564.

Nilson, J.H., Nejedlik, M.T., Virgin, J.B., Crowder, M.E. & Nett, T.M. (1983) Expression of α subunit and luteinizing hormone β genes in the ovine anterior pituitary. Estradiol suppresses accumulation of mRNAs for both α subunit and luteinizing hormone β. *J. biol. Chem.* **258**, 12087–12090.

Padmanabhan, V., Kesner, J.S. & Convey, E.M. (1978) Effects of estradiol on basal and luteinizing hormone-releasing hormone (LHRH)-induced release of luteinizing hormone (LH) from bovine pituitary cells in culture. *Biol. Reprod.* **18**, 603–613.

Peters, A.R., Lamming, G.E. & Fisher, M.W. (1981) A comparison of plasma LH concentrations in milked and suckling post-partum cows. *J. Reprod. Fert.* **62**, 567–573.

Randel, R.D., Short, R.E. & Bellows, R.A. (1976) Suckling effect on LH and progesterone in beef cows. *J. Anim. Sci.* **42**, 267, Abstr.

Restall, B.J. & Starr, B.G. (1977) The influence of season of lambing and lactation on reproduction activity and plasma LH concentrations in Merino ewes. *J. Reprod. Fert.* **49**, 297–303.

Rippel, R.H., Moyer, R.H., Johnson, S.E. & Mauer, R.E. (1974) Response of the ewe to synthetic gonadotropin releasing hormone. *J. Anim. Sci.* **38**, 605–612.

Saiddudin, S., Riesen, W., Tyler, W.J. & Casida, L.E. (1968) Studies on the postpartum cow. 5.0.0 Relation of the postpartum interval to pituitary gonadotropins, ovarian follicular development and fertility in dairy cows. *Bull. Wisconsin Agr. Exp. Stn* **270**, 15–22.

Sarkar, D.K. & Fink, G. (1979) Effect of gonadal steroids on output of luteinizing hormone releasing factor into pituitary stalk blood in the female rat. *J. Endocr.* **80**, 303–313.

Short, R.E., Bellows, R.A., Moody, E.L. & Howland, B.E. (1972) Effects of suckling and mastectomy on bovine postpartum reproduction. *J. Anim. Sci.* **34**, 70–74.

Stabenfeldt, G.H., Drost, M. & Franti, C.E. (1972) Peripheral plasma progesterone levels in the ewe during pregnancy and parturition. *Endocrinology* **90**, 144–150.

Tamanini, C., Crowder, M.E. & Nett, T.M. (1986) Effect of oestradiol and progesterone on pulsatile secretion of luteinizing hormone in ovariectomized ewes. *Acta endocr., Copenh.* **111**, 172–178.

Walters, D.L., Kaltenbach, C.C., Dunn, T.G. & Short, R.E. (1982a) Pituitary and ovarian function in post-partum beef cows. I. Effect of suckling on serum and follicular fluid hormones and follicular gonadotropin receptors. *Biol. Reprod.* **26**, 640–646.

Walters, D.L., Short, R.E., Convey, E.M., Staigmiller, R.B., Dunn, T.G. & Kaltenbach, C.C. (1982b) Pituitary and ovarian function in postpartum beef cows. III. Induction of estrus, ovulation and luteal function with intermittent small-dose injections of GnRH. *Biol. Reprod.* **26**, 655–662.

Whisnant, C.S., Kiser, T.E., Thompson, F.N. & Barb, C.R. (1985) Influence of calf removal on the serum luteinizing hormone response to naloxone. *J. Anim. Sci.* **61** (Suppl. 1), 42, Abstr.

Wise, M.E., Nilson, J.H., Nejedlik, M.T. & Nett, T.M. (1985) Measurement of messenger RNA for luteinizing hormone β-subunit and α-subunit during gestation and the postpartum period in ewes. *Biol. Reprod.* **33**, 1009–1015.

Wise, M.E., Glass, J.D. & Nett, T.M. (1986a) Changes in the concentration of hypothalamic and hypophyseal receptors for estradiol in pregnant and postpartum ewes. *J. Anim. Sci.* **62**, 1021–1026.

Wise, M.E., Sawyer, H.R., Jr & Nett, T.M. (1986b) Functional changes in luteinizing hormone-secreting cells from pre- and postpartum ewes. *Am. J. Physiol.* **250**, E282–E287.

Wright, P.J., Geytenbeek, P.E., Clarke, I.J. & Findlay, J.K. (1980) Pituitary responsiveness to LH-RH, the occurrence of oestradiol-17β-induced LH-positive feedback and the resumption of oestrous cycles in ewes *post partum. J. Reprod. Fert.* **60**, 171–176.

Wright, P.J., Geytenbeek, P.E., Clarke, I.J. & Findlay, J.K. (1983) LH release and luteal function in postpartum acyclic ewes after the pulsatile administration of LH-RH. *J. Reprod. Fert.* **67**, 257–262.

Zawadowsky, M.M., Eskin, I.A. & Ovsjannikov, G.F. (1935) The regulation of the sexual cycle in cows. *Trudy-Din. Razv.* **9**, 75–79.

J. Reprod. Fert., Suppl. **34** (1987), 215–226
Printed in Great Britain
© 1987 Journals of Reproduction & Fertility Ltd

Short light cycles induce persistent reproductive activity in Ile-de-France rams

J. Pelletier and G. Almeida

Institut National de la Recherche Agronomique, Station de Physiologie de la Reproduction, 37380 Nouzilly, France

Summary. European breeds of rams appear to be responsive to photoperiodic changes even though there are large differences between breeds in the timing and amplitude of endocrine (LH and testosterone) and gametogenetic variations before the sexual season. Light regimens such as 6-month light cycles or alternations of constant short and long days every 12–16 weeks are able to entrain the parameters of sexual activity. In these regimens in which the period of the light cycle is shortened, LH release is markedly stimulated during decreasing daylength and evidence is presented, from the relationship of LH and testosterone patterns, that the dampening of LH stimulation could simply result from the effect of steroid feedback. However, there is a gap of several weeks between the maximum LH and testosterone concentrations during which testis growth occurs. Experiments were conducted with Ile-de-France rams, markedly seasonal breeders, in which the period of the light cycle was decreased, in different groups of animals, from 6 to 4, 3, 2 or 1 month. Rams submitted to the three light regimens with the longest periods presented testicular variations which paralleled those of the photoperiod, but those kept in the two regimens with the shortest periods had a progressive increase in testicular weight up to the maximum value (300–350 g) with no further major changes. Therefore, in rams kept in 2-month light cycles, testicular weight remained constant for twelve successive cycles (2 years). LH and testosterone plasma measurements indicated that LH was sufficiently stimulated to maintain testicular development during each decreasing daylength phase but that the stimulation was shifted before testosterone could reach levels at which feedback effects could be exerted. However, all the measures of sperm production were at values characteristic of the sexual season. Similar testicular weight maintenance was also obtained in rams submitted to a regimen in which short days (8L:16D) alternated every month with a split photoperiod interpreted as a long day (7L:8D:1L:8D). It is concluded that short light cycles are able to induce persistent reproductive activity in Ile-de-France rams, which may have practical applications in sheep production systems.

Introduction

Rams exhibit seasonal variations in both behaviour and testicular endocrine and gametogenic activity. Generally both parameters are high at the end of summer and in autumn and low at the end of winter and in spring, thus suggesting a role of photoperiod. In the ram, it has been proposed that light acts on gonadotrophin release, which in turn controls reproduction according to (i) a gonad-independent mechanism and (ii) a change in the magnitude of feedback in relation to the increase or decrease of daylength (Pelletier & Ortavant, 1975a, b). However, in view of the existence of different mechanisms, it is important to distinguish induction of seasonal activity from suppression. The first aim of this paper is therefore to review our knowledge on both the inductive and inhibitory effects of light on LH release and testicular activity. Seasonality is also however a brake to breeding throughout the year, and the second aim is to present a summary of the attempts made to abolish periodic inactivity.

In the first place it is of interest and importance to comment on the natural seasonal variations of LH and testosterone, and of testicular weight. The latter provides, mainly through changes in spermatogenic activity, a reflection of gonadotrophin release.

Seasonality of LH release and testicular activity in rams of different breeds

Seasonality of LH release

In the great majority of rams plasma LH is low during the winter and early spring and high during the summer months, before the breeding season. However, the results are far from homogeneous, perhaps due to different breeds, latitude or protocol (number of animals involved and frequency of bleeding). In Ile-de-France and Préalpes-du-Sud rams, the frequency of LH pulses increases in spring and, at least in the latter breed, remains high until the beginning of autumn (Pelletier *et al.*, 1982). An increase in the number of LH pulses or of the mean LH concentration is also observed before the summer solstice in Soay rams (Lincoln, 1976) and in Polled Dorset and Romney rams (Barrell & Lapwood, 1978/1979). However, in Finnish Landrace males no increase of LH pulsatility is observed until September (Sanford *et al.*, 1977) and the height of LH pulses increases significantly in July and declines quickly during autumn (Sanford *et al.*, 1984). From these data it is hard to compare with precision the photoperiodic effects on LH release in different breeds. This leads us to consider the seasonal changes of testicular values, even though they can be thought of as more distal markers of the light influence than LH pulsatility. Amongst these measures testosterone concentrations and testicular weight are the best documented.

Seasonality of testicular activity

Testosterone release. A testosterone pulse follows each LH pulse but the amplitude of the testicular response to LH varies greatly through the year and maximal plasma testosterone concentrations are not coincident with maximal plasma LH but occur later. In Ile-de-France and Préalpes-du-Sud rams, mean testosterone concentrations begin to increase around the middle of spring (Pelletier *et al.*, 1982). Similarly, in almost all breeds of sheep plasma testosterone starts to increase before the summer solstice (Suffolk: Schanbacher & Lunstra, 1976; Merino, Polled Dorset and Romney: Barrell & Lapwood, 1978/1979; D'Occhio & Brooks, 1983; Soay: Lincoln, 1976). Following the late LH increase, plasma testosterone augmentation is delayed until September in Finnish Landrace sheep (Sanford *et al.*, 1977), although in another study the initial increase in this breed is found somewhat earlier (Schanbacher & Lunstra, 1976). In all these cases, maximum plasma concentrations are found at the end of summer and autumn. A clearly different phenomenon is seen in Ouled-Djellal rams from Algeria in which testosterone concentrations increase as early as mid-winter, and are maximum in early summer and minimum in mid-autumn (Darbeïda & Brudieux, 1980).

Testicular weight. The testis has the advantage of integrating different facets of LH release and of being the target of both LH and FSH. It could therefore be considered as a representative parameter of male seasonal activity. Generally, testicular weight is highest at the end of summer and the beginning of autumn and lowest at the end of winter. Testis redevelopment begins in early summer in Merino and Finnish Landrace breeds (Islam & Land, 1977). Significant testis growth occurs from the beginning of June onwards in Ile-de-France (Fig. 1) (Pelletier & Ortavant, 1970) and Suffolk or Suffolk crosses (Dufour *et al.*, 1984). Finally, in rams of some breeds such as the Soay (Lincoln & Short, 1980) or Préalpes-du-Sud (Pelletier *et al.*, 1981), testicular growth begins still earlier, clearly before the summer solstice and well before the time of reproduction.

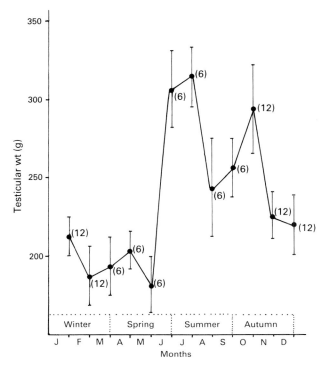

Fig. 1. Seasonal variations of testicular weight in Ile-de-France rams; testicular weights are adjusted by covariance to body weight. Values are the mean ± s.e.m. for the no. of animals in parentheses. (After Pelletier & Ortavant, 1970.)

From the preceding data it is clear that all breeds of rams exhibit seasonal variations of both endocrine and sexual activities. However, many studies underline breed differences with respect to either the time of the seasonal gonadotrophin increase or testicular growth or the duration of sexual activity (Islam & Land, 1977; Barrell & Lapwood, 1978/1979; Dufour *et al.*, 1984). The magnitude of these differences can be marked if one considers the increase of LH pulsatility in early spring for Préalpes-du-Sud (Pelletier *et al.*, 1982) *versus* summer for Finnish Landrace rams (Schanbacher & Lunstra, 1976; Sanford *et al.*, 1977). If light is generally considered as the major cue driving reproduction through the stimulation of the hypothalamo–hypophysial system in European breeds, important differences in the response of the system can be anticipated. Furthermore, some breeds such as the Merino with a weak seasonality, should have a decreased sensitivity to light since seasonal variations were relatively similar when studied at 55°N (Islam & Land, 1977) or 35°S (D'Occhio & Brooks, 1983), i.e. under a very different amplitude of photoperiod change. All these phenomena suggest that sheep of particular breeds retain some intrinsic property of sensitivity to light.

The influence of light is still less evident at lower latitudes. For example, in Ouled-Djellal rams, the shift in seasonal endocrine profiles is sufficiently marked compared to European breeds as to suggest a role of temperature as a predominant cue (Darbeïda & Brudieux, 1980). In the Barbarine or Noire de Thibar rams (35°N, Tunisia), seasonal variations are extremely slight and at the limit of significance (Mehouachi, 1985). However, precise information concerning breeds from tropical or equatorial countries is still required.

In conclusion, in spite of the above mentioned differences, seasonality is thought to be a general phenomenon in European breeds of sheep and all of them appear to be responsive to photoperiod change. This is particularly clear when animals are under artificial light regimens as indicated below. However, in breeds originating from lower latitudes, where photoperiodic changes are moderate, seasonality is much weaker. One can therefore conclude that seasonality is perhaps not a specific trait of sheep but is likely to be related to adaptation to a particular photoperiodic environment.

Light entrainment of LH release and testicular activity

Attempts to drive the activity of the hypothalamo–hypophysial axis, and thus gonadal activity, by artificial light regimens, have been conducted using three basic types of protocol. The first one involves a simple 6-month shift in the annual daylength change in order to induce full activity in males at the time when they are normally quiescent. The second one involves a shortening of the period of the light cycle to multiply the number of sexual seasons during the year. The last procedure includes a decrease of the duration of the light cycle with a simplified photoperiodic schedule limited to abrupt switches from long to short days and *vice versa*.

The annual cycle with a 6-month shift. In Ile-de-France rams, a 6-month shift of the normal annual change in daylength produces an equivalent shift in the pattern of LH release together with testicular development and maximal production of testosterone (Alberio, 1976). Under a similar light schedule, another study also showed that weekly sperm production is maximal in the photoperiodic autumn, whatever the actual time of year at which decreasing daylength occurs (Colas *et al.*, 1985). This protocol is used with success in different artificial insemination centres for sperm production in France (G. Colas, personal communication). However, to study photoperiodic mechanisms, workers have frequently used light regimens with a shortened period.

Influence of sinusoidal light cycles with shortened periods. A light regimen which mimics the normal annual photoperiodic changes in 6 months leads in Ile-de-France rams to two 6-month cycles of testicular weight in 1 year (Ortavant & Thibault, 1956). The testes begin to grow 1·5 months after the shift from long to short days, i.e. when daylength is 12 h. The maximum number of primary spermatocytes at the leptotene stage is observed when decreasing daylength is 10 h (Ortavant, 1961) and finally testis weight is greatest in short days (8 h). Similar effects of 6-month light cycles on sperm production have also been shown in Suffolk rams (Jackson & Williams, 1973).

Measurements of mean plasma LH in intact and castrated Ile-de-France rams indicate that the levels are highest during decreasing daylength (Pelletier & Ortavant, 1975a) as a result of an increase in pulsatility (Lindsay *et al.*, 1984). An increase in testosterone release also occurs in decreasing daylength in Finnish Landrace rams exposed to the same light regimen (Sanford *et al.*, 1978). However, the number of LH pulses is lowest during increasing daylength.

Alternation of constant long and short days. When Soay rams are exposed to a light regimen whereby constant long days (16 h) switch abruptly to short days (8 h), the switch is followed in 6–12 days by an increase in mean plasma LH concentration and LH pulsatility, followed about 3 weeks later by an enlargement of the testes (Lincoln, 1976; Lincoln & Peet, 1977). Alternating short and long days every 16 weeks leads to an entrainment of LH, testosterone and testicular weight. Similar results were observed by D'Occhio *et al.* (1984) for rams of different breeds such as Polled Dorset, Finnish Landrace, Rambouillet and Suffolk when exposed to alternating short and long days every 12 weeks.

In summary, all the results to date obtained from the use of artificial light regimens with shortened periods indicate that photoperiodic cycles are able to entrain the activity of the hypothalamo–hypophysial axis. In all cases stimulation was observed during decreasing daylength. In addition, this indicates that animals are able to measure daylength and to adjust their endocrine activity accordingly. The mechanism of daylength measurement therefore appears to be of crucial importance.

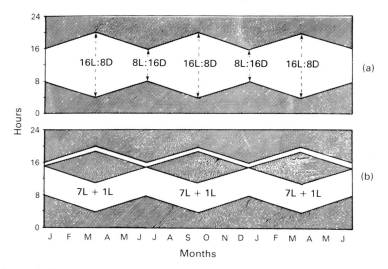

Fig. 2. A 6-month light cycle in which (a) daylength varies from 8 to 16 h and *vice versa* or (b) the light photoperiod is provided in two blocks, one of 7 h adjusted to the dawn of the previous group and the other one of 1 h adjusted to dusk.

Daylength measurement in Ile-de-France rams

Ravault & Ortavant (1977) have shown that 8 h light given in two blocks (7L:9D:1L:7D) are able to mimic a long day and to stimulate prolactin release in Préalpes-du-Sud rams. We have also examined the effect of a split photoperiod on the testicular weight of Ile-de-France rams by using an original protocol in which the second block (or 'light pulse') was mobile.

One group of 12 rams was exposed to a 6-month light cycle in which daylength varied from 8 to 16 h in 3 months and from 16 to 8 h in the other 3 months (Fig. 2a). Another group (N = 12) received light in two blocks, one of 7 h adjusted to 'dawn' of the previous group and a pulse of 1 h adjusted to 'dusk' as indicated in Fig. 2(b). Testicular weight changes were strictly identical in the two groups and similar to the pattern shown in Fig. 4(a). Minimal values, 170–200 g, were observed just after the longest days and maximum values, 300–320 g, were obtained after the shortest days. Furthermore, in the group of rams receiving the light in two blocks, testicular patterns were similar in all individuals. These results confirm previous ones and show that: (1) daylength is measured between two limits even if the interval between them is not always illuminated, (2) the measurement of daylength is homogeneous between animals since apparently none of them took the pulse for the main dawn (Thimonier *et al.*, 1985; J. Pelletier & J. Thimonier, unpublished results). The importance of dawn as a reference mark for LH release is substantiated by the patterns of LH pulse frequency through the day: both in winter and in summer the maximum number of pulses is found 3–4 h after dawn (Ortavant *et al.*, 1982). Furthermore, experiments with split photoperiods show that a 1 h light pulse given each day 16–17 h after dawn stimulates LH release and testicular weight development. Light pulses given at other times of the night are less efficient, or unable to stimulate LH and from these results the existence of a circadian rhythm of sensitivity was postulated (Pelletier *et al.*, 1981). However, the stimulation observed when the photosensitive phase is illuminated tends to vanish in a few weeks. This could be due to a displacement of the photosensitive phase relative to dawn or to the development of a so-called refractoriness.

Is photorefractoriness a phenomenon related to steroid feedback in the ram?

Light regimens and appearance of photorefractoriness

Finnish Landrace and Soay rams kept for several years under a constant light schedule exhibit testicular changes with a period of 35–40 weeks (Howles *et al.*, 1982; Almeida & Lincoln, 1984) as if animals of these normally light entrained breeds have an endogenous rhythm of activity. It therefore appears unrewarding to try to identify a definitive stimulatory photoperiod. Although the mechanism(s) by which the endogenous rhythm is expressed is still unknown, it nevertheless recalls a general phenomenon detectable in almost all light regimens studied.

When Soay rams are transferred in early winter from a natural to an artificial constant 8L:16D regimen, photorefractoriness to initially stimulatory short days is shown by a delayed testicular development (Lincoln, 1980). Similarly, when rams are switched from long to short days, LH release and testicular development occur but both measures begin to decline before the next switch from short to long days (Lincoln & Short, 1980; D'Occhio *et al.*, 1984). A diminution of LH pulsatility is also observed in Ile-de-France rams kept under a 6-month light regimen, when daylength decreases from 10 h 40 min to 8 h (Lindsay *et al.*, 1984). Finally, when rams are kept under natural daylength, regression of testicular weight clearly begins before the winter solstice (see Fig. 1). Thus, whilst LH release and testicular weight are initially stimulated by decreasing daylength, photorefractoriness occurs for both.

Conversely, photorefractoriness also occurs with constant long days or increasing daylength in a sinusoidal 6-month light regimen and leads to an increase in LH pulsatility. It is therefore possible that the increase in LH pulse frequency and the early testicular redevelopment observed in late spring in rams under natural conditions are also due to winter photorefractoriness. This phenomenon is examined below in an experiment using an 8-month light cycle.

Photorefractoriness to constant incremental increases and decreases in daylength and relationship to steroid feedback. In this experiment, the daily light increment and decrement was twice the normal equinoctial values in order to magnify the putative intrinsic properties of increasing and decreasing daylengths. In addition, an 8-month period of the light cycle was chosen so that photorefractoriness could be detected easily. With these parameters the range of daylength variations is 6–20 h (Fig. 3a).

Results indicated that mean LH release and the number of LH pulses increased significantly when daylength was greater than 12 h (Fig. 3b). However, the amplitude of the pulses was low. Although a testosterone pulse normally follows each LH pulse, the increase in mean testosterone level was negligible. As soon as daylength declined from 20 h to 18 h 30 min, there was an abrupt increase in the magnitude of LH pulses (Fig. 3c), resulting in a steep augmentation of the mean plasma concentration. Then, when decreasing daylength reached 12 h, testosterone pulses and the mean plasma value increased in their turn and, at that time, a concomitant decline in LH release occurred.

These results indicate that (a) LH release rises slowly with increasing daylength either by the absence of a strong testosterone feedback since the levels are low at that time, and/or by illumination of the photosensitive phase; conversely, LH release rises quickly in decreasing long daylength, suggesting the existence of at least two mechanisms; (b) the abrupt decline in LH release could essentially be due to high testosterone levels; and (c) there is a gap of about 2 months between the maximum LH and testosterone concentrations in plasma (Pelletier, 1986). This last finding suggests that under light cycles in which the period is greater than 4 months, photorefractoriness is likely to occur. Photorefractoriness to increasing daylength could be equivalent, at least in part, to a reduced steroid feedback, whereas it was the result of strong steroid feedback in decreasing daylength. The opposite patterns of LH and testosterone release are observed in Finnish Landrace (Sanford *et al.*, 1978) and Soay rams (Lincoln & Short, 1980). These conclusions on the whole are in agreement with classical endocrine data and do not preclude the intervention of other

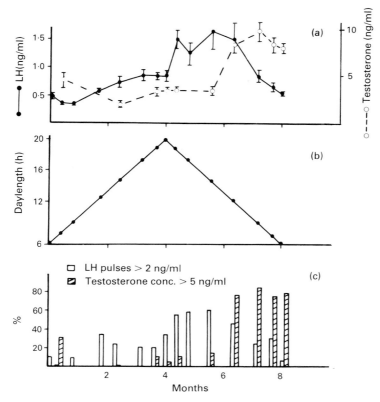

Fig. 3. Changes of plasma LH and testosterone concentrations (a) and LH pulses >2 ng/ml and testosterone pulses >5 ng/ml (c) in 12 Ile-de-France rams according to (b) daylength variations in an 8-month light regimen. In (a) values are mean ± s.e.m. (After Pelletier, 1986.)

mechanisms of LH stimulation when schedules imply constant or sinusoidal light regimens of long duration.

Abolition of seasonal variations in testicular function in Ile-de-France rams

Since the previous results indicated that photorefractoriness requires at least 2 months to develop, it is pertinent to check the effects of shorter and shorter light cycles. Furthermore when the period of the light regimens decreases from 12 to 8 or 6 months the stimulation of LH release tends to be confined to decreasing daylengths whilst increasing daylengths appear not to be stimulatory and could allow 'relaxation' of hypothalamo–hypophysial activity.

Influence of shortened light cycles on testicular weight, LH and testosterone release. The following protocol was used in four groups of 6 Ile-de-France rams submitted to an artificial light regimen. Daylength varied from 8 to 16 h and *vice versa*. The period of the light cycle was 6, 4, 3 or 2 months and testicular weight was measured every 2 weeks for 18, 12, 12 and 18 months respectively. Changes in testicular weight were analysed using an harmonic regression analysis following the model $y(t) = \mu + \alpha \sin (2(\pi t/\tau) + \varphi)$ where μ, α, τ and φ are the mean, amplitude, period and phase respectively (Marquardt, 1963).

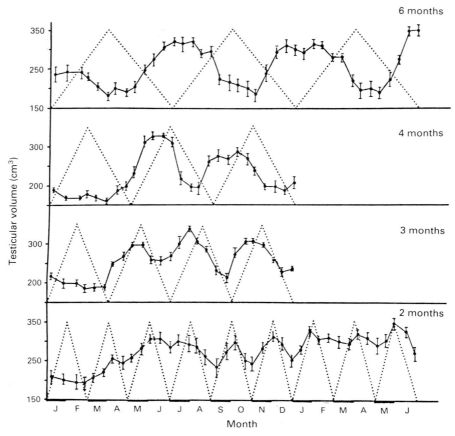

Fig. 4. Variations of testicular weight (—) according to the light regimen (. . .) in which the period of the cycle is 6, 4, 3, or 2 months. There are 6 rams in each group and values are the mean ± s.e.m. (From Pelletier *et al.*, 1985, with permission.)

The results (Fig. 4) clearly indicate that rams kept in the 6-, 4- and 3-month light regimens exhibited seasonal variations in testicular weight. Conversely, rams kept in the 2-month light regimen showed a progressive increase in testicular weight which finally remained steady close to the maximum level. When the period of the light cycle decreases the adjusted mean testis weight increases while the magnitude of variations declines (Pelletier *et al.*, 1985). Results obtained with another group of rams kept for 1 year under a regimen in which the period was 1 month were similar to those obtained in rams under 2-month cycles. In the 6-month group, the changes in testicular weight fit the sinusoidal model ($R^2 = 0.89$) and the magnitude of variations, i.e. twice the computed amplitude, was about 130 g (Table 1). Furthermore, the phase shift between the maximum of daylength and that of testis weight was 105 days. On the other hand, the values obtained in rams kept in the 2-month and 1-month light cycles do not fit the model ($R^2 < 0.20$) and a phase shift with the light regimen cannot be computed. However, it is clear that the amplitude of testicular change was 4 times less than in the group of rams exposed to the 6-month cycle.

Plasma LH and testosterone concentrations were studied on four occasions (daylength: 8 h, 12 h increasing, 16 h and 12 h decreasing) during two successive cycles in the rams kept in the 2-month light cycles. The pooled data for equivalent daylengths are given in Table 2 and show that

Table 1. Variation in testis weight of Ile-de-France rams according to different light regimens: determination of the coefficient between the observed and computed curves, mean, amplitude, period and phase of the computed curves as a function of the period (T) of the light cycle

Period of the light cycle (months)	R^2	Mean (g)	Amplitude* (g)	Period (days)	Phase (days)
6	0·89	262 (256–268)	66 (57–74)	186 (181–191)	−105 (−113 to −97)
2	0·19	297 (289–304)	18 (6–30)	60 (58–61)	(—)
1†	0·15	255 (240–270)	15 (6–36)	54 (42–64)	(—)

Values in parentheses are confidence limits ($P = 0.05$).
*Half of the difference between minimal and maximal testicular weight.
†Animals in this group were 2 years younger than in the two other groups.

Table 2. LH pulsatility, mean plasma LH and testosterone concentrations according to daylength in Ile-de-France rams (6/group) exposed to 2-month light cycles*

Daylength (h)	No. of pulses/8 h	Plasma LH (ng/ml)	Plasma testosterone (ng/ml)
8	2·5	1·43 ± 0·05†	1·75 ± 0·07
12 (increasing)	2·4	1·19 ± 0·03	1·85 ± 0·08
16	2·4	1·08 ± 0·03	1·73 ± 0·09
12 (decreasing)	1·8	1·00 ± 0·02	1·24 ± 0·06

*Mean or mean ± s.e.m. of the pooled data collected during two successive light cycles.
†Variations according to daylength significant ($P < 0.05$).

the number of LH pulses on the one hand, and mean plasma concentrations of LH and testosterone on the other hand, were highest when rams were kept in short daylengths (8 h) resulting from the previous stimulation by decreasing daylength. However, only the mean values differed significantly and the extent of changes, 43 and 29% for LH and testosterone respectively, are far less than those recorded either under natural daylength (Schanbacher *et al.*, 1987) or in an artificial light regimen in which the period is at least 6 months (see Fig. 4). Similarly, the ratio of the highest to the lowest LH pulse frequency during the photoperiodic cycle is only 1·37 as compared to 1·66 when Ile-de-France rams are kept in natural daylength (Pelletier *et al.*, 1982) and 8·00 when they are exposed to a 6-month light cycle (Lindsay *et al.*, 1984). There is therefore a clear decrease in the magnitude of LH and testosterone changes when the period is as short as 2 months.

Most importantly, these results suggest that, in animals exposed to light cycles with a short period, LH pulsatility is sufficient to maintain testicular weight close to its maximum. Furthermore, in the apparent absence of overstimulation of testosterone release, it is likely that there is no strong feedback effect exerted at the level of the hypothalamo–hypophysial system and effectively no regression of the testicular weight occurs. To date, no major decline of testicular weight has been observed in animals kept under a 2-month light cycle for about 2 years. Finally, the efficiency of the

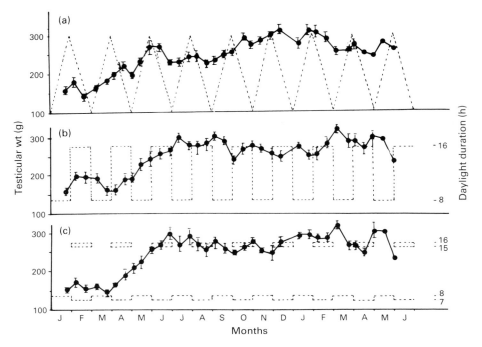

Fig. 5. Variations in testicular weight (mean ± s.e.m.) in rams (N = 6) exposed to short light cycles (period = 2 months) in which the total daylength (– – –) is: (a) 8 h given in two blocks, one of 7 h and one of 1 h, and the interval between the initial dawn and the second dusk varies from 8 to 16 h (as in Fig. 2b); (b) short days (8 h) and long days (16 h) alternating every month; and (c) total daylength is 8 h but every other month the 8th hour is given 15–16 h after the initial dawn.

short light cycles in the maintenance of constant sexual activity in the Ile-de-France ram was ascertained by measurement of different sperm values: the volume, concentration and the number of spermatozoa per ejaculate were quite high, at levels characteristic of the sexual season (G. Almeida, G. Touré, G. Colas, Y. Guérin & J. Pelletier, unpublished data).

Because short light cycles are not easy to use from a practical point of view for farmers, other light schedules have been developed.

Testicular maintenance by a combination of short light cycles and light pulses. Three parallel experiments are presented in Fig. 5 for 3 groups of 6 Ile-de-France rams.

(a) Light was given in two blocks, one of 7 h and one of 1 h so that the interval between the first 'dawn' and the final 'dusk' varied from 8 to 16 h in 1 month and from 16 to 8 h in the next month. The light schedule was therefore similar to that of the previous experiment (see Fig. 2b) except that the period was 2 months.

(b) The period was also 2 months but short days (8 h) and long days (16 h) were alternated every month.

(c) 'Short' days (8 h) were alternated with 'long' days provided by 2 blocks of light, one of 7 h and a pulse of 1 h given 15–16 h after 'dawn'.

In the three groups of rams testicular weight variations occurred in three phases: (i) there was no change during the 2–3 first months corresponding to the previous natural photorefractory phase since the experiment began in January; (ii) during the next 2–3 months there was a regular increase close to maximum values; and (iii) there were steady high values for the next 5 light cycles, with a tendency for the maximum to rise as the animals, still young (2·5 years old), increased in age.

These different results confirm that testicular weight is maintained when rams are exposed to short light cycles and that 'seasonal' regression is abolished under these conditions. Furthermore, as shown previously with the 6-month light cycles, there is no difference in the patterns of testicular weight if light is provided in one or two blocks. In particular, a light pulse given 15–16 h after 'dawn' can mimic long days. Finally, in these short light cycles, there is no difference in testicular weights when the daylength changes are progressive or when the switches are abrupt.

From a practical point of view the alternation of long and short days every month is easier to apply than the daily adjustment of daylength. Similarly, the use of light pulses can provide 'long' days and save energy, particularly in winter.

Conclusions

Seasonal reproduction in sheep breeds indigenous to mid-latitudes is detrimental to productivity. These breeds which have a long period of selection usually have particularly interesting traits and cannot be replaced easily. We have developed, therefore, a new light regimen which effectively prevents male seasonal quiescence. It is based on a working hypothesis according to which at least the early photorefractory stage is mainly dependent on gonadal steroids. Indeed, multiplying the number of light stimulations by short applications of decreasing daylength is able to moderate the amplitude of cyclic hypothalamo–hypophysial activity as suggested by the minor variations of the LH and testosterone releases. However, these cyclic changes are such that they allow gametogenesis to be maintained at the highest level without producing testosterone values at which feedback is strongly inhibitory. The use of light pulses makes our schedule a practical possibility. The next step, at present being investigated, is to verify that rams kept in natural daylength (i.e. kept in normal sheep-folds and not in a light-proof building) and exposed each day for 30 days in a 2-month period to a light pulse corresponding to the time of the summer solstice dusk, are able to maintain their full activity all the year round as they do under successive 2-month light cycles. Finally it is expected that repeated 'long days' will prevent photorefractoriness in late autumn. It is also possible that switches from 'long days' to natural daylength will be able to stimulate LH release in winter or early spring and thus to maintain the testicular weight. The mechanism probably involves changes in melatonin secretion (see Bittman & Karsch, 1984) and will be examined.

The measurement of a long day when a light pulse is given 15–16 h after dawn is not limited to Ile-de-France rams: in the ewe such a treatment is also interpreted as a long day. The work of J. Thimonier (unpublished data) offers an interesting alternative to the use of short light cycles since the alternation of long days and short days is provided by the alternation of light pulse treatment (mimicking long days) with feeding melatonin (mimicking short days) to induce ovarian activity at the anticipated time.

In summary, we have transformed the highly seasonal Ile-de-France ram into a male potentially able to reproduce all the year round.

We thank O. Moulin for assistance.

References

Alberio, R. (1976) *Rôle de la photopériode dans le développement de la fonction de reproduction chez l'agneau Ile-de-France de la naissance à 21 mois.* Thèse Doct. 3° Cycle, Université Paris VI, 57 pp.

Almeida, O.F.X. & Lincoln, G.A. (1984) Reproductive photorefractoriness in rams and accompanying-changes in the patterns of melatonin and prolactin secretion. *Biol. Reprod.* **30**, 143–158.

Barrell, G.K. & Lapwood, K.R. (1978–1979) Seasonality of semen production and plasma luteinizing hormone, testosterone and prolactin levels in Romney, Merino and Polled Dorset rams. *Anim. Reprod. Sci.* **1**, 213–228.

Bittman, E.L. & Karsch, F.J. (1984) Nightly duration of pineal melatonin secretion determines the reproductive response to inhibitory daylength in the ewe. *Biol. Reprod.* **30**, 585–593.

Colas, G., Guérin, Y., Clanet, V. & Solari, A. (1985) Influence de la durée d'éclairement sur la production et la fécondance des spermatozoïdes chez le bélier adulte Ile-de-France. *Reprod. Nutr. Dévelop.* **25**, 101–111.

Darbeïda, H. & Brudieux, R. (1980) Seasonal variations in plasma testosterone and dihydrotestosterone levels and in metabolic clearance rate of testosterone in rams in Algeria. *J. Reprod. Fert.* **59**, 229–235.

D'Occhio, M.J. & Brooks, D.E. (1983) Seasonal changes in plasma testosterone concentration and mating activity in Border Leicester, Poll Dorset, Romney and Suffolk rams. *Aust. J. exp. Agric. Anim. Husb.* **23**, 248–253.

D'Occhio, M.J., Schanbacher, B.D. & Kinder, J.E. (1984) Profiles of luteinizing hormone, follicle-stimulating hormone, testosterone and prolactin in rams of diverse breeds: effects of contrasting short (8L:16D) and long (16L:8D) photoperiods. *Biol. Reprod.* **30**, 1039–1054.

Dufour, J.J., Fahmy, M.H. & Minvielle, F. (1984) Seasonal changes in breeding activity, testicular size, testosterone concentration and seminal characteristics in rams with long or short breeding season. *J. Anim. Sci.* **58**, 416–422.

Howles, C.M., Craigon, J. & Haynes, N.B. (1982) Long-term rhythms of testicular volume and plasma prolactin concentrations in rams reared for 3 years in constant photoperiod. *J. Reprod. Fert.* **65**, 439–446.

Islam, A.B.M.M. & Land, R.B. (1977) Seasonal variation in testis diameter and sperm output of rams of breeds of different prolificacy. *Anim. Prod.* **25**, 311–317.

Jackson, G. & Williams, H.L.L. (1973) The effect of imposed light rhythms on semen production of Suffolk rams. *J. agric. Sci., Camb.* **81**, 179–188.

Lincoln, G.A. (1976) Secretion of LH in rams exposed to two different photoperiods. *J. Reprod. Fert.* **47**, 351–353.

Lincoln, G.A. (1980) Photoperiodic control of seasonal breeding in rams. The significance of short-day refractoriness. *Proc. 6th Int. Congr. Endocrinology, Melbourne,* pp. 283–286. Eds I. A. Cumming, J. W. Funder & F. A. O. Mendelsohn. Austr. Acad. Sci., Canberra.

Lincoln, G.A. & Peet, M.J. (1977) Photoperiodic control of gonadotrophin secretion in the ram: a detailed study of the temporal changes in plasma levels of follicle-stimulating hormone, luteinizing hormone and testosterone following an abrupt switch from long to short days. *J. Endocr.* **74**, 355–367.

Lincoln, G.A. & Short, R.V. (1980) Seasonal breeding: Nature's contraceptive. *Recent Prog. Horm. Res.* **36**, 1–43.

Lindsay, D.R., Pelletier, J., Pisselet, C. & Courot, M. (1984) Changes in photoperiod and nutrition and their effect on testicular growth of rams. *J. Reprod. Fert.* **71**, 351–356.

Marquardt, J. (1963) An algorithm for least squares estimation of non-linear parameters. *J. Soc. indust. appl. Math.* **11**, 431–441.

Mehouachi, M. (1985) *Variations saisonnières de la production spermatique chez les béliers de races Barbarine et Noire de Thibar.* Thèse Univ., Tunis, 134 pp.

Ortavant, R. (1961) Réponses spermatogénétiques du bélier à différentes durées d'éclairement. *Proc. 4th int. Congr. Anim. Reprod. & A.I. The Hague* **2**, 236–242.

Ortavant, R. & Thibault, C. (1956) Influence de la durée d'éclairement sur les productions spermatiques du bélier. *C. r. Séanc. Soc. Biol.* **150**, 358–362.

Ortavant, R., Daveau, A., Garnier, D.H., Pelletier, J., de

Reviers, M.M. & Terqui, M. (1982) Diurnal variation in release of LH and testosterone in the ram. *J. Reprod. Fert.* **64**, 347–353.

Pelletier, J. (1986) Contribution of increasing and decreasing daylength to the photoperiodic control of LH secretion in Ile-de-France rams. *J. Reprod. Fert.* **77**, 505–512.

Pelletier, J. & Ortavant, R. (1970) Influence du photopériodisme sur les activités sexuelle, hypophysaire et hypothalamique du bélier Ile-de-France. In *La Photorégulation de la Reproduction chez les Oiseaux et les Mammifères,* pp. 483–495. Eds J. Benoît & I. Assenmacher. CNRS, Montpellier.

Pelletier, J. & Ortavant, R. (1975a) Photoperiodic control of LH release in the ram. I. Influence of increasing and decreasing light photoperiods. *Acta endocr., Copenh.* **78**, 435–441.

Pelletier, J. & Ortavant, R. (1975b) Photoperiodic control of LH release in the ram. II. Light androgens interaction. *Acta endocr., Copenh.* **78**, 442–450.

Pelletier, J., Blanc, M., Daveau, A., Garnier, D.H., de Reviers, M.M. & Terqui, M. (1981) Mechanism of light action in the ram: a photosensitive phase for LH, FSH, testosterone and testis weight? In *Photoperiodism and Reproduction in Vertebrates,* pp. 117–134. Eds R. Ortavant, J. Pelletier & J. P. Ravault. INRA, Paris.

Pelletier, J., Garnier, D.H., de Reviers, M.M., Terqui, M. & Ortavant, R. (1982) Seasonal variation in the LH and testosterone release in rams of two breeds. *J. Reprod. Fert.* **64**, 341–346.

Pelletier, J., Brieu, V., Chesneau, D., Pisselet, C. & de Reviers, M. (1985) Abolition partielle des variations saisonnières du poids testiculaire chez le bélier par diminution de la période du cycle lumineux. *C. r. hebd. Séanc. Acad. Sci., Paris* **301**, Série III, 665–668.

Ravault, J.P. & Ortavant, R. (1977) Light control of prolactin secretion in sheep. Evidence for a photoinducible phase during a diurnal rhythm. *Annls Biol. anim. Biochim. Biophys.* **17**, 459–473.

Sanford, L.M., Palmer, W.M. & Howland, B.E. (1977) Changes in the profiles of serum LH, FSH, and testosterone and in mating performance and ejaculate volume in the ram during the ovine breeding season. *J. Anim. Sci.* **45**, 1382–1391.

Sanford, L.M., Beaton, D.B., Howland, D.E. & Palmer, W.M. (1978) Photoperiod induced changes in LH, FSH, prolactin and testosterone secretion in the ram. *Can. J. Anim. Sci.* **58**, 123–128.

Sanford, L.M., Howland, B.E. & Palmer, W.M. (1984) Seasonal changes in the endocrine responsiveness of the pituitary and testes of male sheep in relation to their patterns of gonadotropic hormone and testosterone secretion. *Can. J. Physiol. Pharmacol.* **62**, 827–833.

Schanbacher, B.D. & Lunstra, D.D. (1976) Seasonal changes in sexual activity and serum levels of LH and testosterone in Finnish Landrace and Suffolk rams. *J. Anim. Sci.* **43**, 644–650.

Schanbacher, B.D., Orgeur P., Pelletier, J. & Signoret, J.P. (1987) Behavioural and hormonal responses of sexually-experienced Ile-de-France rams to oestrous females. *Anim. Reprod. Sci.* (in press).

Thimonier, J., Brieu, V., Ortavant, R. & Pelletier, J. (1985) Daylength measurement in sheep. *Biol. Reprod.* **32**, Suppl. 1, 55 Abstr.

J. Reprod. Fert., Suppl. **34** (1987), 227–236

Expression of the genes encoding bovine LH in a line of Chinese hamster ovary cells

J. H. Nilson and D. M. Kaetzel

*Department of Pharmacology, School of Medicine, Case Western Reserve University, Cleveland,
OH 44106, U.S.A.*

Summary. Synthesis of biologically active LH is complex, due in part to its hetero-
dimeric subunit structure and to the numerous post-translation modifications of each
subunit. Through the use of mammalian expression vectors we have been able to
·introduce the bovine α subunit and LH-β genes into a Chinese hamster ovary cell line
deficient in dihydrofolate reductase. The bovine genes are actively expressed and the
Chinese hamster ovary cells secrete biologically active LH. The expression vector
containing the bovine α subunit gene also contains a modified mouse gene encoding
dihydrofolate reductase, permitting the use of methotrexate to amplify selectively the
bovine α subunit gene after its integration into the genome of the Chinese hamster cells.
This provides a novel means for assessing the importance of α subunit concentration
with respect to assembly of the heterodimer. In addition, methotrexate selection leads
to the over-production of LH ($10 \, \mu g/10^6$ cells/24 h). Finally, because the bovine LH
produced in the Chinese hamster ovary cells is glycosylated, this transfection system
can be used in conjunction with in-vitro mutagenesis to determine whether site-specific
changes in glycosylation have an effect on subunit assembly and biological activity.
This transfection approach therefore offers multiple avenues to explore further the
molecular mechanisms underlying the complex biosynthetic pathway of bovine LH.

Introduction

Luteinizing hormone (LH) and chorionic gonadotrophin (CG) are two closely related hetero-
dimeric glycoproteins synthesized in the pituitary and placenta, respectively (Pierce & Parsons, 1981;
Chin, 1985). The α subunits of LH and CG are identical, a property of all glycoprotein hormones,
whereas their β subunits are different, sharing an amino acid homology of 80%. Both hormones
bind to the same gonadal receptor and stimulate steroidogenesis and gametogenesis, consistent
with their structural homology (Pierce & Parsons, 1981).

Synthesis of biologically active LH and CG is complex, involving a number of post-
translational events. The initial modification begins with the cleavage of the signal peptide sequence
which occurs as the nascent α and β subunits cross the membrane of the endoplasmic reticulum
(Chin *et al.*, 1978; Daniels-McQueen *et al.*, 1978). Cleavage of the signal sequence is followed
immediately by the addition of *N*-linked, high mannose oligosaccharide cores (Bielinska & Boime,
1979; Magner & Weintraub, 1982; Hoshina & Boime, 1982). Non-covalent subunit combination
occurs shortly thereafter. Following assembly, the heterodimers move through the Golgi complex
where additional post-translational modifications take place, such as O-linked glycosylation and
sulphation (Parsons *et al.*, 1983; Green *et al.*, 1985). Fully modified LH is sequestered in secretory
granules where its release is regulated by hypothalamic-releasing hormones, gonadal steroids and
other known secretagogues (Goverman *et al.*, 1982). In contrast, hCG is released constitutively
after its synthesis due to the lack of secretory granules in placental cells (Hussa, 1980).

Characteristically, intracellular levels of free α subunit exceed the levels of dimeric hormone

(Daniels-McQueen *et al.*, 1978; Workewych & Cheng, 1979) whereas free β subunit is often not detectable (Workewych & Cheng, 1979). A common interpretation of this imbalance is that the extent of heterodimer formation is limited by the concentration of β subunit. However, Peters *et al.* (1984) indicate that almost 50% of both the α and CGβ subunits fail to combine in a human placental cell line. Both free α and β subunits of CG can be detected in the culture medium. Similar results have been reported for the assembly of α and β subunits of TSH in mouse pituitary tumour cells (Weintraub *et al.*, 1980), suggesting that extent of heterodimer formation may not be limited solely by the concentration of the β subunit. If extent of assembly is dependent on the concentration of both subunits, then changes in the extent of assembly should be proportional to the product of the change in concentration of both subunits. Thus, a small change in the concentration of each subunit will lead to a larger or multiplicative change in the concentration of heterodimer. This is an intriguing possibility and might explain the biological significance of the steadily emerging evidence that expression of the α and β subunit genes of gonadotrophin rise and fall together (Nilson *et al.*, 1983; Chin, 1985; Milsted *et al.*, 1985).

While it may be intuitively obvious that extent of heterodimer assembly depends on the concentration of both subunits, there is no direct experimental evidence *in vivo* to support this notion. We have reported that the genes encoding the bovine α and β subunits of LH can be transferred stably into a line of Chinese hamster (*Cricetulus griseus*) ovary cells (Kaetzel *et al.*, 1985). The bovine genes are expressed and the resulting subunits assemble and are secreted as biologically active hormone. By exploiting the properties of the DNA transfection system, we have begun to develop an approach that permits the concentration of one of the gonadotrophin subunits to be changed, while the other remains constant. Thus, by monitoring the extent of assembly within the cell, we can test whether changes in the concentration of α subunit affect assembly of the heterodimer. Such an approach is a prelude to future studies designed to test the importance of both subunits in assembly of the heterodimer, and to address further the structural requirements for assembly and secretion of biologically active LH.

Here we shall review the structural characteristics of the gonadotrophin genes and the properties of the DNA transfection system used to establish permanent cell lines of Chinese hamster ovary cells harbouring functional copies of the bovine LH genes. Then, we shall present preliminary evidence indicating that a selective change in the concentration of α subunit alters the extent of heterodimer assembly—the first step in determining whether assembly is a function of the concentration of both subunits.

Structural features of the genes for α and β subunits of bovine LH

We have isolated and characterized three overlapping lambda clones from a bovine genomic library that contain portions of the α subunit gene (Goodwin *et al.*, 1983). The gene contains 4 exons and 3 introns and spans 16 kilobase pairs (kbp), even though the mature mRNA is only 730 nucleotides long (Fig. 1). The length of the bovine α subunit gene is due primarily to the first intron which is 13 kbp long and is positioned in the 5′-untranslated region, only 90 bp from the start-site of transcription. The other two introns are much smaller and interrupt the coding sequence.

The α subunit gene is present as a single copy in the bovine (Goodwin *et al.*, 1983) and human (Fiddes & Goodman, 1981; Boothby *et al.*, 1981) genome. While expression of the bovine gene is restricted to the pituitary, the human α subunit is expressed in the pituitary and placenta (Pierce & Parsons, 1981; Chin, 1985). The lack of α subunit gene expression in the bovine placenta is consistent with the absence of gonadotrophins in the placenta of ruminants (Pierce & Parsons, 1981).

In contrast to the bovine α subunit gene which spans 16 kbp, the gene subunit for the LH-β subunit spans less than 1·1 kbp (Fig. 1; Virgin *et al.*, 1985). We have determined the nucleotide sequence for the entire gene and 776 bp of 5′-flanking sequence, confirming that it encodes an authentic LH-β subunit (Virgin *et al.*, 1985). The bovine LH-β gene contains three exons and encodes an mRNA of 550 nucleotides, excluding the poly A tail. The mRNA cap site and polyadenylation site have been mapped by primer extension and S1 nuclease protection, respectively (Virgin *et al.*, 1985). Surprisingly, the 5'-untranslated region is only 6–11 nucleotides long. This is unusually short for a eukaryotic mRNA, and stands in contrast to the 5'-untranslated region of the closely related human CG-β gene which is 350 nucleotides long (Talmadge *et al.*, 1984). The functional signifi-cance of this difference is unknown.

Gene quantitation studies reveal that the bovine LH-β gene is unique and that there are no other closely related genes in the genome of cattle (Virgin *et al.*, 1985). A similar result has also been reported for the rat (Jameson *et al.*, 1984). Additional studies indicate that the bovine LH-β gene is not expressed in the placenta. The unique LH-β gene found in cattle and rats contrasts to the human gene family which consists of 7 CG-β genes and 1 LH-β gene sharing greater than 90% homology in nucleotide sequence (Fiddes & Talmadge, 1984). The high level of nucleotide sequence homology between the human LH-β and CG-β genes, and the lack of CG-β genes in the cow and rat, suggest that the CG-β genes have evolved recently from an ancestral LH-β gene via a series of duplications (Fiddes & Talmadge, 1984).

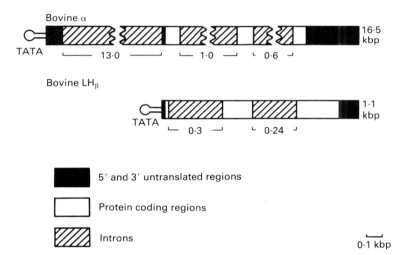

Fig. 1. Schematic comparison of the structure of the bovine α subunit and LH-β genes (see Goodwin *et al.* (1983) and Virgin *et al.* (1985) for further details).

Expression of bovine α subunit and LH-β genes in Chinese hamster ovary cells

Genes for the α and β subunits of bovine LH can be transferred to a line of Chinese hamster ovary cells deficient in dihydrofolate reductase (DHFR−) via DNA-mediated gene transfer (Kaetzel *et al.*, 1985). Because DHFR is required for de-novo synthesis of purines and pyrimidines, Chinese hamster ovary (DHFR−) cells require media supplemented with nucleosides (Kaufman *et al.*, 1985). If the bovine genes are co-transfected with a vector containing a DHFR gene, then clonal cell lines containing the DHFR genes and bovine genes can be selected by growing the transfected cells in media lacking nucleosides.

We have constructed two expression vectors, each containing a gene encoding one of the sub-units of bovine LH (Fig. 2). Both vectors (pDSVα and pSV2LHβ) contain an α or LH-β gene linked to a strong viral promoter (SV40 late and SV40 early, respectively). This is intended to maximize expression of the bovine genes in non-pituitary cells, such as Chinese hamster ovary cells. Due to

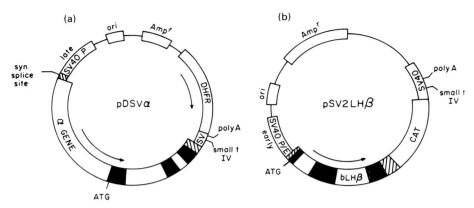

Fig. 2. Construction of two expression vectors containing the bovine α and LH-β subunit genes.
(a) The 8·7 kbp α subunit gene fragment (intron sequences in white, exons in black, 5' and 3'
untranslated regions hatched) is represented, with its initiator methionine (ATG) codon and
direction of transcription indicated. Also shown are the late SV40 promoter element (SV40 P);
a synthetic (syn) fragment containing a consensus splice–donor sequence; a SV40 small tumour
antigen (t)-gene intron (IV) and polyadenylation signal, and the mouse DHFR minigene with
its direction of transcription (arrow). ori, Origin of replication; Amp^r, ampicillin-resistance
gene. (b) A 1·8 kbp Pst I genomic fragment containing the entire LH-β subunit gene was incorporated into pSV2LHβ. Still remaining in the final construct, at the 3' end of the LH-β gene, is
the bacterial chloramphenicol acetyltransferase (CAT) gene from the parent vector, pSV2CAT.
P/E, promoter/enhancer. Reprinted from Kaetzel et al. (1985) with permission of the publisher.

the large size of the α subunit gene, it was convenient to relocate only the portion of the gene which
contained the complete coding sequence. As shown in Fig. 2, this fragment contains the 3' half of
the first intervening sequence and exons 2, 3 and 4; the ATG initiation codon is located in exon 2.
To ensure the correct removal of the 3' half of the first intron, pDSVα contains a synthetic fragment
carrying a consensus splice donor site positioned between the SV40 late promoter and the truncated
α subunit gene. The vector also contains a mini-gene encoding mouse dihydrofolate reductase
(Kaufman & Sharp, 1982). Linkage of the DHFR and α subunit genes should increase the percentage of transfected cells capable of growth in selective media and that retain the α subunit gene.

The bovine LH-β expression vector, pSV2LHβ, contains the entire LH-β subunit gene
(including the RNA cap site and TATAA sequence) located on a 1·8-kbp Pst I genomic fragment
(Virgin et al., 1985). The gene is juxtaposed between the SV40 early promoter/origin of replication
and the bacterial chloramphenicol acetyltransferase (CAT) gene (Gorman et al., 1982). The CAT
gene is a remnant from the SV2CAT parent vector (Gorman et al., 1982) and is non-functional.
The LH-β expression vector can be transferred to Chinese hamster ovary cells (DHFR−) by co-
transfection with the α subunit expression vector. Even though LH-β gene is not linked to the
DHFR gene, a sufficient number of clones capable of growth in nucleoside-free media should
contain the LH-β gene.

To date, we have obtained several clonal lines of Chinese hamster ovary cells capable of growth
in nucleoside-free media by using the expression vectors described above. Analysis of these clonal
lines by RNA blot hybridization, RIA, electrophoresis of immunoprecipitated ^{35}S-labelled proteins, and LH-specific bioassay, reveals that the clones can be divided into three classes: some
synthesize and secrete only α subunit, others only LH-β, while others produce both subunits
(Kaetzel et al., 1985). Intact and biologically active LH is found only in the clones which synthesize
both subunits (Kaetzel et al., 1985). Together, these results suggest that the bovine α and LH-β
genes can be expressed in Chinese hamster ovary cells and that the resulting subunits assemble and

are secreted as biologically active LH. These findings also suggest that the LH of Chinese hamster ovary cells is glycosylated because biological activity appears to depend on glycosylation (Pierce & Parsons, 1981).

Methotrexate selectively increases synthesis of α subunit and secretion of biologically active LH

When cells are selected for growth in the presence of increasing concentrations of methotrexate, resistant subpopulations arise which contain amplified copies of the endogenous DHFR gene (Alt *et al.*, 1978). Furthermore, if Chinese hamster ovary cells (DHFR −) are co-transfected with a DHFR gene and a non-selectable gene, then methotrexate selection commonly results in the co-amplification of both transfected genes (Kaufman *et al.*, 1985). Having established that Chinese hamster ovary cells support expression of the bovine gonadotrophin genes, we wanted to ascertain whether methotrexate selection can affect LH synthesis. For this purpose, we selected a cell line that expresses approximately equal amounts of α and LH-β mRNA (CHO-LH20 or simply LH20 cells; see Table 1). Initial selection was performed at a concentration of 3 nM-methotrexate. After about 3 weeks, the surviving subpopulation of cells was either maintained in 3 nM-methotrexate (LH20-3 cells), or exposed to a stepwise increase in methotrexate (10 nM) for an additional 3-week period. By repeatedly increasing the concentrations of methotrexate in the media, we have isolated a number of subpopulations of LH20 cells, each of which is resistant to a defined concentration of the drug.

To determine whether methotrexate had an effect on LH synthesis, we incubated LH20 cells, and a stable population of LH20 cells resistant to 100 nM-methotrexate (LH20-100), for 18 h with 500 μCi [^{35}S]methionine/ml and 150 μCi [^{35}S]cysteine/ml. Media were collected and then subjected to quantitative immunoprecipitation with rabbit antiserum specific for bovine α or LH-β subunits (Kaetzel *et al.*, 1985). The immunoprecipitates were analysed by electrophoresis through SDS-polyacrylamide gels (Fetherston & Boime, 1982) followed by autoradiography. Two specific polypeptides were precipitated from LH20 and LH20-100 media samples by the LH-β-specific antibody (Fig. 3). Their molecular weights of 20 500 and 16 000 were slightly larger than those reported for bovine α and LH-β subunits (Pierce & Parsons, 1981; Kaetzel *et al.*, 1985). The specificity of

Table 1. Secretion of biologically active bovine LH after methotrexate selection of transfected Chinese hamster ovary cells

Cell line	Methotrexate (nM)	mRNA (relative level)* alpha	beta	LH bioassay (μg LH/10⁶ cell/24 h)†
CHO (DHFR −)	0	0	0	<0·008
LH20	0	1	1·4	0·40 ± 0·01
LH20-3	3	1·3	2·2	0·62 ± 0·04
LH20-100	100	2·6	1·7	2·8 ± 0·03
LH20-1000	1000	5·8	1·6	10·0‡
LHβ25	0	0	0·60	<0·008
LHβ25-1000	1000	0	0·65	<0·008

*mRNA levels were determined by northern blot analysis followed by densitometry of the resulting autoradiographs. ^{32}P-labelled, single-stranded DNA probes specific for bovine α and LH-β mRNA were prepared as previously described (Nilson *et al.*, 1983).
†LH was calculated from the amount of progesterone secreted from luteal cells after treatment with 3 different dilutions of culture medium. Samples were assayed in triplicate. Values are expressed as mean ± s.d. of the dilution nearest the midpoint of the LH standard dose–response curve.
‡Single determination.

the antibody for the LH-β subunit was revealed by the addition of excessive amounts of non-radioactive α or LH-β subunits during immunoprecipitation because only LH-β displaced the radiolabelled bands. Both the presumptive α and LH-β bands were displaced by the unlabelled antigen, suggesting that the antiserum brings down intact heterodimer through recognition of the LH-β subunit. The possibility that the antibody was recognizing LH-β subunits with different molecular weights was ruled out because α-specific antiserum also precipitates the same two labelled proteins (data not shown). Furthermore, the molecular weights of the two labelled proteins were identical to the α and LH-β subunits secreted by other Chinese hamster ovary cell lines which have been transfected with only one of the two gonadotrophin subunit genes.

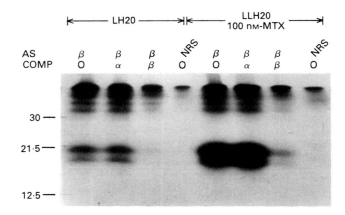

Fig. 3. Expression of bovine LH in Chinese hamster ovary cells. LH20 cells, and a subpopulation of LH20 cells selected for growth in the presence of 100 nM-methotrexate (LH20-100), were labelled with [³⁵S]methionine and [³⁵S]cysteine for about 18 h. Medium was subjected to immunoprecipitation, NaDodSO₄/PAGE, and autoradiography (Kaetzel *et al.*, 1985). Immunoprecipitation was carried out with 4 μl antiserum (AS) directed against the β subunit of LH, in the presence or absence of 10 μg unlabelled α subunit or LH-β competitor (COMP). NRS, normal rabbit serum. Numbers at left represent $M_r \times 10^{-3}$ of marker proteins run in parallel.

The stable population of LH20 cells capable of growth in 100 nM-methotrexate (LH20-100) secreted approximately 8-fold more LH than did the LH20 parent cell line as indicated by the increased intensity of both α and LH-β bands (Fig. 3). Presumably, methotrexate selection caused an increase in both subunits through gene amplification (Kaufman & Sharp, 1982). Alternatively, it is possible that only one gene encoding an LH subunit increased, but that this increase is sufficient to change the extent of assembly. Indirect evidence bearing on this latter point is presented below.

    We have also examined the effects of methotrexate selection by subjecting media samples from several unlabelled cell lines to electrophoresis in SDS-polyacrylamide gels. For these determinations, β-mercaptoethanol was omitted from the sample buffer and electrophoresis was performed at 4°C. Several laboratories have reported that LH will not dissociate under these conditions (Chin *et al.*, 1981; Strickland & Puett, 1982; Strickland & Pierce, 1983). To visualize LH and any free subunit, the proteins were transferred from the gel to nitrocellulose by electrophoresis and then incubated successively with rabbit antiserum specific for bovine α subunit and goat anti-rabbit IgG conjugated to alkaline phosphatase. Colour was developed by further incubation with a chromogenic alkaline phosphatase substrate. As judged from the reaction products formed with the standards, LH was clearly separated from free α subunit after electrophoresis (Fig. 4). From the differential staining intensity of the reaction products associated with equivalent amounts of the α subunit and LH standards, it is apparent that the antibody had more affinity for the free subunit rather than the intact heterodimer (Fig. 4). This was expected because purified α subunit was used to elicit antibody formation, and the finding is consistent with our previous estimates of cross-reactivity (Kaetzel *et al.*, 1985). It is difficult to determine whether the free α subunit band detected in LH standard is a contaminant or represents partial denaturation of the dimer. The same is true for α subunit detected in all of the samples from the subpopulations of LH20 cells capable of growth in increasing concentrations of methotrexate. However, the antibody and the gel system can be used to provide a minimum estimate of LH concentration in media samples. Therefore, the data from the subpopulations of LH20 cells suggests that resistance to increasing concentrations of methotrexate is strongly associated with increased secretion of intact LH.

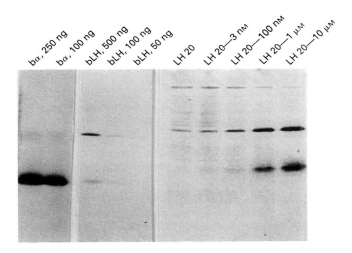

Fig. 4. Western blot analysis of secreted bovine LH and free α subunit from methotrexate-treated LH20 cells. Purified α subunit, LH, or media samples from LH20 cells and subpopulations of LH20 cells treated with the indicated concentrations of methotrexate were diluted with equal volumes of 2 × NaDodSO$_4$-PAGE sample buffer, except for the omission of β-mercaptoethanol and heat treatment (Strickland & Puett, 1982). Electrophoresis was performed at 4°C in 15% polyacrylamide gels containing 0·1% NaDodSO$_4$ (Strickland & Puett, 1982). After electrophoresis, the proteins were transferred from the gel to nitrocellulose and subjected to immunostaining as described in the Bio-Rad Immuno-blot manual (Bio-Rad, Richmond, CA).

Is the effect of methotrexate on LH secretion caused by an increase in the amounts of both LH subunits? We have begun to address this question by quantitating the relative amounts of the mRNAs encoding the α and LH-β subunits by northern blot hybridization and scanning densitometry. In addition, we have measured the amount of biologically active LH through the use of an ovarian luteal cell bioassay (Hoyer *et al.*, 1984; Kaetzel *et al.*, 1985). Results from these experiments are summarized in Table 1. When LH20 cells were selected for growth against increasing concentrations of methotrexate, LH-β mRNAs remained relatively constant while α subunit mRNAs increased with increasing concentrations of the drug. This suggests that the α subunit gene was amplified in response to methotrexate selection, while the LH-β gene remained unaffected. This is not unexpected because the α subunit gene was directly linked to the mouse DHFR gene in the vector. LH-β mRNAs remained unchanged even in the LH-β25-1000 cell line which was resistant to 1 μM-methotrexate. This selective effect of methotrexate is consistent with a report that amplification in response to methotrexate occurs along a gradient, with genes nearest the DHFR gene amplified to a greater extent than genes farther away (Kaufman et al., 1985). Further verification of selective amplification in our system requires measurement of gene copy number by DNA 'dot-blot' hybridization (Kafatos et al., 1979).

Bioassay measurements of media from the subpopulations of LH20 cells indicate that methotrexate selection correlated positively with an increase in secreted levels of LH, from a nadir of $0.4\,\mu g/10^6$ LH20 cells to a peak of $10\,\mu g/10^6$ LH20-1000 cells during a 24-h collection period. This increase correlated with similar increases in α subunit mRNA and suggests that heterodimer assembly is not complete in the parent LH20 cell line even though levels of α and LH-β mRNAs were essentially equal. Although preliminary, these results also suggest that extent of heterodimer assembly may be related to the concentrations of both subunits because an increase in α subunit leads to increased secretion of biologically active LH. Further substantiation will require measurement of intracellular levels of both α and LH-β proteins to confirm that α subunit protein levels indeed increase whereas LH-β protein levels remain constant. Because the methotrexate effect appears to be a function of linkage to the DHFR gene, the LH-β gene can be linked to the DHFR gene to test whether selective increases in LH-β protein levels also lead to an increase in secreted LH. Perhaps the most definitive test will be to link both LH genes to DHFR and determine whether the methotrexate-induced increase in LH synthesis and secretion is a function of the product of the concentration of both subunits.

The bioassay data indicate indirectly that LH produced by the Chinese hamster ovary cells is glycosylated because only glycosylated LH is hormonally active (Pierce & Parsons, 1981). We have been able to label CHO-LH by incubation of LH20 cells with [³H]glucosamine and [³H]mannose, further confirming that CHO-LH is glycosylated. Because the asparagine-linked oligosaccharides of CHO cells have been extensively characterized, CHO-LH is likely to contain a biantennary oligosaccharide structure with terminal sialic acid residues linked to galactose which in turn is linked to N-acetyl glucosamine (Hubbard & Ivatt, 1981). If verified, CHO-LH would have a different type of a complex Asn-linked oligosaccharide from that found normally in pituitary LH (Green et al., 1985). Such a difference may indicate that the biological activity of LH is not strictly dependent on the type of complex oligosaccharide attached to the polypeptide backbone. The subpopulations of LH20 cells selected with methotrexate produce more than enough hormone to permit direct assessment of this possibility.

Conclusions

Bovine LH genes can be expressed in Chinese hamster ovary cells and their expression leads to the appearance of biologically active LH. Through the use of expression vectors which contain the bovine α subunit gene linked to a modified mouse DHFR gene, we have been able to use methotrexate selection to increase the concentration of α subunit while maintaining the concentration of LH-β. This approach provides a novel means to assess the importance of α subunit concentration with respect to assembly of the heterodimer. Our data indicate that heterodimer assembly is incomplete before methotrexate selection and that the amount of assembled LH can be increased by selectively increasing the concentration of α subunit. This suggests that the concentration of LH-β subunit may not be the sole determinant of the extent of heterodimer formation. The transfection approach described herein can be used to verify that changes in extent of assembly are a function of the change in the product of the concentration of both subunits. In addition, because

CHO-LH is glycosylated and biologically active, the transfection system can also be used with in-vitro mutagenesis to determine whether site-specific changes in glycosylation have an effect on subunit assembly and biological activity. The DHFR-based expression vector and the DHFR-deficient cell line from Chinese hamster ovaries offer several unique avenues to explore further the molecular mechanisms underlying the complex biosynthetic pathway of bovine LH.

We thank Dr R. Goodwin, Dr J. Virgin, Dr A. Thomason and Dr T. Nett for contributions to various phases of this work; and Dr L. Webster, Jr, for numerous and helpful suggestions regarding the manuscript. This work has been supported by grants from the NIH (AM28559), NSF (PCM-8309164) and AmGen, Inc. In addition, the authors acknowledge support by a National Institutes of Health Research Career Development Award (AM-01316; J.H.N.) and a National Institutes of Health Postdoctoral Fellowship (AM-06981; D.M.K.).

References

Alt, F.W., Kellems, R.E., Bertino, J.R. & Schimke, R.T. (1978) Selective amplification of dihydrofolate reductase genes in methotrexate-resistant variants of cultured murine cells. *J. biol. Chem.* **253**, 1357–1370.

Bielinska, M. & Boime, I. (1979) Glycosylation of human chorionic gonadotropin in mRNA-dependent cell-free extracts: post-translational processing of an asparagine-linked mannose-rich oligosaccharide. *Proc. natn. Acad. Sci. U.S.A.* **76**, 1208–1212.

Boothby, M., Ruddon, R.W., Anderson, C., McWilliams, D. & Boime, I. (1981) A single gonadotropin α-subunit gene in normal tissue and tumor-derived cell-lines. *J. biol. Chem.* **256**, 5121–5127.

Chin, W.W. (1985) Organization and expression of glycoprotein hormone genes. In *The Pituitary Gland*, pp. 103–125. Ed. H. Imura. Raven Press, New York.

Chin, W.W., Habener, J.F., Kieffer, J.D. & Maloof, F. (1978) Cell-free translation of the messenger RNA coding for the α subunit. *J. biol. Chem.* **253**, 7985–7988.

Chin, W.W., Maloof, F. & Habener, J.F. (1981) Thyroid-stimulating hormone biosynthesis: Cellular processing, assembly, and release of subunits. *J. biol. Chem.* **256**, 3059–3066.

Daniels-McQueen, S., McWilliams, D., Birken, S., Canfield, R., Landefeld, T. & Boime, I. (1978) Identification of mRNAs encoding the α and β subunits of human chorionic gonadotropin. *J. biol. Chem.* **253**, 7109–7114.

Fetherston, J. & Boime, I. (1982) Synthesis of bovine lutropin in cell-free lysates containing pituitary microsomes. *J. biol. Chem.* **257**, 8143–8147.

Fiddes, J.C. & Goodman, H.M. (1981) The gene encoding the common alpha subunit of the four human glycoprotein hormones. *J. molec. appl. Genet.* **1**, 3–18.

Fiddes, J.C. & Talmadge, K. (1984) Structure, expression, and evolution of the genes for the human glycoprotein hormones. *Recent Prog. Hormone Res.* **40**, 43–79.

Goodwin, R.G., Moncman, C.L., Rottman, F.M. & Nilson, J.H. (1983) Characterization and nucleotide sequence of the gene for the common α subunit of the bovine pituitary glycoprotein hormones. *Nucleic Acids Res.* **11**, 6873–6882.

Gorman, C., Moffat, L.F. & Howard, B.H. (1982) Recombinant genomes which express chloramphenicol acetyltransferase in mammalian cells. *Molec. cell. Biol.* **2**, 1044–1051.

Goverman, J.M., Parsons, T.F. & Pierce, J.G. (1982) Enzymatic deglycosylation of the subunits of chorionic gonadotropin: effects on formation of tertiary structure and biological activity. *J. biol. Chem.* **257**, 15059–15064.

Green, E.D., van Halbeek, H., Boime, I. & Baenziger, J.U. (1985) Structural elucidation of the disulfated oligosaccharide from bovine lutropin. *J. biol. Chem.* **260**, 15623–15630.

Hoshina, H. & Boime, I. (1982) Combination of rat lutropin subunits occurs early in the secretory pathway. *Proc. natn. Acad. Sci. U.S.A.* **79**, 7649–7653.

Hoyer, P.B., Fitz, T.A. & Niswender, G.D. (1984) Hormone-independent activation of adenylate cyclase in large steroidogenic ovine luteal cells does not result in increased progesterone secretion. *Endocrinology* **114**, 604–608.

Hubbard, S.C. & Ivatt, R.J. (1981) Synthesis and processing of asparagine-linked oligosaccharides. *Ann. Rev. Biochem.* **50**, 555–583.

Hussa, R.O. (1980) Biosynthesis of human chorionic gonadotropin. *Endocrine Rev.* **3**, 268–294.

Jameson, J.L., Chin, W.W., Hollenberg, A.N., Chang, A.C. & Habener, J.F. (1984) The gene encoding the β subunit of rat luteinizing hormone: analysis of gene structure and evolution of nucleotide sequence. *J. biol. Chem.* **259**, 15474–15480.

Kaetzel, D.M., Browne, J.K., Wondisford, F., Nett, T.M., Thomason, A.R. & Nilson, J.H. (1985) Expression of biologically active bovine luteinizing hormone in Chinese hamster ovary cells. *Proc. natn. Acad. Sci. U.S.A.* **82**, 7280–7283.

Kafatos, F.C., Jones, C.W. & Efstratiadis, A. (1979) Determination of nucleic acid sequence homologies and relative concentrations by a dot hybridization procedure. *Nucleic Acids Res.* **7**, 1541–1551.

Kaufman, R.J. & Sharp, P.A. (1982) Amplification and expression of sequences cotransfected with a modular dihydrofolate reductase gene. *J. molec. Biol.* **159**, 601–621.

Kaufman, R.J., Wasley, L.C., Spiliotes, A., Gossels, S.D., Latt, S.A., Larsen, G.R. & Kay, R.M. (1985) Co-amplification and coexpression of human tissue-type plasminogen activator and murine dihydrofolate reductase sequences in Chinese hamster ovary cells. *Molec. cell. Biol.* **5**, 1750–1759.

Magner, J.A. & Weintraub, B.D. (1982) Thyroid-stimulating hormone subunit processing and combination in microsomal subfractions of mouse pituitary tumor. *J. biol. Chem.* **257**, 6709–6715.

Milsted, A., Silver, B.J., Cox, R.P. & Nilson, J.H. (1985) Coordinate regulation of the messenger ribonucleic acids encoding the α- and β-subunits of human chorionic gonadotropin in HeLa cells and butyrate-resistant variants. *Endocrinology* **117**, 2033–2039.

Nilson, J.H., Nejedlik, M.T., Virgin, J.B., Crowder, M.E. & Nett, T.M. (1983) Expression of α subunit and luteinizing hormone β genes in the ovine anterior pituitary: estradiol suppresses accumulation of mRNAs for both α subunit and luteinizing hormone β. *J. biol. Chem.* **258**, 12087–12090.

Parsons, T.F., Bloomfield, G.A. & Pierce, J.G. (1983) Purification of an alternate form of the α subunit of the glycoprotein hormones from bovine pituitaries and identification of its O-linked oligosaccharide. *J. biol. Chem.* **258**, 240–244.

Peters, B.P., Krzesicki, R.F., Hartle, R.J., Perini, F. & Ruddon, R.W. (1984) A kinetic comparison of the processing and secretion of the αβ dimer and the uncombined α and β subunits of chorionic gonadotropin synthesized by human choriocarcinoma cells. *J. biol. Chem.* **259**, 15123–15130.

Pierce, J.G. & Parsons, T.F. (1981) Glycoprotein hormones: Structure and function. *Ann. Rev. Biochem.* **50**, 465–495.

Strickland, T.W. & Pierce, J.G. (1983) The α subunit of pituitary glycoprotein hormones: formation of three-dimensional structure during cell-free biosynthesis. *J. biol. Chem.* **258**, 5927–5932.

Strickland, T.W. & Puett, D. (1982) The kinetic and equilibrium parameters of subunit association and gonadotropin dissociation. *J. biol. Chem.* **257**, 2954–2960.

Talmadge, K., Vamvakopoulos, N.C. & Fiddes, J.C. (1984) Evolution of the genes for the β subunits of human chorionic gonadotropin and luteinizing hormone. *Nature, Lond.* **307**, 37–40.

Virgin, J.B., Silver, J.B., Thomason, A.R. & Nilson, J.H. (1985) The gene for the β subunit of bovine luteinizing hormone encodes a gonadotropin mRNA with an unusually short 5'-untranslated region. *J. biol. Chem.* **260**, 7072–7077.

Weintraub, B.D., Stannard, B.S., Linnekin, R. & Marshall, M. (1980) Relationship of glycosylation to *de novo* thyroid-stimulating hormone biosynthesis and secretion by mouse pituitary tumor cells. *J. biol. Chem.* **255**, 5715–5723.

Workewych, J. & Cheng, K.W. (1979) Development of glycoprotein hormones and their α- and β-subunits in bovine fetal pituitary glands. II. Quantitation of free α- and β-subunits by radioimmunoassays and correlation of free α-subunit with thyrotropin, follicle-stimulating hormone, and luteinizing hormone. *Endocrinology* **104**, 1075–1082.

J. Reprod. Fert., Suppl. **34** (1987), 237–250

Printed in Great Britain
© 1987 Journals of Reproduction & Fertility Ltd

Transgenic livestock

J. P. Simons and R. B. Land

AFRC Institute of Animal Physiology and Genetics Research*, King's Buildings, West Mains Road, Edinburgh EH9 3JQ

Summary. Single genes can now be added routinely to the genome of mice by molecular manipulation as simple Mendelian dominants; this complements the normal process of reproduction to give 'transgenic' animals. Success in ruminants is limited to a few examples in sheep and although gene expression has yet to be documented, there is every reason to expect that it will be achieved. The application of this technology to livestock improvement depends on the identification of circumstances in which the phenotype is limited by the deficiency of a single protein. While there is little evidence to indicate that single dominant genes are in general likely to have favourable effects, it is argued that there are likely to be exceptions. These include particular combinations of promoter and structural gene sequences to alter feedback control, for example through a change in tissue specificity, and the alteration of definitive proteins such as those of milk. A mouse model has been established to study the molecular manipulation of sheep milk proteins. The sheep beta lactoglobulin gene has been incorporated and the sheep whey protein is secreted by the mammary gland of transgenic mice.

For the future, means to delete or reduce the expression of existing genes are likely to be important, as are more effective means of incorporation such as retroviral based methods and the incorporation of multigene constructs. The resources required to test transgenic livestock will, however, be greater than those required to create them.

Introduction

Reproduction in domestic ruminants is sexual. Both parents transmit one half of their genes to each offspring and it is these genes inherited by the new individual which are largely responsible for the determination of its characteristics. The independent segregation of chromosomes at meiosis and their recombination in the new individual ensures genetic variation among individuals within a species, even within full sib families. Sexual reproduction therefore accommodates the need for both resemblance and variation. Natural selection among this limited variation is the basis of evolution; artificial selection enables the characteristics of livestock to be changed to better meet the needs of society.

The rate of response to selection is affected directly by the genetic variation in the population in question. This is illustrated dramatically in general terms by the much greater response to artificial selection in populations of *Drosophila* with transposing elements than in those without (McKay, 1985). Cytoplasmic inhibitors prevent the movement of P transposable elements in P but not M type stocks. New variation is induced in a cross in which the female parent is M type and the male parent is P type. The result is a 2-fold greater response to selection in this, dysgenic, population over the reciprocal, non-dysgenic, cross (Fig. 1). By contrast, asexual reproduction (cloning) would

---

*Formerly Animal Breeding Research Organisation.

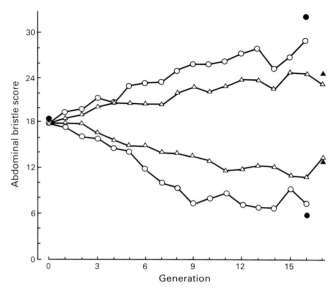

Fig. 1. Increased response to selection on dysgenic *Drosophila melanogaster*. Generation means of abdominal bristle score in lines selected for bristle number. ○, Lines derived from dysgenic flies (M♀♀ × P♂♂); □, non-dysgenic (P♀♀ × M♂♂) control lines. ●, ■, end points of the second replicates of the dysgenic and non-dysgenic selection lines respectively. (After Mackay, 1985.)

remove the genetic variation among individuals. This would both prevent further change and enable favourable existing genotypes to be replicated, as with plants such as fruit trees and potatoes.

Several procedures to introduce genetic variation have been considered. There is evidence from plants that genetic material from irradiated male gametes may be introduced at fertilization with a normal gamete following pollination with a mixture of normal and irradiated pollen (Powell *et al.,* 1983). The evidence for success in animals, however, is equivocal (McKay & Wishart, 1984) and furthermore there is no control over which genes are introduced.

Recombinant DNA techniques and the technology of embryo manipulation and transfer opened the way for specific genes to be introduced into animals (Gordon *et al.,* 1980) such that they are expressed (Brinster *et al.,* 1981; Wagner *et al.,* 1981a, b) and transmitted through the germ line (Costantini & Lacy, 1981; Gordon & Ruddle, 1981; Wagner *et al.,* 1981a); animals carrying newly introduced genes are termed 'transgenic'. Recent advances in the understanding of gene structure and expression, together with the ability to restructure genes, make it possible to target the additional variation to meet particular requirements. Such transgenic reproduction would cause animals to be quantitatively or qualitatively different, increasing the rate or extent to which livestock can adapt to meet the requirements of the community. Before considering transgenic reproduction, the subject of this paper, it is relevant to put this new technology in the context of other reproductive methods.

Multiple ovulation and embryo transfer would affect the rate of response to artificial selection of beef cattle dramatically and for dairy cattle would enable a small closed herd to achieve similar rates of response to those which might be attained in programmes based on progeny testing (Land & Hill, 1975; Nicholas & Smith, 1983). The importance of achieving similar rates of response to that with progeny testing should not be underestimated. With a single herd it is possible to focus decision making to ensure that selection is conducted to achieve breeding objectives. Even for sheep and goats the technique would have a favourable effect (Smith, 1986). Cloning, by contrast, would

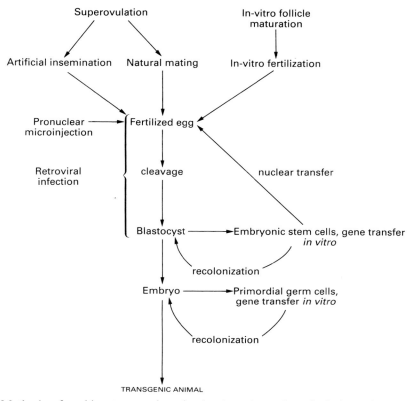

Fig. 2. Methods of making transgenic animals. A variety of methods have been used to make transgenic animals. These and others (not yet proven) are shown together with allied reproductive technologies.

have little impact (Gibson & Smith, 1986). It is proposed to discuss the present and likely future opportunities to transfer genes and then to consider the implications of transgenic reproduction.

Methods of gene transfer

The techniques for the transfer of genes fall into three broad classes: microinjection, infection by retroviral vectors and stem cell transfer. They are not all independent and the relationships between these and other relevant technologies are outlined in Fig. 2. Indeed it might be expected that the combination of these approaches may lead to significant increases in the rate of incorporation. Nevertheless, it is useful to consider them one by one.

Microinjection

The most widely used method for the generation of transgenic mammals is microinjection of DNA into one pronucleus of fertilized eggs. The eggs are usually obtained by superovulation and natural mating of the donor animals. They are immobilized by suction onto a blunt holding pipette, a fine injection pipette is inserted into one of the pronuclei and about 2 pl of the solution containing a few tens to a few hundred molecules of DNA are injected (Fig. 3). Surviving eggs are implanted into foster mothers, in which their development continues. A proportion of the animals which

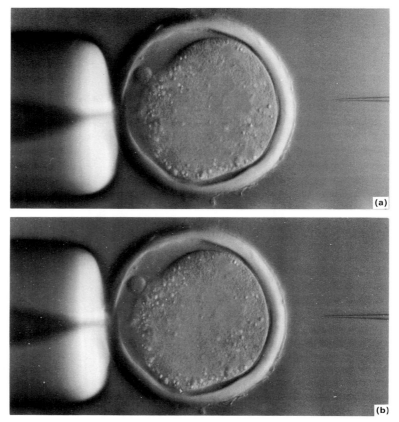

Fig. 3. A sheep egg (a) before and (b) after microinjection of DNA into a pronucleus, visualized by differential interference-contrast microscopy. The egg was held on the holding pipette (left) and injected using the injection pipette (right). The pronuclei can be seen to the right of centre and above and right of centre.

derive from the injected eggs will carry the new gene. These transgenic animals usually contain the injected gene in all cells, integrated into a chromosome. The integrated DNA is usually present in multiple copies, as a tandem repeat, integrated into a single site.

The presence of the gene in all cells is presumably due to integration before the first cleavage. The practical consequence of this is that transmission to the next generation is efficient, whereas if the recipients of the gene are chimaeric, the gene would be inherited by only a proportion of the offspring. This approach has been extremely successful for mice for which an average of 20% of eggs survive injection, transfer and development into young, and about 25% of the young carry the new gene (Brinster *et al.*, 1985). Microinjection has also been used to generate transgenic rabbits, pigs and sheep (Hammer *et al.*, 1985c), but the proportion of injected embryos which develop as transgenic young may be lower (Table 1). The success rates in mice are in general not dependent on the gene introduced. The loss of embryos is likely to arise from lethal physical damage at the time of injection, inefficiency of egg transfer or possibly lethal genetic damage due to the site of incorporation.

Microinjection is technically demanding, especially for farm animals which have almost opaque eggs, making visualization of nuclei difficult. In sheep, careful microscopy with differential interference-contrast optics is required (Hammer *et al.*, 1985c; J. P. Simons & I. Wilmut,

Table 1. Comparison of mice and sheep for efficiency of
transgenic animal production by microinjection

	Mice[a]	Sheep[b]	Sheep[c]
No. of eggs injected and transferred	299	92	1032
No. of young	78[d]	23	73[e]
No. transgenic	23	1	1

(a) J. P. Simons, unpublished; sheep beta lactoglobulin gene.
(b) J. P. Simons, unpublished; pMK injected into 1- and 2-cell eggs.
(c) Hammer *et al.* (1985c); MT-hGH injected into 1-, 2- and 4-cell eggs.
(d) Surviving to weaning.
(e) Fetuses and lambs.

unpublished), and pig pronuclei may be seen after centrifugation of eggs (Wall *et al.*, 1985). Transfer of genes by microinjection of eggs requires large numbers of eggs at a precise stage. Such large numbers of eggs with pronuclei are easy to obtain from mice, but very expensive and time consuming to collect from large animals. For these reasons other possible routes which might increase the efficiency of incorporation would be very advantageous.

Retroviral vectors

Retroviruses have a genome of RNA which, on infection of a cell, is copied into DNA which is then integrated into a chromosome. Infection and integration can be extremely efficient (King *et al.*, 1985). Retroviral vectors have been constructed which carry non-viral sequences and are capable of infecting cells and integrating, but are defective in the production of new virus. Such vectors have the potential to carry new genes into the germ line of animals with minimal manipulation of embryos and with simultaneous treatment of large numbers of embryos. Experiments with mice have shown that this approach does work, but is inefficient as yet (van der Putten *et al.*, 1985; Jaehner *et al.*, 1985). The proportion of animals produced which are transgenic was no better than by microinjection and even this required co-infection with infectious helper virus; integration was less frequent in the absence of helper. An additional problem is that these experiments used eggs in the cleavage stages; transgenic animals derived in this way would be mosaic, the germ line would be chimaeric and genetic transmission less efficient. By contrast with microinjection, genes transferred using retroviral vectors will be integrated as single copies.

Embryo-derived stem cells

Embryo-derived stem cells are pluripotential cells which have been isolated from the inner cell masses of blastocysts (Evans & Kaufman, 1981; Martin, 1981). These cells may be grown *in vitro* for long periods and subsequently reintroduced into blastocysts. The introduced cells efficiently colonize the inner cell mass and contribute to the developing animal, which will be a mosaic of cells derived from the blastocyst and the reintroduced stem cells (Bradley *et al.*, 1984). The introduced stem cells contribute both to the soma and the germ line. A particular advantage of this route is that the fusion gene may be introduced into the stem cells *in vitro*. The genetically manipulated cells can then be selected and characterized directly, and only those with desirable characteristics subsequently introduced into blastocysts. Mosaic transgenic animals have been produced by this route (Lovell-Badge *et al.*, 1985; Wagner *et al.*, 1985). Expression of genes transferred by this route has been demonstrated (Lovell-Badge *et al.*, 1985), although germ-line transmission has not yet been reported. To date, there have been no reports of embryo-derived stem-cell lines isolated from species other than mice.

Achievements

The potential of gene transfer for the gross manipulation of animal phenotypes was dramatically demonstrated by the 'giant' mice of Palmiter *et al.* (1982). These animals carried a hybrid metallothionein/rat growth hormone gene, and were up to 1·87 times the weight of control non-transgenic littermates at 74 days of age. In these mice, rat growth hormone was synthesized under the control of the metallothionein promoter in the liver, an ectopic site for growth hormone synthesis. This ectopic expression of the growth hormone gene is insensitive to the feedback mechanisms which normally operate to determine the concentration of circulating growth hormone. Consequently, the levels of growth hormone in these mice were up to 700 times those of control animals. Similar results have been obtained with mice expressing high levels of human (Palmiter *et al.*, 1983) or bovine (Hammer *et al.*, 1985a) growth hormone. One secondary effect of abnormally high levels of growth hormone is frequent female infertility (Hammer *et al.*, 1984, 1985a). Mice have been generated which express human growth hormone-releasing factor under the control of the metallothionein promoter (Hammer *et al.*, 1985b). They also grow faster, although not to the same extent. The elevated concentrations of growth hormone-releasing factor stimulate increased growth hormone synthesis by the pituitary gland, the normal site of growth hormone synthesis. This manipulation did not lead to female sterility.

Another clear example of the impact of an introduced gene on the phenotype is the restoration of immune competence to respond to particular polypeptides. Some strains of mice are unable to make the I-E antigen, one of the class II major histocompatibility antigens, which are involved in the immune response. This deficiency results in an inability of these mice to mount an immune response to a synthetic polypeptide: poly(glutamic acid-lysine-phenylalanine). The deficient mice have no functional gene for the alpha subunit of the I–E antigen; introduction of functional genes for E alpha has been shown to restore the response to poly(glu-lys-phe) (Le Meur *et al.*, 1985; Yamamura *et al.*, 1985).

In addition to the examples given above, which have clear relevance to applications in ruminants, a great deal of information of less direct relevance has been obtained from the use of transgenic mice. Areas to which transgenic animal studies have been particularly fruitful are regulation of gene expression, oncogenesis and the regulation of immunoglobulin gene rearrangements. The field has been extensively reviewed by Palmiter & Brinster (1986).

The genes of choice

The evidence from large animals, laboratory studies and from theory all indicate that single genes in general are unlikely to have a significant favourable effect on commercial traits and it is concluded that successful genetic manipulation will depend on the identification of exceptions (Land & Wilmut, 1987). Genes code for proteins, additional genes will potentially lead to the synthesis of additional protein. Favourable effects of gene transfer will therefore be dependent upon circumstances in which a single protein is deficient except when the deficiency of that protein arises from an inadequacy of one or more amino acids. The options now available are to introduce genes for novel proteins or for the production of existing proteins at higher levels in different tissues or at different times.

Proteins act as signals (e.g. hormones), enzymes or definitive products such as wool, body structure or milk. The basis for the argument that individual genes are unlikely to have large effects and for the identification of circumstances in which message is likely to be limiting are best considered for each in turn.

Increased levels of signal proteins such as hormones do affect the phenotype. Follicle-stimulating hormone (FSH) is well known to increase the number of eggs shed by the ovary, and growth hormone increases the rate of milk production by the mammary gland. Natural genetic variation in either trait is not, however, necessarily associated with variation in concentrations of the trophic hormone. Equally, the intrinsic synthetic capacity of the trophic system is not limited by the structural gene itself. In the case of FSH, for example, castration of the male and ovariectomy of the female both lead to a several-fold increase in the concentration of the hormone in peripheral plasma. The intrinsic synthesis potential for the hormone is not limiting, plasma concentrations are limited by the feedback control of gene expression. Any application of gene transfer to increase the output of signal proteins such as hormones would therefore require uncoupling of the synthesis of signal protein from the normal feedback control. The previously described giant mice of Palmiter *et al.* (1982) demonstrate well that this approach is effective. A refinement would be to express the structural gene from a promoter which is activated in the desired circumstances. Increased growth hormone synthesis during lactation, for example, would be expected to increase milk yield; such fine control will allow any deleterious effects of the manipulations to be minimized.

Enzymes affect the flux through pathways, but most pathways are made up of several steps, each dependent upon the effects of different enzymes. Kacser & Burns (1979) show theoretically that flux is insensitive to changes in individual enzymes and this is supported by experimental evidence, particularly in micro-organisms. Conversely, selection for the series of enzymes in a pathway does increase the flux and the phenotype, e.g. enzymes in NADP synthesis pathways in which high levels are positively correlative with the thickness of backfat in the pig (Muller, 1986). The implication is that it would be necessary to extend the present technology to constructs coding for several enzymes and this has recently been confirmed experimentally for tryptophan synthesis in yeast. Even many copies of genes for the individual enzymes had little impact on the flux but the introduction of a vector carrying genes for all 5 enzymes in the pathway showed a response in synthesis nearly proportional to the number of plasmids in the cell (P. Niederberger, quoted by Kacser, 1987). The application of transgenic technology to increase the rate of conversion of substrate to product through existing pathways would therefore depend on the development of multi-gene constructs. In addition, as described above, feedback control will often, if not always, need to be subverted.

In the third class of gene products are definitive proteins such as those in muscle, wool and milk. The rate of protein synthesis relates to the level of messenger RNA. The introduction and expression of an additional gene will change the message available for translation and hence, for definitive proteins, gene transfer affects the product directly. Of this group of products, only milk and wool proteins are truly definitive because muscle proteins are continually recycled. (The regulation of protein turnover is likely to have as important an effect on the rate of accretion of muscle as the expression of the structural genes for the muscle proteins themselves.) The keratin genes have been studied extensively, but the evidence indicates that the keratin messenger RNA is abundant in wool follicles and that the rate of wool production is limited more by the availability of the required amino acids than the availability of message (Ward *et al.,* 1986). The qualitative nature and the proportions of wool and milk proteins could, however, potentially be changed by introducing genes expressed in the wool follicle and mammary gland respectively, and the molecular biology programme in our laboratory has focussed on alteration of the composition of milk. In 1982 Palmiter *et al.*, on finding very high levels of circulating growth hormone in some of their transgenic mice, suggested that transgenic animals may be useful as production systems for valuable proteins. Lathe *et al.* (1986) suggested that the mammary gland of the lactating ewe would be the system of choice for such applications, being so prodigious in protein synthesis and secretion, and milk being so easily harvested. There is the opportunity to introduce genes coding for novel proteins of high value such as medicines or industrial enzymes, and to modify the composition and hence the nutritional value of milk as a food.

The molecular manipulation of milk

Milk is composed largely of water, fats, lactose and proteins. The most abundant proteins in the milk of ruminants are the caseins, beta lactoglobulin and alpha lactalbumin. The most abundant single protein is beta casein, at up to 16 g/litre. Each of these proteins is synthesized in the mammary gland; other more minor components, including proteins such as albumin, enter the milk from the blood stream.

Various considerations, however, make it desirable to have a model system for the molecular manipulation of ruminant milk. First, as already outlined, the production of transgenic ruminants is time consuming, expensive and not yet routine. Second, the timescale for experiments with ruminants is very long: in sheep, it may take 2–3 years from the time of injection of the DNA into an egg until milk can be harvested from a transgenic ewe. Two models were considered, primary mammary culture systems in which the effectiveness of alternative constructs could be evaluated after transfection, and transgenic mice. Given that fully differentiated mammary tissue cultures are not available and that transgenic mice will be a closer model, the latter was the option of choice: for a transgenic mouse the interval from egg injection to milking can in principle be 3 months.

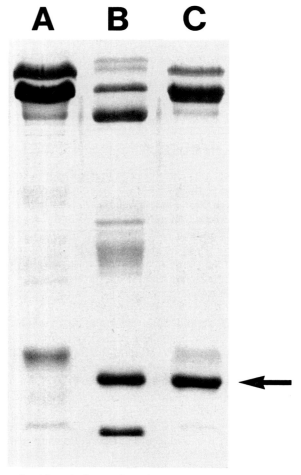

Fig. 4. SDS polyacrylamide gel electrophoresis of milk proteins to show sheep beta lactoglobulin in milk of a transgenic mouse: (A) control mouse whey, (B) control sheep whey, (C) whey from a mouse transgenic for the sheep beta lactoglobulin gene. The arrow shows the position of sheep beta lactoglobulin.

In our first experiments, the sheep gene which codes for beta lactoglobulin was isolated and introduced into mice, which do not normally make beta lactoglobulin. Figure 4 shows an electrophoretic analysis of the whey proteins from one such mouse. A major band which comigrates with sheep beta lactoglobulin is present, and this has now been shown to be genuine sheep beta lactoglobulin by western blotting. The concentration of beta lactoglobulin in this milk is approximately 20 mg/ml. Several other lines of mice transgenic for the sheep beta lactoglobulin gene similarly secrete this protein.

These results demonstrate, first, that the sheep milk protein gene can be incorporated into the genome of the mouse in such a way that it is expressed in the homologous tissue. Breeding experiments also show that the gene and its expression are inherited. The next questions are whether the expression is limited to the mammary gland and whether the milk protein genes can be used as vehicles for the control of structural genes for specific non-milk proteins. A schematic construct and its expression are illustrated in Fig. 5. The mouse model system will allow the assessment of a variety of combinations of components of constructs designed for the production of foreign proteins, allowing relatively rapid optimization of construct design. Having shown that beta lactoglobulin expression is not species specific, the results obtained in mice should be directly relevant to ruminants.

The production of high value proteins such as medicines and industrial enzymes is a simple application of transgenic technology in ruminants, whether they be cattle, goats or sheep. The vigorous expression of the sheep beta lactoglobulin gene in the mouse also indicates that it might well be possible to increase the proportion of protein in ruminant milk by the introduction of additional structural genes for the milk protein required. Such manipulations could be aimed simply to improve the nutritional value of milk (for human consumption or to increase the growth

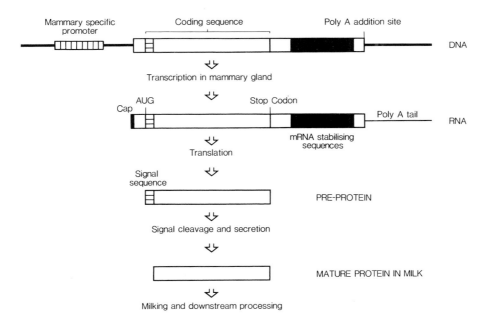

Fig. 5. Diagram of a generalized gene designed to direct synthesis of foreign proteins in the milk of transgenic animals, and steps in its expression. The mammary-specific promoter directs the synthesis of a messenger RNA. Sequences in the RNA ensure that the mRNA is stable in the mammary gland. The coding sequence is translated into a precursor protein from which the signal peptide is cleaved when the protein is secreted. The mature protein is harvested in the milk.

rate of suckled young), or to alter milk qualitatively to make it more suitable for processing by the dairy industry.

Other options would be to reduce the lactose content of milk. Not only is lactose the least valuable of the milk solids, the vast majority of humans lose the ability to digest lactose at weaning. These people cannot ingest milk or dairy products without severe discomfort resulting. The production of lactose-free milk is thus an attractive possibility. One route to this end would be to construct animals which secrete lactase, the enzyme which hydrolyses lactose, in their milk. The lactose would be broken down to glucose and galactose *in situ*. A second method of producing lactose-free milk would be to prevent the synthesis of lactose, which occurs within the mammary gland and which requires alpha lactalbumin as a factor in the combination of glucose and galactose to form the disaccharide. This, however, would require the extinction of alpha lactalbumin expression. While this is presently beyond existing transgenic technology, possible methods are discussed later.

Application to livestock breeding

Novel products are an intermediate term application of new molecular biological knowledge to the livestock industry. The product can be readily identified and harvested. Even if the fertility or another component of normal performance were depressed, the high value of the product would outweigh such a disadvantage. In conventional agriculture, however, it would be necessary for the transgenic process to increase the economic merit of the strain in which it is practised. Therefore, not only must the addition of a single gene have a desirable effect on a biological component of merit, but the benefits of this effect must be greater than any disadvantageous side effects.

Possible strategies for the use of transgenic stock are discussed by Smith *et al.* (1987) who draw attention to two points of particular importance. First, there is at present no control of the site of incorporation; while the direct effects of a particular fusion gene might be similar each time it is introduced, the indirect effects would be expected to be unique through the effects of incorporation on the function of adjacent genes. Each transgenic introduction would therefore have to be evaluated separately. Second, transgenics cannot be assessed accurately as individuals and it will be necessary to establish populations both hemi- and homozygous for the new gene for the effects to be measured. Indeed, the difficulty of assessment is such that it might well be more cost effective simply to introduce potentially useful transgenics to the population and allow the genes to find their own frequency under the effects of selection. In any case, but with such an approach in particular, it would be important to work with the best populations.

To put the question of testing into perspective, resources have never been allocated for the accurate comparison of existing populations so it is not reasonable to assume that they will be available for new ones. It must be remembered that the numbers required are considerable; 50 offspring would be required to give a 95% chance of detecting a gene which increased the mean by one standard deviation; 16 would be required even when the effect was as large as 2 standard deviations. Traditional selection would be expected to change the mean by 2% per generation so that the mean could be increased by 10% in the duration of the 5 or so generations needed to introduce and evaluate a transgenic. A threshold of useful effects might then be 10% of the mean. For typical growth traits with a coefficient of variation of 10%, the 50 offspring given above would be required for testing. For reproductive traits with coefficients of variation of 25%, 250 animals would be required!

On occasion, integration of a new gene disrupts an endogenous gene, termed 'insertional mutation' (see Palmiter & Brinster, 1986). One consequence of disruption of endogenous genes at the site of incorporation is that the adverse side effects of transgenesis may be fully recessive and hence it is important to test homozygous as well as hemizygous populations. If hemizygous favourable transgenics are introduced, half of the population will be hemizygous before the proportion of homozygotes would be expected to reach 10%. Most insertional mutations are recessive lethals, and so retrospective culling could have to be very severe.

As with the need to consider exceptions in the identification of genes which affect the biological components of economic merit, it is also perhaps relevant to consider exceptions in the identification of target biological traits for transgenic improvement. The addition of growth hormone genes could have some beneficial effects, but might also have both direct and indirect adverse effects. In other cases, the only adverse effects which might reasonably be anticipated are those resulting from insertional mutation. One example is disease resistance for which additional MHC loci could be introduced. More specifically, the expression of viral coat protein genes in animals normally susceptible to virus infection would competitively reduce the binding of infecting viruses to the receptor proteins on the cell surface and so confer resistance to infection. Exactly the same path to resistance has recently been demonstrated in plants: the introduction into tobacco plants of a gene which codes for the coat protein of the tobacco mosaic virus interfered with the normal progression of the viral infection (Abel *et al.*, 1986).

Knowledge for the future

The knowledge that gene products are in general unlikely to be limiting indicated the need to look for exceptions. Equally, the same knowledge indicates that the phenotype is susceptible to a reduction in the product of a single gene. The manipulation of single loci could therefore be highly relevant in circumstances such as those in which decreased levels of the product decrease negative feedback on the target system. One example would be the feedback control of reproduction where the inhibition of the production of ovarian hormones controlling the release of FSH would be likely to increase the ovulation rate. Active and passive immunization against ovarian steroids increases the ovulation of sheep (Scaramuzzi & Hoskinson, 1984; Webb *et al.*, 1984), but this technology has not been extended to cattle. The discovery of key feedback hormones would identify targets for immunization, and knowledge of steps in the biosynthesis of these hormones would indicate targets for gene neutralization.

Two potential methods for gene neutralization can be envisaged: anti-sense gene expression and site-directed mutation. The first step in expression of a gene is transcription, i.e. synthesis of a messenger RNA (mRNA) using one of the gene's strands, the coding strand, as a template. The mRNA is a 'sense' transcript. In normal circumstances the complementary strand is not transcribed. By rearranging the gene, however, expression of the complementary strand can be obtained. This transcribed anti-sense RNA and the mRNA from the coding strand are complementary with each other and so can hybridize. This will form a double-stranded RNA, interfering with one or more of the subsequent steps of gene expression: RNA processing, export from the nucleus and translation (Fig. 6). Experimental reduction of gene expression by anti-sense gene expression has been demonstrated (Izant & Weintraub, 1984; Kim & Wold, 1985; McGarry & Lindquist, 1986). One potential problem with this method of gene neutralization is that, due to the kinetics of hybridization, a large excess of anti-sense RNA over mRNA is required. In addition, if the gene being neutralized is subject to feedback control, the result may simply be to increase the rate of transcription. It is probable that anti-sense methods will result in a reduction of expression rather than total extinction; this may well prove to be a useful feature.

A second method for the removal of gene function may be in-vivo site-directed mutagenesis. It has been shown that genes introduced into cultured cells can integrate by homologous recombination (Smithies *et al.*, 1985; Thomas *et al.*, 1986). This could be used to extinguish gene expression by inserting DNA which disrupts the gene (Fig. 7). Such homologous recombination is infrequent, and as yet would not be applicable in transgenic animals for this reason. It is likely, however, that, as the mechanism of homologous recombination becomes elucidated, conditions which increase the frequency will be determined. The combination of the embryo-derived stem-cell route and

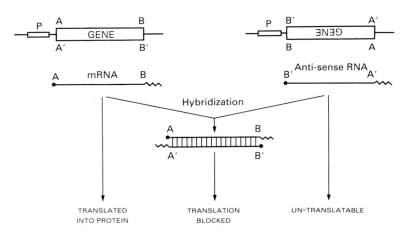

Fig. 6. Neutralization of gene expression by anti-sense gene expression. An anti-sense gene directs the synthesis of anti-sense RNA. This RNA hybridizes to the mRNA transcript of the gene to be neutralized. The hybridization removes the mRNA from the pool available for translation into protein resulting in reduced levels of the protein product.

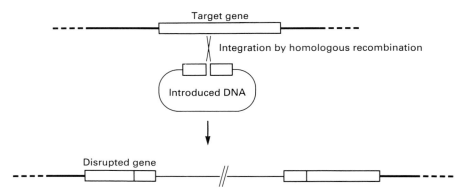

Fig. 7. Gene disruption. The DNA introduced has homology with the target gene (open boxed regions). On occasion, integration will be by homologous recombination; this results in a partial duplication of the gene sequences and an interruption of the gene. The interruption will prevent normal expression of the gene.

homologous recombination is attractive: it would be possible to select those cell lines which have the desired integration *in vitro*, before entering the more expensive phase of work with animals.

It should be pointed out that anti-sense extinction of expression will be a dominant effect since it acts on the RNA transcribed from both alleles of the gene in question. Gene disruption by homologous recombination will in most cases be a recessive mutation, and the effects can only be assessed by breeding to homozygosity. In the longer term we should hope to be able to change single nucleotides in a predetermined way; this would allow, in addition to complete extinction, very fine manipulation of genes potentially in both regulatory and coding regions.

Transgenic biology has clear implications for animal improvement. Although there are many potential pitfalls, the application to livestock improvement has considerable, widespread potential. Simple traits will be the easiest to manipulate. We have chosen to focus on manipulation of milk composition, and have demonstrated, in mice, that this is feasible. Success in these endeavours will depend on careful design of the genes transferred. This design will be aided by an understanding of regulation at all levels: genetic, biochemical and physiological. In addition, the widespread successful application of transgenic technology will require improvement of existing and development of new techniques to increase the efficiency of generation of transgenic animals and to allow new types of manipulation.

The application of the technology will, however, depend on the establishment of a suitable social framework. The new gene is neither a drug nor an infectious agent and falls outside the legislation for either. The basic advantages of genetic improvement over direct manipulation of stock are strengthened by current moves against the use of hormones in commerce and the progressively strengthening preference for natural products. Transgenic practice could enhance these advantages by increasing the rate at which the characteristics of stock could be changed to meet the requirements of the community.

We thank many colleagues who have contributed to the establishment of molecular biology research at ABRO, particularly Dr Bishop, Dr Clark, Dr Lathe and Dr Wilmut and the farm staff who have made it possible.

References

Abel, P.P., Nelson, R.S., De, B., Hoffmann, N., Rogers, S.G., Fraley, R.T. & Beachy, R.N. (1986) Delay of disease development in transgenic plants that express the tobacco mosaic virus coat protein gene. *Science, N.Y.* 232, 738–743.

Bradley, A., Evans, M., Kaufman, M.H. & Robertson, E. (1984) Formation of germ-line chimaeras from embryo-derived teratocarcinoma cell lines. *Nature, Lond.* 309, 255–256.

Brinster, R.L., Chen, H.Y. & Trumbauer, M. (1981) Somatic expression of herpes thymidine kinase in mice following injection of a fusion gene into eggs. *Cell* 27, 223–231.

Brinster, R.L., Chen, H.Y., Trumbauer, M.E., Yagle, M.K. & Palmiter, R.D. (1985) Factors affecting the efficiency of introducing foreign DNA into mice by microinjecting eggs. *Proc. nat. Acad. Sci. U.S.A.* 82, 4438–4442.

Costantini, F. & Lacy, E. (1981) Introduction of a rabbit β-globin gene into the mouse germ line. *Nature, Lond.* 294, 92–94.

Evans, M.J. & Kaufman, M.H. (1981) Establishment in culture of pluripotential cells from mouse embryos. *Nature, Lond.* 292, 154–156.

Gibson, J.P. & Smith, C. (1986) Technology and animal breeding applications in livestock improvement. *Proc. 3rd Wld Congr. Genet. Appl. Livest. Prod., Lincoln, Nebraska,* Vol. 12, 96–105.

Gordon, J.W. & Ruddle, F.H. (1981) Integration and stable germ line transmission of genes injected into mouse pronuclei. *Science, N.Y.* 214, 1244–1246.

Gordon, J.W., Scangos, G.A., Plotkin, D., Barbosa, J.A. & Ruddle, F.H. (1980) Genetic transformation of mouse embryos by microinjection of purified DNA. *Proc. natn. Acad. Sci. U.S.A.* 77, 77380–77384.

Hammer, R.E., Palmiter, R.D. & Brinster, R.L. (1984) Partial correction of murine hereditary growth disorder by germ-line incorporation of a new gene. *Nature, Lond.* 311, 65–67.

Hammer, R.E., Brinster, R.L. & Palmiter, R.D. (1985a) Use of gene transfer to increase animal growth. *Cold Spring Harbor Symp. Quant. Biol.* 50, 379–387.

Hammer, R.E., Brinster, R.L., Rosenfeld, M.G., Evans, R.M. & Mayo, K.E. (1985b) Expression of human growth hormone-releasing factor in transgenic mice results in increased somatic growth. *Nature, Lond.* 315, 413–416.

Hammer, R.E., Pursel, V.G., Rexroad, C.E., Jr, Wall, R.J., Bolt, D.J., Ebert, K.M., Palmiter, R.D. & Brinster, R.L. (1985c) Production of transgenic rabbits, sheep and pigs by microinjection. *Nature, Lond.* 315, 680–683.

Izant, J.G. & Weintraub, H. (1984) Inhibition of thymidine kinase gene expression by anti-sense RNA: a molecular approach to genetic analysis. *Cell* 36, 1007–1015.

Jaehner, D., Kirsten, H., Mulligan, R. & Jaenisch, R. (1985) Insertion of the bacterial *gpt* gene into the germ line of mice by retroviral infection. *Proc. natn. Acad. Sci. U.S.A.* 82, 6927–6931.

Kacser, H. (1987) Biochemical pathways and their control. In *Mathematical Models in Microbiology: 3. Physiological Models* (in press). Eds M. J. Bazin & J. I. Prosser. CRC Press, Boca Raton.

Kacser, H. & Burns, J.A. (1979) Molecular democracy: who shares the controls? *Biochem. Soc. Trans.* 7, 1149–1160.

Kim, S.K. & Wold, B.J. (1985) Stable reduction of thymidine kinase activity in cells expressing high levels of anti-sense RNA. *Cell* 42, 129–138.

King, W., Patel, M.D., Lobel, L.I., Goff, S.P. & Chi Nguyen-Huu, M. (1985) Insertion mutagenesis of embryonal carcinoma cells by retroviruses. *Science, N.Y.* 228, 554–558.

Land, R.B. & Hill, W.G. (1975) The possible use of superovulation and embryo transfer in cattle to increase response to selection. *Anim. Prod.* 21, 1–12.

Land, R.B. & Wilmut, I. (1987) Gene transfer and animal breeding. *Theriogenology* **27**, 169–179.

Lathe, R., Clark, A.J., Archibald, A.L., Bishop, J.O., Simons, P. & Wilmut, I. (1986) Novel products from livestock. In *Exploiting New Technologies in Animal Breeding: Genetic Developments*, pp. 91–102. Eds C. Smith, J. W. B. King & J. C. McKay. Oxford University Press, Oxford.

Le Meur, M., Gerlinger, P., Benoist, C. & Mathis, D. (1985) Correcting an immune-response deficiency by creating E$_a$ gene transgenic mice. *Nature, Lond.* **316**, 38–42.

Lovell-Badge, R.H., Bygrave, A.E., Bradley, A., Robertson, E., Evans, M.J. & Cheah, K.S.E. (1985) Transformation of embryonic stem cells with the human type-II collagen gene and its expression in chimeric mice. *Cold Spring Harbor Symp. Quant. Biol.* **50**, 707–711.

Mackay, T.F.C. (1985) Transposable element-induced response to artificial selection in drosophila melanogaster. *Genetics, Princeton* **111**, 351–374.

Martin, G.R. (1981) Isolation of a pluripotent cell line from early mouse embryos cultured in medium conditioned by teratocarcinoma stem cells. *Proc. natn. Acad. Sci. U.S.A.* **78**, 7634–7638.

McGarry, T.J. & Lindquist, S. (1986) Inhibition of heat shock protein synthesis by heat-inducible antisense RNA. *Proc. natn. Acad. Sci. U.S.A.* **83**, 399–403.

McKay, J.C. & Wishart, G.J.H. (1984) Genetic and physiological effects of irradiation of fowl semen. *Proc. 26th British Poultry Breeders' Roundtable, 1984, Norwich*.

Muller, E. (1986) Physiological and biochemical indicators of growth and composition. In *Exploiting New Technologies in Animal Breeding*, pp. 132–139. Eds C. Smith, J. W. B. King & J. C. McKay. Oxford University Press, Oxford.

Nicholas, F.W. & Smith, C. (1983) Increased rate of genetic change in dairy cattle by embryo transfer and splitting. *Anim. Prod.* **36**, 341–353.

Palmiter, R.D. & Brinster, R.L. (1986) Germline transformation of mice. *Ann. Rev. Genetics* 465–499.

Palmiter, R.D., Brinster, R.L., Hammer, R.E., Trumbauer, M.E., Rosenfeld, M.G., Birnberg, N.C. & Evans, R.M. (1982) Dramatic growth of mice that develop from eggs microinjected with metallothionein-growth hormone fusion genes. *Nature, Lond.* **300**, 611–615.

Palmiter, R.D., Norstedt, G., Gellnas, R.E., Hammer, R.E. & Brinster, R.L. (1983) Metallothionein-human GH fusion genes stimulate growth of mice. *Science, N.Y.* **222**, 809–814.

Powell, W., Caligari, P.D.S. & Hayter, A.M. (1983) The use of pollen irradiation in barley breeding. *Theor. appl. Gen.* **65**, 73–76.

Scaramuzzi, R.J. & Hoskinson, R.M. (1984) Active immunisation against steroid hormones for increasing fecundity. In *Immunological Aspects of Reproduction in Mammals*, pp. 445–474. Ed. D. B. Crighton. Butterworths, London.

Smith, C. (1986) Use of embryo transfer in genetic inprovement of sheep. *Anim. Prod.* **42**, 81–88.

Smith, C., Meuwissen, T. & Gibson, J.P. (1987) On the use of transgenes in livestock improvement. *Anim. Breed. Abstr.* **55**, 1–10.

Smithies, O., Gregg, R.G., Boggs, S.S., Koralewski, M.A. & Kucherlapati, R.S. (1985) Insertion of DNA sequences into the human chromosomal β-globin locus by homologous recombination. *Nature, Lond.* **317**, 230–234.

Thomas, K.R., Folger, K.R. & Capecchi, M.R. (1986) High frequency targeting of genes to specific sites in the mammalian genome. *Cell* **44**, 419–428.

van der Putten, H., Botteri, F.M., Miller, A.D., Rosenfeld, M.G., Fan, H., Evans, R.M. & Verma, I.M. (1985) Efficient insertion of genes into the mouse germ line via retroviral vectors. *Proc. natn. Acad. Sci. U.S.A.* **82**, 6148–6152.

Wagner, T.E., Hoppe, P.C., Jollick, J.D., School, D.R., Hodinka, R.L. & Gault, J.B. (1981a) Microinjection of a rabbit β-globin gene into zygotes and its subsequent expression in adult mice and their offspring. *Proc. natn. Acad. Sci. U.S.A.* **78**, 6376–6380.

Wagner, E.F., Stewart, T.A. & Mintz, B. (1981b) The human β-globin gene and a functional viral thymidine kinase gene in developing mice. *Proc. natn. Acad. Sci. U.S.A.* **78**, 5016–5020.

Wagner, E.F., Keller, G., Gilboa, E., Ruther, U. & Stewart, C. (1985) Gene transfer into murine stem cells and mice using retroviral vectors. *Cold Spring Harbor Symp. Quant. Biol.* **50**, 691–700.

Wall, R.J., Pursel, V.G., Hammer, R.E. & Brinster, R.L. (1985) Development of porcine ova that were centrifuged to permit visualization of pronuclei and nuclei. *Biol. Reprod.* **32**, 645–651.

Ward, K.A., Franklin, I.R., Murray, J.D., Nancarrow, C.D., Raphael, K.A., Rigby, N.W., Byrne, C.R., Wilson, B.W. & Hunt, C.L. (1986) The direct transfer of DNA by embryo microinjection. *Proc. 3rd Wld Congr. Genet. Appl. Livest. Prod., Lincoln, Nebraska*, Vol. 12, 6–21.

Webb, R., Land, R.B. & Pathiraja, N. (1984) Passive immunisation against steroid hormones in the female. In *Immunological Aspects of Reproduction in Mammals*, pp. 475–499. Ed. D. B. Crighton. Butterworths, London.

Yamamura, K., Kikutani, H., Folsom, V., Clayton, L.K., Kimoto, M., Akira, S., Kashiwamura, S., Tonegawa, S. & Kishimoto, T. (1985) Functional expression of a microinjected E$_a^d$ gene in C57BL/6 transgenic mice. *Nature, Lond.* **316**, 67–69.

Note added in proof. Since submission of this paper, we have obtained a further 5 transgenic sheep.

Germ-line transmission of genes transferred into mice via embryo-derived stem-cells has been reported. The gene transfer into stem-cells was performed by repeated retroviral vector infection, without selection (Robertson *et al.*, 1986), or by calcium phosphate-mediated transfection with biochemical selection for the introduced gene (Gossler *et al.*, 1986).

Gossler, A., Doetschman, T., Korn, R., Serfling, E. & Kemler, R. (1986) Transgenesis by means of blastocyst-derived embryonic stem cell lines. *Proc. natn. Acad. Sci U.S.A.* **83**, 9065–9069.

Robertson, E., Bradley, A., Kuehn, M. & Evans, M. (1986) Germ-line transmission of genes introduced into cultured pluripotential cells by retroviral vector. *Nature, Lond.* **323**, 445–448.

J. Reprod. Fert., Suppl. **34** (1987), 251–259

Printed in Great Britain

Use of chimaeras to study development

G. B. Anderson

Department of Animal Science, University of California, Davis, California 95616, U.S.A.

Introduction

In Greek mythology a chimaera was a fire-breathing she-monster having a lion's head, a goat's body and a serpent's tail. Chimaeras have been used extensively as models for research in developmental biology under the more general definition of a composite animal or plant in which different cell populations are derived from more than one fertilized egg, or the union of more than two gametes (McLaren, 1976). This paper is limited to chimaeras produced by combination of cells from two or more mammalian embryos. Characteristics of chimaeras, methods for production and uses in research are described. Effort has been made to include results of direct relevance to domestic animals. Excellent reviews on mammalian chimaeras and their uses in research are available from McLaren (1976) and Le Douarin & McLaren (1984).

Production of mammalian chimaeras

Manipulations to produce mammalian experimental chimaeras are commonly carried out early in embryonic development, which can lead to extensive chimaerism throughout the body. In some non-mammalian vertebrates, combination of embryos or embryonic cells can result in duplication of body parts while failure to replace completely embryonic cells that have been excised can lead to truncation of body parts. The early mammalian embryo has the ability to regulate its development in such a manner that foreign embryonic cells can be incorporated to produce a morphologically normal individual with two cell lines.

Mammalian experimental chimaeras are usually produced during preimplantation stages by aggregation of two or more embryos (aggregation chimaera) or injection of cells into a blastocyst (injection chimaera). Methods for production of aggregation chimaeras were first described by Tarkowski (1961) and Mintz (1962); these procedures have since been adapted for use in many laboratories. Aggregation procedures have been used to produce chimaeras from 2-cell through to morula-stage embryos. Eight-cell stage embryos are often used, however, because their geometric configuration allows greater contact between embryos than is achieved with earlier stage embryos, and by the morula stage tight junctions between blastomeres reduce the likelihood that aggregation will occur. Zona-free embryos are either placed in contact with one another and pushed together or placed in medium containing phytohaemagglutinin (Mintz *et al.*, 1973) or antibodies (Palmer & Dewey, 1983) to facilitate adhesion. Composite embryos are usually left in culture until they reach the blastocyst stage and then are transferred to the reproductive tracts of recipients for development to term. In domestic animals in-vitro culture systems for cleavage-stage embryos are not well developed and aggregated embryonic cells may be incubated for several days in a ligated oviduct of a temporary host before transfer to a recipient for development to term (Fehilly *et al.*, 1984a) or transferred immediately to a recipient (Brem *et al.*, 1984). Blastocysts produced by aggregation of embryos are larger than normal, but regulation of size occurs during gestation so that chimaeras have normal birthweights.

Production of injection chimaeras was first described by Gardner (1968). Simplified procedures for production of injection chimaeras have been published for use with embryos of the mouse

(Moustafa & Brinster, 1972), rabbit (Babinet & Bordenave, 1980) and sheep (Butler, 1986). Cells injected into blastocysts for production of chimaeras most often originate from the inner cell mass (ICM) of another blastocyst. Solter & Knowles (1975) developed an immunosurgical procedure for isolation of relatively intact ICMs. Other cell types that have been used to produce injection chimaeras include cleavage-stage blastomeres (Tucker *et al.*, 1974) and embryonal carcinoma cells (Brinster, 1974). A modification of the procedure for blastocyst injection has been described in which the ICM of one blastocyst is completely replaced with that of another blastocyst (Gardner *et al.*, 1973; Papaioannou, 1982), a procedure referred to as reconstitution of blastocysts. This procedure has been used for induction of successful interspecific pregnancy in which the fetus is surrounded by placental membranes of a different species (Rossant *et al.*, 1983a).

The laboratory mouse has been the species of choice for production of experimental chimaeras, but they also have been produced in the rat (Mayer & Fritz, 1974; Weinberg *et al.*, 1985), rabbit (Gardner & Munro, 1974; Babinet & Bordenave, 1980), sheep (Tucker *et al.*, 1974; Fehilly *et al.*, 1984a; Butler *et al.*, 1985), and cow (Willadsen, 1982; Brem *et al.*, 1984). Viable chimaeras have been produced also when embryos were combined from different species. *Mus musculus*←→*Mus caroli* chimaeras, produced from the domestic laboratory mouse and a South East Asian wild mouse (Rossant & Frels, 1980), will be discussed later. Embryos from *Bos taurus* and *Bos indicus*, two species of cattle that readily hybridize and produce fertile young, have been combined; although no overt chimaeras were produced, one individual had evidence of chimaerism of internal tissue (Summers *et al.*, 1983). Combination of embryos from sheep and goats, *Ovis aries*←→*Capra hircus*, has also lead to viable chimaeras (Fehilly *et al.*, 1984b; Polzin *et al.*, 1986). Chimaeric embryos have been produced from the mouse and bank vole (Mystkowska, 1975) and the mouse and rat (Gardner & Johnson, 1973; Stern, 1973; Zeilmaker, 1973). For these latter combinations normal post-implantation development failed when chimaeric embryos were transferred to the reproductive tracts of one of the parent species. Gardner & Johnson (1975) reported the birth of rat←→mouse chimaeras, all of which were dead at birth or died soon thereafter.

Characteristics of chimaeras

Perhaps the most consistent feature of experimental chimaeras produced from preimplantation mammalian embryos is their variability. Each chimaera is unique, which can be an advantage or a disadvantage for their use in research. Although some degree of control can be exercised over the genotype of the placenta relative to that of the fetus when embryonic cells are combined (Rossant *et al.*, 1982; Fehilly *et al.*, 1984b; Meinecke-Tillman & Meinecke, 1984), cell mixing that occurs during early development and allocation of cells to development of different tissues make it impossible to predict or control the exact cellular distribution of components in a chimaera. Falconer & Avery (1978) described a nearly flat distribution in the frequencies of chimaeras having various proportions of cells from two component lines. Furthermore, genotype of the component lines may influence the extent to which each line is represented in a chimaera (Mullen & Whitten, 1971), from relatively equal representation of the two lines over the population of chimaeras (balanced chimaeras) to the predominance of one line (unbalanced chimaeras). Even in balanced combinations, individual chimaeras may vary from an almost complete exclusion of one line to an almost complete exclusion of the other line. Chimaerism can occur in any and all tissues and organs of the body. In one series of chimaeras, Falconer *et al.* (1981) correlated the degree of chimaerism observed between organ pairs and found correlation coefficients that ranged from 0·37 to 0·89 and an average correlation coefficient of 0·73. For these chimaeras, there was a relatively strong relationship between the degree of chimaerism observed in different organs, even though each chimaeric individual may have differed markedly from another.

The sex ratio of a population of experimental chimaeras often deviates from 1:1. A combination of two female embryos produces a female chimaera; two male embryos yield a male chimaera. The combination of a male and a female embryo, which is expected to occur approximately 50% of the time, is responsible for deviation from the normal sex ratio. In general, these sex chimaeras are phenotypically normal males, which results in a sex ratio in chimaeras of about 75% males and 25% females in 'balanced' strain combinations. However, a more normal 1:1 sex ratio has been reported to occur in 'unbalanced' strain combinations (Mullen & Whitten, 1971). Some XX/XY chimaeras do not differentiate as normal males, but rather as hermaphrodites or normal females (McLaren, 1984b). McLaren (1984b) summarized data on sex of known XX/XY chimaeras from various studies and reported that 22% were females, 7% were hermaphrodites and 71% were males.

Just as chimaerism can occur readily in somatic tissues, germ cell chimaerism is often observed in XX/XX and XY/XY chimaeras. The use of genetic markers in breeding trials has demonstrated that XX/XX female chimaeras are capable of ovulating oocytes of both genetic lines; likewise, XY/XY males can produce both types of spermatozoa (reviewed by McLaren, 1984a). Germ cell chimaerism in XX/XY mammalian sex chimaeras usually does not occur. In some non-mammalian vertebrates differentiation of germ cells depends not on their genetic sex, but rather on the sex of the gonad into which they migrate (Blackler, 1965). In mammals, however, only XY germ cells appear to be capable of developing into spermatozoa; XX/XY chimaeric males are fertile, but produce spermatozoa only from the XY line. Whether or not XY germ cells can develop into viable oocytes is less clear. Offspring of XX/XY females usually develop from XX germ cells. One exception has been reported in which a mouse was born from an XY germ cell in an XX/XY chimaeric female (Ford *et al.*, 1975). A complete discussion of development of germ cells in chimaeras is presented by McLaren (1984a).

Uses of experimental chimaeras in research

The value of chimaeras in studying development and tissue interactions is based on the ability to distinguish between the two genetically different cell populations within the animal. McLaren (1976) described the ideal cell marker to distinguish one chimaeric component from the other as being "cell-localized, cell-autonomous, stable, distributed universally among both the internal and external tissues of the body, and easy to detect, both grossly and in histological sections, without elaborate processing. No such marker exists." Despite the lack of an ideal marker, many useful markers are available. For a detailed review of available markers for use with chimaeras the reader is referred to West (1984).

Studies of normal development

Chimaeras have been used extensively to study the course of normal embryogenesis and several examples are presented here. Genetically marked cells introduced into an embryo and followed through development have been used in the analysis of cell lineage. For example, the fates of ICM and trophectoderm have been examined using reconstituted murine blastocysts in which an ICM was injected into a trophoblastic vesicle (Papaioannou, 1982), rat←→mouse chimaeras (Gardner & Johnson, 1975; Rossant, 1976) and *M. musculus*←→*M. caroli* chimaeras (Rossant *et al.*, 1983b). Chimaeras have also been used to determine the minimum number of clones responsible for development of various tissues. In 1976, McLaren (p. 33) stated, "No instance has been found of a tissue or organ, however small, which in a chimaeric animal is always formed of one component only. This implies that it is groups of cells, rather than single cells, which are directed towards particular developmental pathways and this gives rise to particular tissues and organs." Chimaerism has been found in essentially all tissues studied, which demonstrates that at least two embryonic cells contribute to the formation of most tissues and organs of the body. Analysis of results from chimaeras has also led to the conclusion that only a few cells of the blastocyst contribute to formation of the embryo proper; all other cells contribute to extraembryonic structures. Circumstantial evidence to support this hypothesis comes from observations that injection of only a single cell into another embryo can produce extensive chimaerism in the resulting offspring (Ford et al., 1975; Illmensee & Mintz, 1976). Mintz (1970) proposed that the number of cells in the blastocyst that form the embryo is only three. This hypothesis was based on the observation that, in about 25% of the individuals produced by aggregation of two embryos, only one component cell line is expressed. Markert & Petters (1978) demonstrated that at least 3 cells are allocated to produce the embryo when they produced a triply chimaeric mouse by aggregation of 3 embryos. They pointed out that their results set a lower but not an upper limit on the number of cells that originally contribute to the embryo. The same authors (Petters & Markert, 1980) subsequently reported the production of an aggregation chimaera that expressed four different genotypes, which might argue that 4, not 3, cells of the blastocyst contribute to the embryo. It may be argued also, however, that allocation of cells in an embryo that is four times normal size may be different from that in a single embryo. Recent results have set at 8 the maximum number of cells from which the embryo is derived (Soriano & Jaenisch, 1986).

Another interesting question raised from research with experimental chimaeras is the origin of primordial germ cells that colonize the gonad during early development and ultimately give rise to gametes. From work with non-mammalian species primordial germ cells have been shown to develop from yolk sac endoderm and a similar origin has been assumed for mammalian primordial germ cells. Two lines of evidence indicate that these cells are derived from ectoderm in mammals rather than from endoderm. Falconer & Avery (1978) showed that germ cell chimaerism is positively correlated with chimaerism of somatic tissues, which suggests a similar origin. More conclusively, Gardner & Rossant (1979) showed that ectoderm but not endoderm that is injected into blastocysts contributes to germ cell lineage of the resulting chimaeras.

Study of abnormal development

Experimental chimaeras have also been useful in the study of certain developmental anomalies. Again, several examples will be given to illustrate how the model is used. When mouse chimaeras were produced from normal embryos and embryos from a strain affected by muscular dystrophy, their phenotypes were normal despite the presence of defective genes within multinucleate muscle cells (Peterson, 1974). Similar results have been obtained with bovine chimaeras in which one embryo was homozygous for the double-muscling gene (R. B. Church, personal communication). Normal embryonic cells are also capable of directing the development of parthenogenetically activated ova. While parthenogenesis can be readily induced in mammals, in no case has development proceeded to term. Normal murine aggregation and injection chimaeras containing cells of parthenogenetic origin have been produced, however, when parthenogenetically activated embryos were combined with normally fertilized embryos (Surani et al., 1977; Stevens, 1978). Chimaeras produced by Stevens (1978) produced germ cells from both the normal and parthenogenetic line, which demonstrated that parthenogenetically activated ova retain the ability to contribute to both somatic and germ cells.

An extreme example of the ability of normal cells to direct development of abnormal cells in a chimaera is the reversal of malignancy of embryonal carcinoma cells, which are stem cells derived from a teratocarcinoma. Embryonal carcinoma cells show morphological and biochemical similarities to pluripotent embryonic cells, possess the ability to proliferate indefinitely in the undifferentiated state and, under the appropriate conditions, will differentiate in a more or less orderly fashion (Rossant & Papaioannou, 1984). The most dramatic examples of their differentiation are injection into blastocysts (Brinster, 1974; Mintz & Illmensee, 1975; Papaioannou et al., 1975; Stewart & Mintz, 1981) and aggregation with embryos (Stewart, 1982). Under these conditions embryonal carcinoma cells have the ability to contribute to development of normal tissues and organs of a chimaeric individual. Furthermore, colonization of the germ cell line can occur so that the resulting chimaeras produce gametes that contain haploid genotypes from the embryonal carcinoma cell line. Recently, pluripotential cell lines have been derived in culture from embryos (Martin, 1981; Magnuson *et al.*, 1982; Kaufman *et al.*, 1983; Axelrod, 1984; Wobus *et al.*, 1984a). These cell lines are usually referred to as embryonic stem cells and resemble stem cells derived from teratocarcinomas, but in many respects may be more like pluripotential cells from morulae or early inner cell masses than they are like teratocarcinoma stem cells (Rossant & Papaioannou, 1984). Embryonic stem cells will readily be incorporated into chimaeras and will colonize the germ line at substantially greater frequency than will teratocarcinoma stem cells (Bradley *et al.*, 1984). Of particular interest in domestic livestock is development of embryonic stem cells for incorporation of foreign DNA. The feasibility of DNA transformation of embryonic stem cells has been demonstrated (Wobus *et al.*, 1984b). Theoretically, it may be possible to insert foreign DNA into a stem cell, perhaps screen the transformed cell line for appropriate incorporation of the gene, and introduce the desired gene into the germ cell line of a chimaera (reviewed by Stewart, 1984).

Study of expression of quantitative genetic traits

Most production traits of economic importance in livestock are under the control of many genes, the contribution of each being difficult to identify. Improvement of production traits has resulted from selective breeding, but the physiological basis for genetic differences is not always clear. Experimental chimaeras have been used on only a limited basis in the study of expression of quantitative production traits. Falconer *et al.* (1981) produced aggregation chimaeras from strains of mice selected for large and small body size and unselected controls. Body weight was linearly related to the mean cell proportions of each cell line within an individual chimaera. Correlation coefficients of body weight to chimaerism in blood, liver, lung, spleen, spinal cord, brain, pituitary gland, kidney, adrenal gland and testis indicated that none of these organs had a predominant influence on growth. In our laboratory male aggregation chimaeras from a line of mice selected for rapid post-weaning growth and a line of mice with a normal rate of growth exhibited a linear relationship between rate of growth and proportion of cells from the rapid-growth line (estimated from coat-colour chimaerism) only up to 50% contribution of the rapid-growth line. Animals with greater than 50% contribution of the rapid-growth line had a mean growth rate similar to that of the selected rapid-growth line (unpublished observations), suggesting that diffusible gene product(s) from the rapid-growth line cells affected growth of tissues of both genotypes. Reproductive performance of experimental chimaeras produced from genetic lines of mice that differ in litter size has also been studied (Craig-Veit & Anderson, 1985). Reproductive characteristics were compared for line crosses having half the genetic information of each line in all cells and chimaeras having the full genetic complement of the two lines in different cells of the body. For most traits, means for crossbreds and chimaeras were similar, regardless of whether means were at or above the mid-parent average. In contrast, for ovulation rate and body weight, genetic crossbreds and chimaeras clearly differed, with chimaeric females being similar to the high litter size line and crossbred females exhibiting additive inheritance. Results from studies of the cell types found in ovarian follicles of chimaeras produced from selected lines that differ in ovulation rate suggested that follicles composed of cells of a high-ovulation line may be preferentially recruited for growth and ovulation over follicles composed of cells of the unselected line (unpublished observations). These results may explain why chimaeric females produced from two lines that differ in ovulation rate have an ovulation rate characteristic of the high-ovulation line (Craig-Veit & Anderson, 1985) and may further suggest that increased ovulation rate resulting from selection for large litter size was due to changes at the level of the ovary.

Barriers to interspecies reproduction

An extensively studied model for interspecific pregnancy has been the *M. musculus* ← → *M. caroli* chimaera (reviewed by Rossant *et al.*, 1983a). *M. musculus* and *M. caroli* do not readily produce viable hybrids (West *et al.*, 1977, 1978; Frels *et al.*, 1980). Likewise, *M. caroli* embryos fail to develop in the *M. musculus* uterus, even when accompanied by *M. musculus* embryos that develop normally (Croy *et al.*, 1982, 1985). Immunization of *M. musculus* recipients with *M. caroli* lymphocytes promoted failure of interspecific pregnancy, but transfer of *M. caroli* embryos to T cell-deficient or NK cell-deficient *M. musculus* recipients or treatment of *M. musculus* recipients with cyclosporin A or anti-Ia antiserum failed to prolong survival (Croy *et al.*, 1985). *M. caroli* embryos can be protected from the *M. musculus* uterus, however, by construction of *M. musculus*←→*M. caroli* chimaeric embryos. When *M. caroli* ICM were injected into *M. musculus* blastocysts that were then transferred to the uteri of *M. musculus* recipients, viable interspecific chimaeras were produced (Rossant & Frels, 1980). *M. musculus* ICM transplanted to *M. caroli* blastocysts failed to survive in the *M. musculus* uterus (Rossant *et al.*, 1982). Final proof of the protection of *M. caroli* fetuses by *M. musculus* trophoblast was provided by Rossant *et al.* (1983a) who transferred *M. caroli* ICM into *M. musculus* trophoblastic vesicles that contained no *M. musculus* ICM cells; a *M. caroli* fetus was carried to term in a *M. musculus* uterus. The importance of trophoblast genotype in maintenance of pregnancy has also been demonstrated in production of sheep←→goat chimaeras and sheep/goat interspecific pregnancies. Aggregation chimaeras have been used to construct ovine fetuses with caprine placentas carried to term in does (Fehilly *et al.*, 1984b) and caprine fetuses with ovine placentas carried to term in ewes (Fehilly *et al.*, 1984b; Meinecke-Tillman & Meinecke, 1984). Injection chimaeras produced by transplanting the ICM of one species into the blastocyst of the other species have also resulted in successful interspecific pregnancies and chimaeras between sheep and goats when genotypes of the trophoblast and recipient uterus were the same (Fehilly *et al.*, 1984b; Polzin *et al.*, 1986). Pregnancies of horse←→donkey chimaeras have also been established (R. L. Pashen, personal communication), but none survived to term.

The importance of genotype of the placenta in maintenance of successful interspecific and chimaeric pregnancies has been clearly established, but the degree of chimaerism that will be tolerated in the placenta is not known. Rossant *et al.* (1982) reported that aggregation *M. musculus*←→*M. caroli* chimaeras, which are expected to have chimaeric trophoblast, survive in the *M. musculus* uterus. Furthermore, participation by *M. caroli* cells in formation of the placenta was confirmed in viable Day-9·5 conceptuses. Unfortunately, chimaeric trophoblast in term pregnancies was not identified.

Germ cell chimaerism has been confirmed in *M. musculus*←→*M. caroli* chimaeras (Rossant & Chapman, 1983). When these chimaeras were mated to *M. musculus* males, some litters contained both *M. musculus* and hybrid offspring, which confirmed that chimaeric females were ovulating both *M. musculus* and *M. caroli* oocytes. Intraspecific chimaeras are known to have allogeneic tolerance of the component lines (reviewed by Matsunaga, 1984) and since *M. musculus*←→*M. caroli* and sheep←→goat chimaeras remain generally healthy, this tolerance apparently extends to interspecific chimaeras. (An interesting exception appears to be the chick←→quail spinal cord chimaera, in which rejection of the grafted tissue later extends to that of the host (Kinutani *et al.*, 1986). Allografts of spinal cord performed between two non-histocompatible strains of chickens, however, produced viable chimaeras in which the foreign neural tissue was permanently tolerated (Kinutani & Le Douarin, 1985). Chick←→quail chimaeras represent more localized grafts of foreign tissue with less subsequent cell mixing than occurs in mammalian aggregation and injection chimaeras.) One might suggest that interspecific mammalian chimaeras are able to carry to term pregnancies of either species. As already mentioned, *M. musculus*←→*M. caroli* chimaeric females mated to *M. musculus* males produced litters of both *M. musculus* and hybrid offspring (Rossant & Chapman, 1983). Somewhat surprisingly, pregnancies were not carried to term when chimaeric females were inseminated with *M. caroli* spermatozoa, suggesting that tolerance may be unidirectional. When embryos of the two species were transferred to *M. musculus*←→*M. caroli* chimaeric recipients, *M. musculus* embryos survived but *M. caroli* embryos did not. Further complicating interpretation of these results is the observation that *M. musculus* embryos can survive in the *M. caroli* uterus. The authors concluded that the primary event in *M. caroli* embryo failure in the *M. musculus* uterus is incorrect interaction between *M. caroli* trophoblast and *M. musculus* uterine tissue. Support for this hypothesis comes from results in our laboratory with transfer of ovine and caprine embryos to a female sheep← →goat chimaera. Pregnancy was confirmed by ultrasonography at approximately 70 days of gestation and two fetuses were thought to exist. At about 4 months gestation, however, a single ovine fetus was aborted. The fetus appeared normal except for being small for age. Only about 15 cotyledons were present on the allantochorion (unpublished observations). Abnormal placental formation may have resulted from inappropriate interaction between ovine allantochorion and caprine uterine cells. A similar observation was made when *Bos gaurus* embryos were transferred to *B. taurus* recipients (Stover *et al.*, 1981). Two fetuses that went to term in the domestic recipients were reported to have small placentomes and a reduced number of cotyledons. Horse embryos develop to term in donkey recipients more frequently than the reciprocal transfer in which the donkey chorionic girdle fails to invade the endometrium of the mare to form endometrial cups (Allen, 1982). On the other hand, failure of sheep/goat hybrid pregnancies in ewes has been attributed to maternal cell-mediated attack against trophoblastic tissue in the caruncular areas (Dent *et al.*, 1971). Chimaeras with complete genotypes of two different species will be useful for studies of maintenance of fetal allografts.

References

Allen, W.R. (1982) Immunological aspects of the endometrial cup reaction and the effect of xenogeneic pregnancy in horses and donkeys. *J. Reprod. Fert., Suppl.* **31**, 57–94.

Axelrod, H.R. (1984) Embryonic stem cell lines derived from blastocysts by a simplified technique. *Devl Biol.* **101**, 225–228.

Babinet, C. & Bordenave, G.R. (1980) Chimeric rabbits from immunosurgically-prepared inner-cell-mass transplantation. *J. Embryol. exp. Morph.* **60**, 429–440.

Blackler, A.W. (1965) Germ-cell transfer and sex ratio in *Xenopus laevis*. *J. Embryol. exp. Morph.* **13**, 51–61.

Bradley, A., Evans, M., Kaufman, M.H. & Robertson, E. (1984) Formation of germ-line chimera from embryo-derived teratocarcinoma cell lines. *Nature, Lond.* **309**, 255–256.

Brem, B., Tenhumberg, H. & Kraußlich, H. (1984) Chimerism in cattle through microsurgical aggregation of morula. *Theriogenology* **22**, 609–613.

Brinster, R.L. (1974) The effect of cells transferred into the blastocyst on subsequent development. *J. exp. Med.* **140**, 1049–1056.

Butler, J.E. (1986) Production of experimental chimeras in livestock by blastocyst injection. In *Genetic Engineering of Animals*, pp. 175–185. Eds J. W. Evans & A. Hollender. Plenum Press, New York.

Butler, J.E., Anderson, G.B., BonDurant, R.H. & Pashen, R.L. (1985) Production of ovine chimeras. *Theriogenology* **23**, 183, Abstr.

Craig-Veit, C. & Anderson, G.B. (1985) Reproductive performance of chimeric mice produced from genetic lines that differ in litter size. *J. Anim. Sci.* **61**, 1527–1538.

Croy, B.A., Rossant, J. & Clark, D.A. (1982) Histological and immunological studies of post-implantation death of *Mus caroli* embryos in the *Mus musculus* uterus. *J. Reprod. Immun.* **4**, 277–293.

Croy, B.A., Rossant, J. & Clark, D.A. (1985) Effect of alterations in the immunocompetent status of *Mus musculus* females on the survival of transferred *Mus caroli* embryos. *J. Reprod. Fert.* **74**, 479–489.

Dent, J., McGovern, P.T. & Hancock, J.L. (1971) Immunological implications of ultrastructural studies of goat–sheep hybrid placentae. *Nature, Lond.* **231**, 116–117.

Falconer, D.S. & Avery, P.J. (1978) Variability of chimaeras and mosaics. *J. Embryol. exp. Morph.* **43**, 195–219.

Falconer, D.S., Gauld, I.K., Roberts, R.C. & Williams, D.A. (1981) The control of body size in mouse chimaeras. *Genet, Res., Camb.* **38**, 25–46.

Fehilly, C.B., Willadsen, S.M. & Tucker, E.M. (1984a) Experimental chimaerism in sheep. *J. Reprod. Fert.* **70**, 347–351.

Fehilly, C.B., Willadsen, S.M. & Tucker, E.M. (1984b) Interspecific chimaerism between sheep and goat. *Nature, Lond.* **307**, 634–636.

Ford, C.E., Evans, E.P., Burtenshaw, M.D., Clegg, H.M., Tuffrey, M. & Barnes, R.D. (1975) A functional 'sex-reversed oocyte in the mouse. *Proc. R. Soc. B* **190**, 187–197.

Frels, W.I., Rossant, J. & Chapman, V.M. (1980) Intrinsic and extrinsic factors affecting the development of hybrids between *Mus musculus* and *Mus caroli*. *J. Reprod. Fert.* **59**, 387–392.

Gardner, R.L. (1968) Mouse chimaeras obtained by the injection of cells into the blastocyst. *Nature, Lond.* **220**, 596–597.

Gardner, R.L. & Johnson, M.H. (1973) Investigation of early mammalian development using interspecific chimaeras between rat and mouse. *Nature, New Biol.* **246**, 86–89.

Gardner, R.L. & Johnson, M.H. (1975) Investigation of cellular interaction and deployment in the early mammalian embryo using interspecific chimaeras between the rat and mouse. In *Cell Patterning* (Ciba Found. Symp. 29), pp. 183–200. Associated Scientific Publishers, Amsterdam.

Gardner, R.L. & Munro, A.J. (1974) Successful construction of chimaeric rabbits. *Nature, Lond.* **250**, 146–147.

Gardner, R.L. & Rossant, J. (1979) Investigation of the fate of 4·5 day *post-coitum* mouse inner cell mass cells by blastocyst injection. *J. Embryol. exp. Morph.* **52**, 141–152.

Gardner, R.L., Papaioannou, V.E. & Barton, S.C. (1973) Origin of the ectoplacental cone and secondary giant cells in mouse blastocysts reconstituted from isolated trophoblast and inner cell mass. *J. Embryol. exp. Morph.* **30**, 561–572.

Illmensee, K. & Mintz, B. (1976) Totipotency and normal differentiation of single teratocarcinoma cells cloned by injection into blastocysts. *Proc. natn. Acad. Sci. U.S.A.* **73**, 549–553.

Kaufman, M.H., Robertson, E.J., Handyside, A.H. & Evans, M.J. (1983) Establishment of pluripotential cell lines from haploid mouse embryos. *J. Embryol. exp. Morph.* **73**, 249–261.

Kinutani, M. & Le Douarin, N.M. (1985) Avian spinal cord chimaeras: I. Hatching ability and post hatching survival in homo- and heterospecific chimaeras. *Devl Biol.* **111**, 243–255.

Kinutani, M., Coltey, M. & Le Douarin, N. (1986) Postnatal development of a demyelinating disease in avian spinal cord chimeras. *Cell* **45**, 307–314.

Le Douarin, N. & McLaren, A. (1984) *Chimeras in Developmental Biology*, 456 pp. Academic Press, London.

Magnuson, T., Epstein, C.J., Silver, L.M. & Martin, G.R. (1982) Pluripotent embryonic stem cell lines can be derived from tw5/tw5 blastocysts. *Nature, Lond.* **298**, 750–753.

Markert, C.L. & Petters, R.M. (1978) Manufactured hexaparental mice show that adults are derived from three embryonic cells. *Science, N.Y.* **202**, 56–58.

Martin, G.R. (1981) Isolation of a pluripotential cell line from early mouse embryos cultured in medium conditioned by teratocarcinoma stem cells. *Proc. natn. Acad. Sci. U.S.A.* **78**, 7634–7638.

Matsunaga, T. (1984) The use of chimeric mice in immunology: How does the immune system know self and non-self? In *Chimeras in Developmental Biology*, pp. 217–238. Eds N. Le Douarin & A. McLaren. Academic Press, London.

Mayer, .J.F. & Fritz, H.J. (1974) The culture of preimplantation rat embryos and the production of allophenic rats. *J. Reprod. Fert.* **39**, 1–9.

McLaren, A. (1976) *Mammalian Chimaeras*, 154 pp. Cambridge University Press.

McLaren, A. (1984a) Chimeras and sexual differentiation. In *Chimeras in Developmental Biology*, pp. 381–399. Eds N. Le Douarin & A. McLaren. Academic Press, London.

McLaren, A. (1984b) Germ cell lineages. In *Chimeras in Developmental Biology*, pp. 111–129. Eds N. Le Douarin & A. McLaren. Academic Press, London.

Meinecke-Tillman, S. & Meinecke, B. (1984) Experimental chimaeras—removal of reproductive barrier between sheep and goat. *Nature, Lond.* **307**, 637–638.

Mintz, B. (1962) Formation of genotypically mosaic mouse embryos. *Amer. Zool.* **2**, 432, Abstr. 310.

Mintz, B. (1970) Gene expression in allophenic mice. *Symp. Int. Soc. Cell Biol.* **9**, 15–42.

Mintz, B. & Illmensee, K. (1975) Normal genetically mosaic mice produced from malignant teratocarcinoma cells. *Proc. natn. Acad. Sci. U.S.A.* **75**, 3585–3589.

Mintz, B., Gearhart, J.D. & Guymont, A.O. (1973) Phyto-

hemagglutinin mediated blastomere aggregation and development of allophenic mice. *Devl Biol.* **31**, 195–199.

Moustafa, L.A. & Brinster, R.L. (1972) Induced chimerism by transplanting embryonic cells into mouse blastocysts. *J. exp. Zool.* **181**, 193–202.

Mullen, R.J. & Whitten, W.K. (1971) Relationship of genotype and degree of chimerism in coat color to sex ratio and gametogenesis in chimeric mice. *J. exp. Zool.* **178**, 165–176.

Mystkowska, E.T. (1975) Development of mouse-bank vole interspecific chimaeric embryos. *J. Embryol. exp. Morph.* **33**, 731–734.

Palmer, J. & Dewey, M.J. (1983) Allophenic mice produced from embryos aggregated with antibody. *Experientia* **39**, 196–198.

Papaioannou, V.E. (1982) Lineage analysis of inner cell mass and trophectoderm using microsurgically reconstituted mouse blastocysts. *J. Embryol. exp. Morph.* **68**, 199–209.

Papaioannou, V.E., McBurney, M.W., Gardner, R.L. & Evans, M.J. (1975) Fate of teratocarcinoma cells injected into early mouse embryos. *Nature, Lond.* **258**, 70–73.

Peterson, A.C. (1974) Chimaera mouse study shows absence of disease in genetically dystrophic muscle. *Nature, Lond.* **248**, 561–564.

Petters, R.M. & Markert, C.L. (1980) Production and reproductive performance of hexaparental and octaparental mice. *J. Hered.* **71**, 70–74.

Polzin, V.J., Anderson, D.L., Anderson, G.B., BonDurant, R.H., Butler, J.E., Pashen, R.L., Penedo, M.C.T. & Rowe, J.D. (1986) Production of sheep←→goat chimeras by blastocyst injection. *Theriogenology* **25**, 183, Abstr.

Rossant, J. (1976) Investigation of inner cell mass determination by aggregation of isolated rat inner cell mass with mouse morulae. *J. Embryol. exp. Morph.* **36**, 163–174.·

Rossant, J. & Chapman, V.M. (1983) Somatic and germline mosaicism in interspecific chimaeras between *Mus musculus* and *Mus caroli*. *J. Embryol. exp. Morph.* **73**, 193–205.

Rossant, J. & Frels, W.I. (1980) Interspecific chimeras in mammals: successful production of live chimeras between *Mus musculus* and *Mus caroli*. *Science, N.Y.* **208**, 419–421.

Rossant, J. & Papaioannou, V.E. (1984) The relationship between embryonic, embryonal carcinoma and embryo-derived stem cells. *Differentiation* **15**, 155–162.

Rossant, J., Mauro, V.M. & Croy, B.A. (1982) Importance of trophoblast genotype for survival of interspecific murine chimaeras. *J. Embryol. exp. Morph.* **69**, 141–149.

Rossant, J., Croy, B.A., Clark, D.A. & Chapman, V.M. (1983a) Interspecific hybrids and chimeras in mice. *J. exp. Zool.* **228**, 223–233.

Rossant, J., Vijh, M., Siracusa, L.D. & Chapman, V.M. (1983b) Identification of embryonic cell lineages in histological sections of *M. musculus*←→*M. caroli* chimaeras. *J. Embryol. exp. Morph.* **73**, 179–191.

Solter, D. & Knowles, B.B. (1975) Immunosurgery of mouse blastocyst. *Proc. natn. Acad. Sci. U.S.A.* **72**, 5099–5102.

Soriano, P. & Jaenisch, R. (1986) Retroviruses as probes for mammalian development: allocation of cells to the somatic and germ cell lineages. *Cell* **46,** 19–29.

Stern, M.S. (1973) Chimaeras obtained by aggregation of mouse eggs with rat eggs. *Nature, Lond.* **243,** 472–473.

Stevens, L.C. (1978) Totipotent cells of parthenogenetic origin in a chimaeric mouse. *Nature, Lond.* **276,** 266–267.

Stewart, L.C. (1982) Formation of viable chimaeras by aggregation between teratocarcinomas and pre-implantation mouse embryos. *J. Embryol. exp. Morph.* **67,** 167–179.

Stewart, L.C. (1984) Teratocarcinoma chimeras and gene expression. In *Chimeras in Developmental Biology*, pp. 409–427. Eds N. Le Douarin & A. McLaren. Academic Press, London.

Stewart, T.A. & Mintz, B. (1981) Succesive generations of mice produced from an established culture line of euploid teratocarcinoma cell. *Proc. natn. Acad. Sci. U.S.A.* **78,** 6314–6318.

Stover, J., Evans, J. & Dolensek, E.P. (1981) Inter-species embryo transfer from the gaur to domestic Holstein. *Proc. Am. Ass. Zoo. Vet.* pp. 122–124.

Summers, P.M., Shelton, J.N. & Bell, D. (1983) Synthesis of primary *Bos taurus–Bos indicus* chimaeric calves. *Anim. Reprod. Sci.* **6,** 91–102.

Surani, M.A.H., Barton, S.C. & Kaufman, M.H. (1977) Development to term of chimaeras between diploid parthenogenetic and fertilized embryos. *Nature, Lond.* **270,** 601–603.

Tarkowski, A. (1961) Mouse chimaeras developed from fused eggs. *Nature, Lond.* **190,** 857–860.

Tucker, E.M., Moor, R.M. & Rowson, L.E.A. (1974) Tetraparental sheep chimaeras induced by blastomere transplantation. Changes in blood type with age. *Immunology* **26,** 613–621.

Weinberg, W.C., Howard, J.C. & Iannaccone, P.M. (1985) Histological demonstration of mosaicism in a series of chimeric rats produced between congenic strains. *Science, N.Y.* **227,** 524–527.

West, J.D. (1984) Cell marks. In *Chimeras in Developmental Biology*, pp. 39–67. Eds N. Le Douarin & A. McLaren. Academic Press, London.

West, J.D., Frels, W.I., Papaioannou, V.E., Karr, J.P. & Chapman, V.M. (1977) Development of interspecific hybrids of *Mus. J. Embryol. exp. Morph.* **41,** 233–243.

West, J.D., Frels, W.I. & Chapman, V.M. (1978) *Mus musculus* × *Mus caroli* hybrids: mouse mules. *J. Hered.* **69,** 321–326.

Willadsen, S.M. (1982) Micromanipulation of embryos of the large domestic species. In *Mammalian Egg Transfer*, pp. 185–210. Ed. C. E. Adams. CRC Press, Boca Raton.

Wobus, A.M., Holzhausen, H., Jakel, P. & Schoneich, J. (1984a) Characterization of a pluripotent stem cell line derived from a mouse embryo. *Expl Cell Res.* **152,** 212–219.

Wobus, A.M., Kiessling, U., Strauss, M., Holzhausen, H. & Schoneich, J. (1984b) DNA transformation of a pluripotent mouse embryonal stem cell line with a dominant selective marker. *Cell Differentiation* **15,** 93–97.

Zeilmaker, G.H. (1973) Fusion of rat and mouse morulae and formation of chimaeric blastocysts. *Nature, Lond.* **242,** 115–116.

J. Reprod. Fert., Suppl. **34** (1987), 261–271

Printed in Great Britain
© 1987 Journals of Reproductivity and Fertility Ltd

Detection of early pregnancy in domestic ruminants

R. G. Sasser and C. A. Ruder

Department of Animal Science, University of Idaho, Moscow, Idaho 83843, U.S.A.

Summary. Tests for the detection of pregnancy early after insemination have not yet reached their full potential. Currently, the milk progesterone assay provides the earliest possible test, at an interval of one oestrous cycle after insemination, i.e. 17, 21 and 21 days in sheep, goats and cows respectively. This assay is pregnancy non-specific and rate of detection of pregnant animals is acceptable but less than desirable.

Detection of activity of early pregnancy factor may develop into an excellent early test for many species, but the rosette inhibition test which is currently required has limited development and use. Pregnancy-specific protein B has been developed as a radioimmunoassay and is reliable under laboratory situations for ruminants. It can be used after 24 days of gestation in the cow. Application to field testing awaits development. Other pregnancy-associated or specific substances which are found in maternal body fluids might develop as pregnancy markers. Ultrasonic devices might provide very early detection in cattle but the expense of a test will limit application.

All tests for pregnancy early after insemination have an inherent inaccuracy. Presence of an embryo at the time the test is applied will not assure pregnancy at the time of a confirmatory test, such as birth of live young or rectal examination in cows after 35 days of gestation. Therefore, no matter how early the test, a follow-up examination might be desirable in intensively managed herds or flocks.

The animal industry is on the verge of new biotechnological approaches to reproductive management. The potential seems as great as the imagination.

Introduction

Pregnancy is routinely detected in cows by inserting the hand into the rectum and palpating through the rectal and uterine walls for fetal membranes, the amniotic vesicle or cotyledons within the uterus (Wisnicky & Casida, 1948). This can be accomplished by well-trained technicians by 35 days after insemination but accuracy is considerably improved by waiting until 40–50 days. In small ruminants, older methods have included rectal–abdominal palpation in sheep (Hulet, 1972) and ultrasound detection in sheep and goats (Hulet, 1969). These methods have not been sensitive enough to be accurate much earlier than 60–80 days after insemination.

Proteins produced by the placenta have been used to detect pregnancy for many years in certain species. For example, human chorionic gonadotrophin (hCG) was discovered by Aschheim & Zondek (1928) in urine of pregnant women and can be assayed in urine and blood as early as 8–10 days after conception (Marshall *et al.*, 1968). A host of other pregnancy proteins in the human have been found, some of which reach maternal serum and may be useful as pregnancy markers (Bohn, 1985). Pregnant mare serum gonadotrophin (PMSG) was first reported by Cole & Hart (1930) and measurement in serum was shown to be useful for detection of pregnancy (Cole & Hart, 1942). Other than in equids, reliable pregnancy markers have not been used to detect pregnancy in farm animals.

Several laboratories have reported the presence of pregnancy proteins in cattle and sheep. Rowson & Moor (1967) found that sheep embryos of 13 days of age produced a heat-labile substance which appeared to be responsible for maintenance of the corpus luteum. Similarly, the embryo of the cow affects luteal maintenance (Northey & French, 1980; Humblot & Dalla Porta, 1984). This effect is probably antiluteolytic in action and is caused, at least in part, by trophoblastin (Martal *et al.*, 1979) or ovine trophoblast protein-1 (OTP-1) or other conceptus secretory proteins in sheep (Godkin *et al.*, 1984). Uterine flushings from ewes more than 14 days pregnant contained three proteins not found in serum (Roberts *et al.*, 1976). A heat-labile protein with a molecular weight greater than 8000 was found in serum of pregnant ewes (Cerini *et al.*, 1976a, b). These studies were extended by Staples *et al.* (1978). When first found, this protein showed promise of being useful for pregnancy detection.

In the cow, Roberts & Parker (1976) detected three proteins which appeared in uterine flushings beginning on Day 7 of gestation. Laster (1977) found a pregnancy-specific protein with a molecular weight of 50 000–60 000 in the cow uterus at Day 15. Butler *et al.* (1982) reported a protein of similar molecular weight which could be measured in sera of pregnant cattle (Sasser *et al.*, 1986). This protein is termed pregnancy-specific protein B (PSPB). Several proteins, including bovine trophoblast protein 1 produced by cow conceptuses in culture, have also been reported (Bartol *et al.*, 1984; Knickerbocker *et al.*, 1984) and are known to extend luteal life-span, probably through an antiluteolytic role, when infused into the uterus of cyclic cows. Reviews on these proteins have been published (Bazer *et al.*, 1985; Thatcher *et al.*, 1985). Future work may show that several of these secretory products can enter the circulation of the mother and can be measured in body fluids, thus permitting the detection of pregnancy.

For a test of pregnancy to be useful in livestock management it should detect pregnancy before the time of next expected oestrus after insemination. Unfortunately, the classical established techniques that were mentioned above for cattle and sheep cannot be used that soon after insemination. The only test in practice that approaches this is the assay for progesterone in milk or serum. This can be used to confirm the absence of a corpus luteum and therefore non-pregnancy near the time of next expected oestrus. Other tests such as real-time ultrasound, early pregnancy factor (EPF) in serum (Morton *et al.*, 1979a) and pregnancy-specific protein B (Sasser *et al.*, 1986) are not routinely used in practice but may develop into appropriate early tests. Measurement of oestrone sulphate can probably provide a convenient and reliable test later in pregnancy (Holdsworth *et al.*, 1982). This review will consider only the ultrasonic and more recently described biochemical tests for early pregnancy detection.

Ultrasonic detection of pregnancy

Ultrasonic detection of pregnancy has been used with various degrees of success depending on the type of instruments that are used. A-mode is a one-dimensional display of echo amplitude versus distance. B-mode produces a two-dimensional image of soft-tissue cross-section. Real-time ultrasound scanning is a modification of B-mode which utilizes a number of crystals within the transducer head. The result is a moving two-dimensional image that can be displayed on a video monitor (Reeves *et al.*, 1984). This method was tested by Memon & Ott (1980) for accuracy in ewes and goats. In general, externally applied transducers that did not employ real-time were able to detect pregnancy accurately in the last half of gestation. Intrarectal probes permitted slightly earlier detection but were more labour intensive.

The image provided by real-time scanning allows for a more precise and accurate decision by the operator as to the state of pregnancy. This has resulted in an overall accuracy of 97·1% (5370 correct scans/5530 total scans) in sheep. Ewes that were less than 40 days pregnant were too early to diagnose and were not included in the analysis (Fowler & Wilkins, 1984). Earlier detection has been reported (Tainturier *et al.*, 1983) but more time per animal is required and numbers of animals in the study were limited.

Probes for transrectal use have been applied in pregnancy detection in horses as early as 14 days (Chevalier & Palmer, 1982; Simpson *et al.*, 1982; Rantanen *et al.*, 1982). The same instrument that was used by Rantanen *et al.* (1982) was used by Reeves *et al.* (1984) for transrectal probing for pregnancy in cattle. Pregnancy was detected as early as 28 days after insemination and they attributed the difference in time of detection to a difference in anatomy of the reproductive tracts. Pierson & Ginther (1984) were able to detect the bovine embryonic vesicle by 12–14 days, and the fetus and fetal heartbeat at 26–29 days. In 32 cows scanned at regular intervals between 20 and 140 days of gestation, pregnancy was detected with confidence at 30 days (White *et al.*, 1985).

The use of real-time scanning in practical farm situations is limited by the cost of the instrument. It is likely that less expensive equipment will become available for use in cows, sheep and goats.

Biochemical detection of pregnancy

Biochemical detection of pregnancy relies upon detection of substances that arise in the maternal system when an embryo is present in the uterus. Pregnancy-specific substances are produced by the conceptus. A qualitative assay in body fluids may be sufficient for pregnancy detection. Pregnancy-associated substances are produced by the maternal system but generally are increased in quantity when a conceptus is present. In this case a quantitative assay may be more important. The pregnancy-specific substances would include hCG and PMSG and a pregnancy-associated substance would be milk or serum progesterone. Several such proteins have been listed by Chard (1985) for humans.

Biochemical markers for pregnancy in farm animals are either steroidal or proteinaceous. Those which may be used effectively are progesterone and oestrone sulphate in the former category and early pregnancy factor (EPF) and pregnancy specific protein B (PSPB) in the latter. These will be discussed in more detail.

Steroids

Progesterone. Detection of pregnancy by analysis of plasma (Shemesh *et al.*, 1968) or milk (Laing & Heap, 1971; Lamming & Bulman, 1976; Shemesh *et al.*, 1978) for progesterone content relies upon the decline in concentration at the time of the next expected oestrus if conception does not occur after insemination. A low progesterone concentration suggests that luteolysis has occurred (non-pregnant) and a high content suggests that luteal maintenance had occurred in response to stimuli of the conceptus. Low progesterone concentrations predict oestrus or non-pregnancy (Booth, 1979; Shemesh *et al.*, 1978) with a high degree of accuracy. However, high progesterone concentration is not as reliable a predictor of pregnancy since luteal life-span can vary and samples may be collected too early or too late with respect to luteolysis in non-pregnant animals. Early embryonic death after the sample is collected and before rectal examination at about 50 days after insemination also contribute up to 8% error in the test (Cavestany & Foote, 1985).

Radioimmunoassay (RIA) for milk progesterone for pregnancy detection has been reported several times. Such assays are in use today but more recently enzyme immunoassays have been developed. The accuracy of certain tests in cattle and goats are presented in Table 1. The day of milk sampling for cattle ranged from 20 to 25 days after insemination and is 21 days for the goat. In general the optimum time is at 23, 24 or 25 days for the cow. Two studies (Cavestany & Foote, 1985; Marcus & Hockett, 1986) showed improvement in accuracy of predicting pregnancy by sampling milk at the time of insemination and again in 3 weeks; both values are then used in the evaluation of pregnancy status.

All assays have been adequate in detecting non-pregnancy. Most values approached 100% and so low milk progesterone concentration is a good indicator that luteolysis has occurred and that the conceptus was absent. Detection of pregnant cows is generally less than 85% accurate and ranges from 60 to 100%. The exceptional RIA with 100% accuracy (Shemesh *et al.*, 1981) utilized progestagen-impregnated vaginal sponges for the purpose of reducing variability in time of return

Table 1. Accuracy of the milk progesterone test (radioimmunoassay (RIA) or enzyme immunoassay (EIA)) for pregnancy detection

Reference	Type of assay	No. of samples	Day of sampling after insemination	Detected pregnant correctly (%)	Detected not pregnant correctly (%)
Hoffman *et al.* (1974)	RIA	168	20	77	100
Heap *et al.* (1976)	RIA	176	24	85	100
Pennington *et al.* (1976)	RIA	116	21	73	98
Shemesh *et al.* (1981)	RIA				
Group A		96	21–24	79	100
Group B*		50	21–24	100	100
Montigny *et al.* (1982)	RIA	275 (goats)	21	91	97
Nakao *et al.* (1982)	EIA	268	20, 21 or 22	60	100
Chang & Estergreen (1983)	EIA	115	22	79	98
Cavestany & Foote (1985)	RIA				
Group A		313	23 or 25	73	99
Group B		313	0 & 23 or 25†	84	100
Marcus & Hockett (1986)	EIA	152	0 & 21†	96	90
Sauer *et al.* (1986)	EIA	110	24	94	100

*Cows were treated with progestagen-impregnated vaginal sponges from 6 to 17 days after insemination.
†Assays at insemination and at an interval of one oestrous cycle later.

to oestrus and so if a conceptus was absent low milk progesterone would be more likely on the day of milk sampling.

The enzyme immunoassays (EIA) have been as successful as RIAs with the exception of one recent assay by Sauer *et al.* (1986) which detected 93·5% of the pregnant animals. The other EIA, which reported 96% accuracy, required two samplings, one at time of insemination and the other 21 days later (Marcus & Hockett, 1986). A value was not reported for success based upon only the 21-day sample but the authors did not believe it was adequate.

Accuracy of overall detection of pregnancy by RIA of progesterone in milkfat of goats was 90·5% while that on whole milk was 85·4%. Concentration of progesterone in whole milk and in milkfat of pregnant goats was 5 times and over 10 times greater respectively than that of non-pregnant goats on Day 21 after insemination. Milkfat is the preferred medium for analysis in goats (Montigny *et al.*, 1982). Similar success in detecting pregnancy in sheep with the serum progesterone test has been reported (Shemesh *et al.*, 1973). An improvement of accuracy was obtained with RIA for serum progesterone by taking multiple samples in animals without known breeding dates (Döbeli & Schwander, 1985). Three samples were collected at 6-day intervals beginning 9 days after the removal of rams from the flock. Accuracy in predicting pregnancy or non-pregnancy was 98·8% and 100% respectively.

The milk or serum progesterone assay has come into greater use with the availability of reliable commercial assay kits. EIAs permit safe, simple-to-conduct, laboratory or on-the-farm evaluation of milk progesterone for pregnancy detection. Disadvantages are that insemination dates must be known to obtain results from a single sample and, since progesterone is not pregnancy-specific, incorrect positive detection can occur.

Oestrone sulphate. This substance is primarily produced by the fetal–placental unit and enters the circulation of the mother. Oestrone sulphate in sera of pregnant pigs was elevated over that of non-pregnant pigs between 20 and 29 days of gestation (Robertson & King, 1974) and can provide an early test for pregnancy. However, in sheep the increase in serum does not occur until 70–90 days of gestation and can be measured at 100 days as a late test for pregnancy (Thimonier *et al.*,

1977). It has been used for late pregnancy detection in goats after 60 days of gestation and was 96% accurate in pregnant (N = 260) and 98% accurate in non-pregnant (N = 60) does (P. Humblot, unpublished). In the pregnant cow, concentrations of oestrone sulphate in milk were increased above non-pregnancy levels by 100 days of gestation and remained high throughout the remainder of gestation (Holdsworth *et al.*, 1982). An EIA in microtitre plates has been developed for rapid solid-phase measurement of oestrone sulphate in bovine milk without extraction. The assay was 99 and 86% accurate in predicting pregnant and non-pregnant animals respectively (Power *et al.*, 1985). Assay of milk oestrone sulphate is of significance and is mentioned herein since it would provide an easy, late, reliable test for confirmation of pregnancy. A serum test might be particularly advantageous for small ruminants for which insemination dates are not known.

Proteins

Early pregnancy factors. EPF was first identified in pregnant mice by its ability to inhibit rosette formation between T lymphocytes and heterologous red blood cells. A rosette inhibition test is a useful assessment of lymphocyte activity (Morton *et al.*, 1976). When antilymphocyte serum is added to lymphocytes before the addition of red blood cells, the inhibition of rosette formation can reach 25%. The titre (rosette inhibition titre, RIT) at which an antilymphocyte serum can inhibit rosette formation does not vary significantly when using lymphocytes from normal animals, but if lymphocytes are subjected to in-vivo immunosuppressive agents the formation of rosettes occurs at a much higher dilution of antilymphocyte sera (Shaw & Morton, 1980).

The RIT has been used to detect early pregnancy in sheep (Morton *et al.*, 1979a, b) and cows (Nancarrow *et al.*, 1981). A positive test was achieved for sera of sheep from 72 h after insemination to 16 weeks of gestation and then it was no longer detectable (Morton *et al.*, 1979a). Only in pigs has EPF activity been found throughout pregnancy (Morton *et al.*, 1983).

An excellent review of use of EPF in detecting pregnancy in farm animals has been presented by Koch & Ellendorff (1985). In this detailed study to evaluate the RIT assay for sera of pigs, they showed that it was possibile to detect non-pregnancy in 91·4% and pregnancy in 55·6% of pigs. Using data of three tests on the same samples of sera, the respective values were 91·4% for non-pregnancy and 87·8% for pregnancy detection.

The RIT test is time consuming and difficult to maintain. Development of a RIA or EIA for EPF would probably provide a more reliable test. This will first require chemical isolation of EPF. Progress on this has been reported (Wilson *et al.*, 1984; Clark & Wilson, 1985). Until the assay is improved one must withhold judgement as to the potential use of EPF in detection of pregnancy in farm animals (see Ellendorf & Koch, 1985).

Pregnancy-specific protein B. This protein was isolated and partly purified by us at the University of Idaho (Butler *et al.*, 1982). The protein was found after immunization of a rabbit with homogenates of placenta, collection of rabbit antisera and adsorption of antisera with somatic tissues to remove antibodies to proteins not specific to the placenta. Remaining antibodies were against alpha-fetoprotein and PSPB of the placenta. Immunoelectrophoresis and immunodiffusion techniques using the adsorbed antisera were used as markers in chemical isolation of PSPB. Placental extracts were subjected to ion-exchange and gel-filtration chromatography and isoelectric focussing. A preparation of PSPB (R-37) was used to immunize rabbits to obtain an antiserum for development of a double-antibody radioimmunoassay (Sasser *et al.*, 1986). It was established that the assay was specific for bovine PSPB with minimal cross-reactivity (<0·05%) with bovine LH and ovine FSH and partial cross-reactivity with ovine placental lactogen (<0·05%). There was no cross-reactivity with hCG, PMSG, bovine PRL, bovine TSH or bovine growth hormone (Fig. 1). Pools of sera from rams, wethers and steers and non-pregnant and pregnant ewes were assayed in

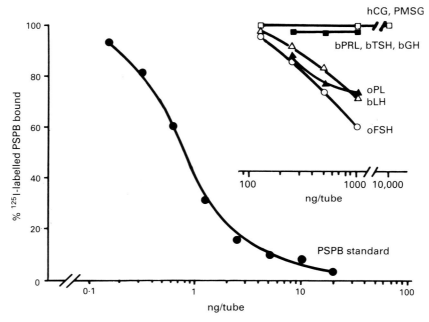

Fig. 1. Cross-reactivity of various pituitary and placental hormones with antiserum to the PSPB standard. The amount of radiolabelled PSPB bound to antiserum in tubes containing no PSPB was designated as 100% ^{125}I bound. Note the location of the continued *x* axis. (From Sasser *et al.*, 1986.)

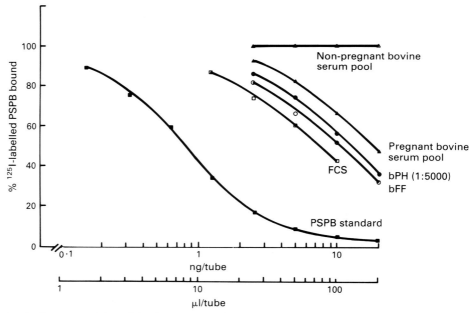

Fig. 2. Cross-reactivity of fetal calf serum (FCS), bovine fetal fluids (bFF), bovine placental homogenate (bPH, diluted 1:5000) and pools of sera with antiserum to the PSPB standard. (From Sasser *et al.*, 1986.)

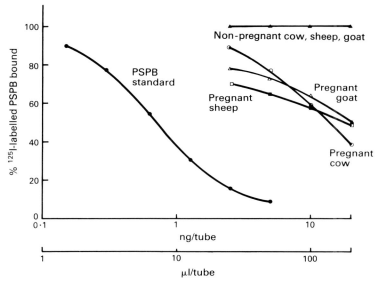

Fig. 3. Cross-reactivity of sheep and goat sera with antiserum to the PSPB standard.

the PSPB RIA. There was no PSPB in ram, wether, steer or non-pregnant ewe sera. Pools of sera from pregnant cows inhibited binding in a manner parallel to the standard curve, whereas serum pools from non-pregnant cows did not. Also, bovine fetal fluids, supernatant from placental homogenate and fetal calf serum resulted in an inhibition curve that was parallel to that of the standard. This indicated that the isolation procedure resulted in a protein similar to the native protein (Fig. 2). Pregnant ewe and goat sera contained cross-reacting antigens but the inhibition curve was not parallel to the standard curve (Fig. 3). These data suggest that not only could PSPB be measured in serum of pregnant cows, but that it was pregnancy-specific and, therefore, the RIA might be useful for pregnancy detection.

For PSPB to be considered as a pregnancy marker it would be desirable that it be present throughout pregnancy. To determine this, 5 dairy cows were bled twice weekly from mating until 21 days *post partum*. PSPB was detectable from about 24 days after insemination until 21 days *post partum* (Fig. 4). To confirm the time that PSPB first appeared in maternal serum, 21 cows were bled daily from 15 to 30 days after insemination. PSPB was present in sera of 3 cows at 15 days and by 24 days 12 had PSPB in the sera; 3 more acquired PSPB by Day 28 but were non-pregnant by rectal palpation at 45 days after insemination and had long interoestrous intervals while the former 12 cows were confirmed pregnant. PSPB could not be detected in the serum of the remaining cows and these animals were confirmed as non-pregnant at 45 days after insemination.

We have found that PSPB remains in maternal serum for several weeks after parturition. To determine whether there was continued secretion of PSPB after parturition (Ruder & Sasser, 1986), 5 cows were hysterectomized on Day 21 after parturition and 5 other cows served as intact controls. Blood was collected from parturition until Day 53 *post partum*. Half-life for PSPB in control and hysterectomized cows was 8·4 days and 7·3 days ($P = 0.04$) respectively from Days 21 to 53 *post partum*. These results showed that PSPB has a long half-life and is released from the uterus after parturition or, on the other hand, the uterus could alter metabolism of PSPB resulting in an increase in half-life. Continued secretion is not likely since immunocytochemical studies have shown that PSPB is localized in (Eckblad *et al.*, 1985) and secreted from (Reimers *et al.*, 1985) the binucleated cells of the trophoblastic ectoderm.

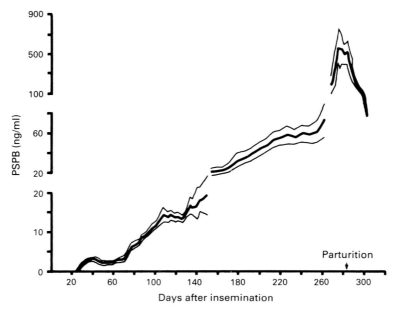

Fig. 4. Concentration (mean ± s.e.; heavy and light lines, respectively), of PSPB in serum of 5 dairy cows from insemination until 21 days after parturition. Note the three scales of the y axis. (From Sasser *et al.*, 1986.)

The long half-life could pose a major limitation on this method of pregnancy detection if cows are to be examined before 80 days *post partum*. This would not usually be the case since cows are rarely inseminated before 55 days *post partum* and PSPB would not be detectable until 24 days later for a total of 79 days *post partum*. A major advantage of this method is that insemination dates are not required because PSPB is pregnancy-specific and is present continually after 24 days of gestation.

Pregnancy detection by RIA of PSPB in sera was highly accurate in a commercial herd of beef cows. At first examination 99 cows were detected pregnant by PSPB assay; 98 calved and 3 of 4 non-pregnant ones were accurately detected. Serum samples were collected on the 27th day after the end of a 75-day breeding period. There was an exceptionally high number of pregnant animals in this study and it was impossible to evaluate accuracy in non-pregnant cows. However, we have shown that RIA of serum for PSPB detected 177 of 187 non-pregnant cows from which uteri were examined for presence of a conceptus at slaughter. In detection of pregnant animals the RIA predicted 189 of 191 cows. Cattle had been inseminated 25–61 days before slaughter (Maurer *et al.*, 1985).

We have shown that assay for PSPB is a reliable serological test for pregnancy in sheep. It was as accurate as abdominal real-time scanning at 35 days of gestation. Real-time ultrasound predicted pregnancy in 163 and non-pregnancy in 17 ewes while RIA predicted 161 and 19 respectively; 159 ewes gave birth and 21 did not (Ruder *et al.*, 1984). Although there are cross-reacting antigens in sera of pregnant goats, we have not conducted a study of the accuracy of assay for PSPB in pregnancy detection for these animals.

These are the first experiments to demonstrate a simple, accurate and specific serological test for pregnancy in cattle and sheep. In cattle pregnancy can be detected as early as 24 days after insemination until parturition. Insemination dates are not needed as they are for measurements of progesterone in milk or serum or EPF in serum. The assay can be applied to other ruminants as well (Sasser *et al.*, 1985; Wood *et al.*, 1986) and it will probably be as versatile as the pregnancy-specific chorionic gonadotrophin tests in humans (Marshall *et al.*, 1968).

Development of an animal-side test

Recent developments in diagnostic testing have made it possible to have reliable, sensitive and accurate assays that do not require sophisticated equipment. Approaches to use of non-radioactive markers in immunological assays are presented by Ngo & Lenhoff (1980). These assays utilize enzymic reactions resulting in development of a colour. Colour is monitored with an instrument (spectrophotometer) or visually. Enzyme immunoassays are of two types. The heterogeneous ones require physical separation of the unbound antigen while homogeneous assays do not (Rubenstein *et al.*, 1972). In the latter, total activity can be measured without prior separation and exploits the ability of the sample antigen to modulate an immunospecific signal.

Enzyme immunoassays can be immobilized on an inert support (Zuk *et al.*, 1985) for testing in less technical laboratories or in the field. Such assays are currently available for analysis of progesterone in milk or serum. Any substance that becomes a pregnancy marker could easily be assayed by this method. In addition, quantitative assay on test-strips (Zuk *et al.*, 1985) may provide an added advantage in predicting the stage of pregnancy if the antigen in question varies considerably throughout pregnancy.

This study was supported by the Idaho Agricultural Experiment Station, the United Dairymen of Idaho, and the National Association of Animal Breeders. We thank NIAMDD for the hormone preparations and I.N.R.A., Jouy-en-Josas, France, for use of library and other facilities in preparation of this manuscript.

References

Aschheim, S. & Zondek, B. (1928) Diagnosis of pregnancy by demonstrating the presence of the hormone of the anterior hypophysis in the urine. *Klin. Wschr.* **7,** 1404–1411.

Bartol, F.F., Roberts, R.M., Bazer, F.W., Lewis, G.S., Godkin, J.D. & Thatcher, W.W. (1984) Characteristics of proteins produced *in vitro* by periattachment bovine conceptuses. *Biol. Reprod.* **32,** 681–694.

Bazer, F.W., Roberts, R.M., Thatcher, W.W. & Sharp, D.C. (1985) Mechanism related to establishment of pregnancy. In *Reproduction and Perinatal Medicine: Early Pregnancy Medicine,* Vol. 1, pp. 13–24. Eds F. Ellendorff & E. Koch. Perinatology Press, Ithaca.

Bohn, H. (1985) Biochemistry of pregnancy proteins–an overview. In *Reproduction and Perinatal Medicine: Early Pregnancy Factors,* Vol. 1, pp. 127–139. Eds F. Ellendorff & E. Koch. Perinatology Press, Ithaca.

Booth, J.M. (1979) Milk progesterone testing: Application to herd management. *J. Dairy Sci.* **62,** 1829–1834.

Butler, J.E., Hamilton, W.C., Sasser, R.G., Ruder, C.A., Hass, G.M. & Williams, R.J. (1982) Detection and partial purification of two bovine pregnancy-specific proteins. *Biol. Reprod.* **26,** 925–933.

Cavestany, D. & Foote, R.H. (1985) The use of milk progesterone and electronic vaginal probes as aids in large dairy herd reproductive management. *Cornell Vet.* **75,** 441–453.

Cerini, M., Cerini, J.C., Findlay, J.K. & Lawson, R.A.S. (1976a) Preliminary characterization of pregnancy-specific antigens in the ewe. *J. Reprod. Fert.* **46,** 534, Abstr.

Cerini, M., Findlay, J.K. & Lawson, R.A.S. (1976b) Pregnancy-specific antigens in sheep: application to the diagnosis of pregnancy. *J. Reprod. Fert.* **46,** 65–69.

Chang, C.F & Estergreen, V.L. (1983) Development of a direct enzyme immunoassay of milk progesterone and its application to pregnancy diagnosis in cows. *Steroids* **41,** 173–195.

Chard, T. (1985) Biological and clinical significance of pregnancy proteins. In *Reproduction and Perinatal Medicine: Early Pregnancy Factor,* Vol. 1, pp. 29–41. Eds F. Ellendorff & E. Koch. Perinatology Press, Ithaca.

Chevalier, F. & Palmer, E. (1982) Ultrasonic echography in the mare. *J. Reprod. Fert., Suppl.* **32,** 423–430.

Clark, F. & Wilson, S. (1985) In search of EPF. In *Reproduction and Perinatal Medicine: Early Pregnancy Factor,* Vol. 1, pp. 165–177. Eds F. Ellendorff & E. Koch. Perinatology Press, Ithaca.

Cole, H.H. & Hart, G.H. (1930) The potency of blood serum of mares in progressive stages of pregnancy in affecting the sexual maturity of the immature rat. *Am. J. Physiol.* **93,** 57–68.

Cole, H.H. & Hart, G.H. (1942) Diagnosis of pregnancy in the mare by hormonal means. *J. Am. vet. med. Ass.* **101,** 124–128.

Döbeli, M. Von & Schwander, B. (1985) Trächtigkeits-diagnose in einer Schafherde anhand dreimaliger Progesteronbestimmung im Blutplasma. *Zucht-hygiene* **20,** 192–199.

Eckblad, W.P., Sasser, R.G., Ruder, C.A., Panlasigui, P. & Kuczynski, T. (1985) Localization of pregnancy-specific protein B (PSPB) in bovine placental cells using a glucose oxidase-anti-glucose oxidase immuno-histochemical stain. *J. Anim. Sci.* **61** (Suppl.), 149–150, Abstr.

Ellendorff, F. & Koch, E. (Eds) (1985) *Reproductive and Perinatal Medicine: Early Pregnancy Factors,* Vol. 1, 276 pp. Perinatology Press, Ithaca.

Fowler, D.G. & Wilkins, J.F. (1984) Diagnosis of preg-nancy and number of foetuses in sheep by real-time ultrasonic imaging. I. Effects of number of foetuses, stage of gestation, operator and breed of ewe on accuracy of diagnosis. *Livestock Prod. Sci.* **11,** 437–450.

Godkin, J.D., Bazer, F.W. & Roberts, R.M. (1984) Ovine trophoblast protein 1, an early secreted blastocyst protein, binds specifically to uterine endometrium and affects protein synthesis. *Endocrinology* **114,** 120–130.

Heap, R.B., Holdsworth, R.J., Sadaby, J.E., Laing, J.A. & Walters, D.E. (1976) Pregnancy diagnosis in the cow from milk progesterone concentration. *Br. vet. J.* **132,** 445–464.

Hoffman, B., Hamburger, R., Günzler, O., Kerndörfer, L. & Lohoff, H. (1974) Determination of progesterone in milk applied for pregnancy diagnosis in the cow. *Theriogenology* **2,** 21–28.

Holdsworth, R.J., Heap, R.B., Booth, J.M. & Hamon, M. (1982) A rapid direct radioimmunoassay for the measurement of oestrone sulphate in the milk of dairy cows and its use in pregnancy diagnosis. *J. Endocr.* **95,** 7–12.

Hulet, C.V. (1969) Pregnancy diagnosis in the ewe using an ultrasonic doppler instrument. *J. Anim. Sci.* **28,** 44–47.

Hulet, C.V. (1972) A rectal abdominal palpation tech-nique for diagnosing pregnancy in the ewe. *J. Anim. Sci.* **35,** 814–819.

Humblot, P. & Dalla Porta, M.A. (1984) Effect of con-ceptus removal and intrauterine administration of conceptus tissue on luteal function in the cow. *Reprod. Nutr. Dével.* **24,** 529–541.

Knickerbocker, J.J., Thatcher, W.W., Bazer, F.W., Drost, M., Barron, D.H., Fincher, K.B. & Roberts, R.M. (1984) Proteins secreted by cultured day 17 conceptuses extend luteal function in cattle. *Proc. 10th Int. Congr. Anim. Reprod. & A.I., Urbana–Champaign* **1,** 88, Abstr.

Koch, E. & Ellendorff, F. (1985) Detection of early preg-nancy factor (EPF) activity in farm animals. In *Reproduction and Perinatal Medicine: Early Preg-nancy Factor,* Vol. 1, pp. 25–28. Eds F. Ellendorff & E. Koch. Perinatology Press, Ithaca.

Laing, J.A., & Heap, R.B. (1971) The concentration of progesterone in the milk of cows during the repro-ductive cycle. *Br. vet. J.* **127,** 19–22.

Lamming, G.E. & Bulman, D.C. (1976) The use of milk progesterone radioimmunoassay in diagnosis and treatment of subfertility in dairy cows. *Br. vet. J.* **132,** 507.

Laster, D.B. (1977) A pregnancy-specific protein in the bovine uterus. *Biol. Reprod.* **16,** 682–690.

Marcus, G.I. & Hockett, A.J. (1986) Use of an enzyme-linked immunosorbent assay for measurement of bovine serum and milk progesterone without extrac-tion. *J. Dairy Sci.* **69,** 818–824.

Marshall, J.R., Hammond, C.B., Ross, G.T., Jacobson, A., Rayford, P. & Odell, W.D. (1968) Plasma and urinary chorionic gonadotrophin during early human pregnancy. *Obstet. Gynec., N.Y.* **32,** 760–764.

Martal, J., Lacroix, M.C., Loudes, C., Saunier, M. & Wintenberger-Torres, S. (1979) Trophoblastin, an antiluteolytic protein present in early pregnancy in sheep. *J. Reprod. Fert.* **56,** 63–67.

Maurer, R.R., Ruder, C.A. & Sasser, R.G. (1985) Effect-iveness of the protein B radioimmunoassay to diag-nose pregnancy in beef cattle. *J. Anim. Sci.* **61** (Suppl. 1), 390, Abstr.

Memon, M.A. & Ott, R.S. (1980) Methods of pregnancy diagnosis in sheep and goats. *Cornell Vet.* **70,** 226–231.

Montigny, G. de, Millerioux, P., Jeanguyot, N., Humblot, P. & Thibier, M. (1982) Milk fat progesterone con-centrations in goats and early pregnancy diagnosis. *Theriogenology* **17,** 423–431.

Morton, H., Hegh, V. & Clunie, G.J.A. (1976) Studies of the rosette inhibition test in pregnant mice: evidence of immunosuppression? *Proc. R. Soc. B* **193,** 413–419.

Morton, H., Clunie, G.J.A. & Shaw, F.D. (1979a) A test for early pregnancy in sheep. *Res. vet. Sci.* **26,** 261–262.

Morton, H., Nancarrow, C.D., Scaramuzzi, R.J., Evison, B.M. & Clunie, G.J.A. (1979b) Detection of early pregnancy in sheep by the rosette inhibition test. *J. Reprod. Fert.* **56,** 75–80.

Morton, H., Morton, D.J. & Ellendorff, F. (1983) The appearance and characterization of early pregnancy factor in the pig. *J. Reprod. Fert.* **68,** 437–446.

Nakao, T., Sugihashi, A., Ishibashi, Y., Tosa, E., Nakagawa, Y., Yuto, H., Nomura, T., Ohe, T., Ishimi, S., Takahashi, H., Koiwa, M., Tsunoda, N. & Kawata, K. (1982) Use of milk progesterone enzyme immuno-assay for early pregnancy diagnosis in cows. *Therio-genology* **18,** 267–274.

Nancarrow, C.D., Wallace, A.L.C. & Grewal, A.S. (1981) The early pregnancy factor of sheep and cattle. *J. Reprod. Fert., Suppl.* **30,** 191–199.

Ngo, T.T. & Lenhoff, H.M. (1980) Enzyme modulators as tools for the development of homogeneous enzyme immunoassays. *FEBS Letters* **116,** 285–288.

Northey, D.L. & French, L.R. (1980) Effect of embryo removal and intrauterine infusion of embryonic homogenates on the lifespan of the bovine corpus luteum. *J. Anim. Sci.* **50,** 298–302.

Pennington, J.A., Spahr, S.L. & Lodge, J.R. (1976) Pro-gesterone diagnosis in dairy cattle by progesterone concentration in milk. *J. Dairy Sci.* **59,** 1528.

Pierson, R.A. & Ginther, O.J. (1984) Ultrasonography for detection of pregnancy and study of embryonic development in heifers. *Theriogenology* **22,** 225–233.

Power, M.J., Cleare, W.F., Gosling, J.M., Fottrell, P.F., Langley, O.H. & Sreenan, J.M. (1985) A direct, high throughput, enzyme immunoassay for oestrone sul-phate in the milk of cows. *Irish Vet. J.* **39,** 18–24.

Rantanen, N.W., Torbeck, R.L. & Dumond, S.S. (1982) Early pregnancy diagnosis in the mare using transrectal ultrasound scanning techniques: a preliminary study. *J. eq. vet. Sci.* **2**, 27–29.

Reeves, J.J., Rantanen, N.W. & Houser, M. (1984) Transrectal real-time scanning of the cow reproductive tract. *Theriogenology* **21**, 485–494.

Reimers, T.J., Sasser, R.G. & Ruder, C.A. (1985) Production of pregnancy-specific protein by bovine binucleate trophoblastic cells. *Biol. Reprod.* **32** (Suppl.), 65, Abstr.

Roberts, G.P. & Parker, J.M. (1976) Fractionation and comparison of proteins from bovine uterine fluid and bovine allantoic fluid. *Biochem. Biophys. Acta* **446**, 69–76.

Roberts, G.P., Parker, J.M. & Symonds, H.W. (1976) Macromolecular components of genital tract fluids from sheep. *J. Reprod. Fert.* **48**, 99–107.

Robertson, H.A. & King, G.J. (1974) Plasma progesterone and oestrogen in the pig. *J. Reprod. Fert.* **40**, 133–141.

Rowson, L.E.A. & Moor, R.M. (1967) The influence of embryonic tissue homogenate infused into the uterus on the life-span of the corpus luteum in sheep. *J. Reprod. Fert.* **13**, 511–516.

Rubenstein, K.E., Schneider, R.S. & Ullman, E.F. (1972) Homogeneous enzyme immunoassay. A new immunochemical technique. *Biochem. Biophys. Res. Commun.* **47**, 846–851.

Ruder, C.A. & Sasser, R.G. (1986) Source of bovine pregnancy-specific protein B (bPSPB) during the postpartum period and estimation of half-life of bPSPB. *J. Anim. Sci.* **63** (Suppl.), 335, Abstr.

Ruder, C.A., Sasser, R.G., Ivani, K.A., Panlasigui, P.M., Dahmen, J.J. & Stellflug, J.N. (1984) Diagnosis of pregnancy in sheep by measurement of a blood antigen that cross-reacts in a radioimmunoassay for pregnancy-specific protein B. *J. Anim. Sci.* **59**, (Suppl.), 367, Abstr.

Sasser, R.G., Ruder, C.A., Ivani, K.A., Short, R.E. & Wood, A.K. (1985) Pregnancy-specific protein B in serum of various species. in *Reproductive and Perinatal Medicine: Early Pregnancy Factors*, Vol. 1, pp. 161–163. Eds F. Ellendorff & E. Koch. Perinatology Press, Ithaca.

Sasser, R.G., Ruder, C.A., Ivani, K.A., Butler, J.E. & Hamilton, W.C. (1986) Detection of pregnancy by radioimmunoassay of a novel pregnancy-specific protein in serum of cows and a profile of serum concentrations during gestation. *Biol. Reprod.* **35**, 936–942.

Sauer, M.J., Foulkes, J.A., Worsfold, A. & Morris, B.A. (1986) Use of progesterone II-glucuronide-alkaline phosphatase conjugate in a sensitive microtitre-plate enzyme immunoassay of progesterone in milk and its application to pregnancy testing in dairy cattle. *J. Reprod. Fert.* **76**, 375–391.

Shaw, F.D. & Morton, H. (1980) An immunological approach to pregnancy diagnosis: a review. *Vet. Record* **106**, 268–269.

Shemesh, M., Ayalon, N. & Lindner, H.R. (1968) Early effect of conceptus on plasma progesterone level in the cow. *J. Reprod. Fert.* **15**, 161–164.

Shemesh, M., Ayalon, N. & Lindner, H.R. (1973) Early pregnancy based upon plasma progesterone levels in the cow and ewe. *J. Anim. Sci.* **36**, 726–729.

Shemesh, M., Ayalon, N., Shalev, E., Nerya, A., Schindler, H. & Milaguir, F. (1978) Milk progesterone measurement in dairy cows: correlation with estrous and pregnancy determination. *Theriogenology* **9**, 343–352.

Shemesh, M., Ayalon, N., Marcus, S., Danielli, Y., Shore, L. & Lavi, S. (1981) Improvement of early pregnancy diagnosis based on milk progesterone by the use of progestin-impregnated vaginal sponges. *Theriogenology* **15**, 459–462.

Simpson, D.J., Greenwood, R.E.S., Ricketts, S.W., Rossdale, P.D., Sanderson, M. & Allen, W.R. (1982) Use of ultrasound echography for early diagnosis of single and twin pregnancy in the mare. *J. Reprod. Fert., Suppl.* **32**, 431–439.

Staples, L.D., Lawson, R.A.S & Findlay, J.K. (1978) The occurrence of an antigen associated with pregnancy in the ewe. *Biol. Reprod.* **19**, 1076–1082.

Tainturier, D., Lijour, L., Charri, M., Sardjana, K.W. & Denis, B. (1983) Diagnostic de la gestation chez la brebis par échographie. *Revue Méd. Vét.* **134**, 523–526.

Thatcher, W.W., Knickerbocker, J.J., Helmer, S.D., Hansen, P.J., Bartol, F.F., Roberts, R.M. & Bazer, F.W. (1985) Characteristics of conceptus secretory proteins and their effect on $PGF_{2\alpha}$ secretion by the maternal endometrium. In *Reproduction and Perinatal Medicine: Early Pregnancy Factors*, Vol. 1, pp. 25–28. Eds F. Ellendorff & E. Koch. Perinatology Press, Ithaca.

Thimonier, J., Bosc, M., Djiane, J., Martal, J. & Terqui, M. (1977) Hormonal diagnosis of pregnancy and number of fetuses in sheep and goats. In *Management of Reproduction in Sheep and Goats*, pp. 78–88. University of Wisconsin, Madison.

White, I.R., Russel, A.J.F., Wright, I.A. & Whyte, T.K. (1985) Real-time ultrasonic scanning in the diagnosis of pregnancy and the estimation of gestational age in cattle. *Vet. Rec.* **117**, 5–8.

Wilson, S., McCarthy, R. & Clark, F. (1984) In search of early pregnancy factor: characterization of active polypeptides isolated from pregnant ewes' sera. *J. Reprod. Immunol.* **6**, 253–260.

Wisnicky, W. & Casida, L.E. (1948) A manual method for the diagnosis of pregnancy in cattle. *Am. vet. med. Ass.* **113**, 451–452.

Wood, A.K., Short, R.E., Darling, A.E., Dusek, G.L., Sasser, R.G. & Ruder, C.A. (1986) Serum assays for detecting pregnancy in mule and white-tailed deer. *J. Wildlife Mgmt* **50**, 684–687.

Zuk, R.F., Ginsberg, V.K., Houts, T., Rabbie, J., Merrick, H., Ullman, E.F., Fisher, M.M., Sizto, C.C., Stiso, S.N. & David, J. (1985) Enzyme immunochromatography. A quantitative immunoassay requiring no instruments. *Clin. Chem.* **31**, 1144–1150.

INDEXES

AUTHORS

Entries in **bold** type indicate citations in the reference sections.

SUBJECT INDEX